Advances in Electrical Power System Quality

Advances in Electrical Power System Quality

Guest Editors

Marcel Nicola
Petre-Marin Nicolae
Dan Selisteanu

MDPI

Basel • Beijing • Wuhan • Barcelona • Belgrade • Novi Sad • Cluj • Manchester

Guest Editors
Marcel Nicola
Research and Development Department
National Institute for Research, Development and Testing in Electrical Engineering
Craiova
Romania

Petre-Marin Nicolae
Department of Electrical Engineering and Energetic Doctoral School
University of Craiova
Craiova
Romania

Dan Selisteanu
Department of Automatic Control and Electronics
University of Craiova
Craiova
Romania

Editorial Office
MDPI AG
Grosspeteranlage 5
4052 Basel, Switzerland

This is a reprint of the Special Issue, published open access by the journal *Energies* (ISSN 1996-1073), freely accessible at: https://www.mdpi.com/journal/energies/special_issues/OE050917LQ.

For citation purposes, cite each article independently as indicated on the article page online and as indicated below:

Lastname, A.A.; Lastname, B.B. Article Title. *Journal Name* **Year**, *Volume Number*, Page Range.

ISBN 978-3-7258-7803-1 (Hbk)
ISBN 978-3-7258-7804-8 (PDF)
https://doi.org/10.3390/books978-3-7258-7804-8

Contents

Preface

The following Special Issue is dedicated to addressing the evolving challenges associated with modern electrical power systems, which are undergoing a profound transformation driven by the large-scale integration of renewable energy sources, the rapid expansion of electric mobility, and the increasing deployment of smart grid technologies. These developments introduce new levels of complexity, variability, and uncertainty, requiring advanced approaches for analysis, control, and operation. Within this context, power quality has emerged as a critical concern. The extensive use of power electronic converters and distributed energy resources has led to increased levels of harmonic distortion, voltage deviations, and other disturbances that can negatively impact both system stability and end-user equipment. Ensuring compliance with power quality standards while maintaining system efficiency and flexibility has therefore become a key objective for researchers and practitioners alike. To address these challenges, modern control and optimization techniques are increasingly being adopted. In particular, intelligent methods based on artificial intelligence—such as neural networks, fuzzy logic, and machine learning—offer powerful tools for managing the nonlinear and dynamic behavior of contemporary power systems. These approaches enable adaptive, data-driven solutions that enhance system robustness and support real-time decision-making in complex operating environments. The contributions included in this Special Issue reflect the interdisciplinary nature of current research in power system quality. They present recent advances in control strategies, optimization methods, signal processing techniques, and energy management solutions, with applications ranging from renewable energy systems and electric drives to industrial and smart grid infrastructures. Both theoretical developments and practical implementations are considered, including experimental validations, hardware-in-the-loop simulations, and real-time deployment aspects, providing valuable insights into current trends and emerging directions. Looking ahead, the ongoing integration of digital technologies and intelligent control is expected to play a central role in the evolution of power systems toward more autonomous, resilient, and efficient operation. Future developments will likely emphasize enhanced interoperability, cybersecurity, standardization frameworks, and data-driven optimization, further strengthening system adaptability, scalability, and reliability. In addition, increased collaboration between academia and industry will be essential to accelerate the transfer of innovative solutions into real-world applications. By fostering innovation and facilitating knowledge exchange, this Special Issue aims to contribute to the development of sustainable energy systems capable of meeting the increasingly complex demands of modern society.

Marcel Nicola, Petre-Marin Nicolae, and Dan Selisteanu

Guest Editors

energies

MDPI

Article

Analysis of Reactive Power in Electrical Networks Supplying Nonlinear Fast-Varying Loads

Yuriy Sayenko [1,*], Ryszard Pawelek [1] and Tetiana Baranenko [2,†]

1 Institute of Electrical Power Engineering, Lodz University of Technology, 20 Stefanowskiego Street, 90-537 Lodz, Poland; ryszard.pawelek@p.lodz.pl
2 Department of Automation Electrical Systems and Electric Drive, Pryazovskyi State Technical University, 7 Universytetska Street, 87555 Mariupol, Ukraine; tbaranenko@gmail.com
* Correspondence: yuriy.sayenko@p.lodz.pl
† Current address: Department of Electric Power Complexes and Systems, Pryazovskyi State Technical University, 49005 Mariupol, Ukraine.

Abstract: This study concerns problems related to the assessment of reactive power in power networks with nonlinear fast-varying loads, such as electric arc furnaces, rolling mill drives, etc. The operation of this type of load is characterized by the introduction of interharmonic currents (including higher harmonics) into the power supply network and a relatively low power factor. Rapid changes in the RMS value of the current also cause voltage fluctuations and the related phenomenon of light-flickering. Therefore, there is a need to evaluate the power selection of compensating devices, taking into account the random nature of load changes and the distortion of current and voltage waveforms, in particular, interharmonic components, the impact of which has not been fully investigated so far. To analyze the random nature of load changes, autocorrelation functions were used, which allowed for the estimation of the expected values of the arc furnace current distortion coefficient (based on the recorded waveforms). In order to determine the parameters of reactive power compensating devices, and in particular capacitor banks, an autocorrelation function in the exponential-cosine-sine form was used, which meets the conditions of differentiation. This study contains comparative results of the reactive power of capacitor banks determined using different methods. The criterion for selecting capacitor bank parameters was the minimization of energy losses in the power supply network. The calculations presented in this study show that by taking into account higher harmonics and interharmonics in the voltage and currents of fast-varying loads, the installed power of the capacitor bank can be reduced by approximately 7%, and energy losses in the power grid can be reduced by 3–5%.

Keywords: reactive power; higher harmonics; interharmonics; random processes; autocorrelation function

1. Introduction

In power networks with different voltage levels, there is a continuous increase in the use of nonlinear loads. Nonlinear loads cause distortions in the waveforms of current and voltage curves and are at the same time reactive power consumers.

With distorted current and voltage waveforms, deformed reactive power appears, which does not allow for the classic approach to reactive power compensation used in networks with sinusoidal waveforms. Therefore, in power networks with nonlinear loads, the problem of reactive power control, and especially its compensation, arises.

Many published works [1–10] have been devoted to the problem of determining reactive power in conditions of distortion and asymmetry of supply voltage, but most often they concern the analysis of network operation with a constant nature of the load.

Connecting nonlinear fast-varying loads (e.g., arc furnaces) to power networks may cause further problems. It should be emphasized here that the phenomena caused by nonlinear fast-varying loads are most often random. Therefore, when calculating the reactive power of such receivers, the random nature of changes in the RMS value of the load current should be taken into account.

Rapid changes the RMS current value cause the appearance of interharmonic components in the frequency spectra of these currents. Fast-varying loads are therefore sources of interharmonic currents. Of particular importance are the interharmonic components associated with the fundamental harmonic. These components, especially subharmonics (components with a frequency lower than the fundamental), may consequently cause voltage fluctuations and the related phenomenon of light-flickering [11]. The method of assessing the frequency spectrum of current waveforms containing harmonics and interharmonics is given in the EN 61000-4-7 standard [12]. According to this standard, the total distortion ratio of current waveform (incl. harmonics and interharmonics) can be derived from the expression

$$TDR_I = \frac{\sqrt{I^2 - I_1^2}}{I_1} = \sqrt{THD_I^2 + TID_I^2} \quad (1)$$

where I is the current RMS value aggregated in the time interval of load fluctuation, I_1 is the current fundamental RMS value aggregated in the time interval of load fluctuation, THD_I is the total harmonic distortion factor of current, and TID_I is the total interharmonic distortion factor of current.

In [11], it was shown that the share of interharmonics in the stationary amplitude-modulated current of a nonlinear receiver increases with the increase in the modulation depth of the waveform.

The characteristics of selected receivers with quickly changing loads are given in [13].

Research related to the analysis of reactive power in power networks supplying nonlinear fast-varying loads is a new issue. The latest research publications related to time-varying loads [14–18] concern load variability during the day, taking into account the impact of renewable energy sources. The load changes analyzed in these works do not cause the generation of interharmonics.

Previous research on the problem of determining, calculating, and compensating the reactive power of nonlinear receivers takes into account the influence of higher harmonic sources. So far, there have been no methods developed for calculating and compensating reactive power in the presence of interharmonics. Therefore, the presented work fills a gap in this research topic.

Solving the problem of reactive power compensation in electrical networks with interharmonic sources is the main goal of this study. Therefore, when analyzing the reactive power consumption and compensation of these loads, the random nature of load changes should be taken into consideration.

2. Determination of Reactive Power in Nonlinear Electrical Circuits

Harmonic analysis is often used in the calculation of nonlinear electrical circuits. Based on the expansion of current and voltage waveforms into a Fourier series, many methods of determining reactive power have been developed.

In [19] it is proposed to define reactive power in the form of the Reimann integral:

$$Q = \frac{1}{2\pi} \int_0^T u(t) \frac{di(t)}{dt} dt \quad (2)$$

Based on expression (2), one can obtain an equation determining the reactive power with distorted voltage and current waveforms, which generally can be presented in the form of the following relationships:

$$u(t) = \sum_{h=1}^{\infty} U_{mh} \sin(h\omega t + \alpha_h), \quad (3)$$

$$i(t) = \sum_{h=1}^{\infty} I_{mh} \sin(h\omega t + \beta_h), \quad (4)$$

where h is the harmonic number, and U_{mh}, I_{mh}, α_h, β_h are the amplitudes and initial phases of harmonic components of the order h of voltage and current, respectively, ω is the fundamental angular frequency.

Substituting these expressions into Equation (2) we obtain, provided that Parseval's theorem [20] is satisfied, the following relationship:

$$\begin{aligned} Q = \frac{1}{2\pi}\int_0^T \sum_{h=1}^{\infty} U_{mh} \sin(h\omega t + \alpha_h) \sum_{h=1}^{\infty} hI_{mh} \cos(h\omega t + \beta_h)dt \\ = \sum_{h=1}^{\infty} hU_h I_h \sin\varphi_h \end{aligned}, \quad (5)$$

where U_h, I_h are the RMS values of the h-th harmonics of voltage and current and $\varphi_h = \alpha_h - \beta_h$ is the phase shift angle between the waveforms of the h-th harmonics of voltage and current.

3. Reactive Power Compensation of Nonlinear Loads

In linear circuits (with sinusoidal voltage and current waveforms), the power of the compensation device Q_k should be equal to the reactive power of the load Q_o with the opposite sign:

$$Q_k = -Q_o = -UI\sin\varphi, \quad (6)$$

where U is the RMS value of voltage, I is the RMS value of load current, φ is the phase shift angle between voltage and current waveforms.

Under conditions of distortion of voltage and current waveforms, i.e., when the network operates with a nonlinear load, determining the reactive power of compensation devices in accordance with Equation (6) is incorrect (inaccurate) [21–25]. To analyze electromagnetic processes related to the exchange of energy between the source and the load in nonlinear systems, the concept of instantaneous reactive power should be used, just as instantaneous voltage and current waveforms (or their frequency spectra) are used, instead of their RMS values [19,26].

When supplying nonlinear loads (e.g., power electronic converters), in addition to the reactive power resulting from the presence of reactance elements (capacitance and inductance), there may be distortion power, i.e., a reactive power component associated with the phase shift of the current in relation to the voltage forced by the converter control systems, or associated with the distortion of current and voltage waveforms [20].

However, when solving compensation problems, there is no need to separate the reactive power into these two components. This is due to the fact that the effects of reactive power are the same, despite the significantly different physical nature of the origin of the two components. The reactive power of the distortion can be compensated by LC elements (passive filters for higher harmonics) and the reactive power resulting from the reactance of the loads by means of power electronics (APF active filters, STATCOM static compensators). In addition, classic solutions such as capacitor banks or synchronous compensators can be used for reactive power compensation. The considerations included in this study apply to any type of compensator, with particular emphasis on the practical application of capacitor banks.

The problem of selecting the parameters of compensation devices, from the point of view of minimizing energy losses in the power supply network, can basically be reduced to determining the minimum RMS value of the network current, which is determined by the sum of the instantaneous currents of the load $i_o(t)$ and the compensation device $i_k(t)$. Therefore, the following condition should be met:

$$\int_0^T (i_o(t) + i_k(t))^2 dt \to \min. \tag{7}$$

The minimum square of the effective value of the current in the supply network can be determined by equating the partial derivatives with respect to the parameters of the compensation devices (expression (7)) to zero. If a capacitor bank is used as a compensation device, then its capacitance C can be determined from the following equation:

$$\frac{d}{dC}\left[\frac{1}{T}\int_0^T \left(i_o(t) + C\frac{du(t)}{dt}\right)^2 dt\right] = 0, \tag{8}$$

the solution of which has the form:

$$C = -\frac{\frac{1}{T}\int_0^T u'(t)i(t)dt}{\frac{1}{T}\int_0^T (u'(t))^2 dt}, \tag{9}$$

where

$$u'(t) = \frac{du(t)}{dt}. \tag{10}$$

Then the power of the capacitor bank, determined at its rated voltage U_N, is

$$Q_k = U_N^2 \omega C = -\frac{U_N^2 \omega \frac{1}{T}\int_0^T u'(t)i(t)dt}{\frac{1}{T}\int_0^T (u'(t))^2 dt}. \tag{11}$$

If the voltage $u(t)$ and the load current $i_o(t)$ are determined by expressions (3) and (4), then the relationship (11) takes the form

$$Q_k = U_N^2 \frac{\sum_{h=1}^{\infty} hU_h I_h \sin\varphi_h}{\sum_{h=1}^{\infty} h^2 U_h^2}. \tag{12}$$

4. Reactive Power Compensation in Networks with Fast-Varying Nonlinear Loads

The analysis of reactive power compensation in a network with interharmonics was carried out for the equivalent network model shown in Figure 1. In this system, a nonlinear fast-changing load is a source of higher harmonics and interharmonic currents.

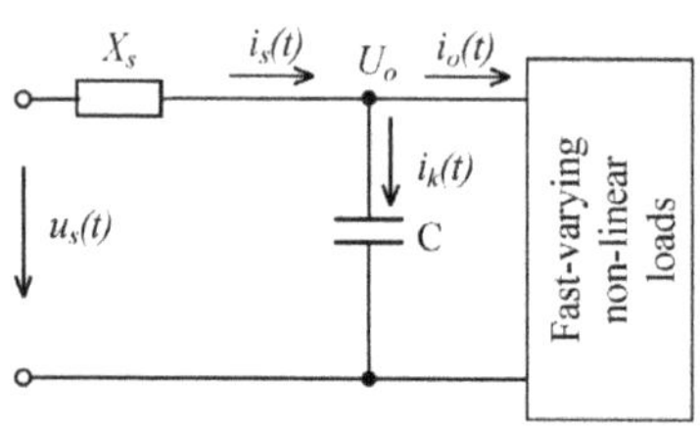

Figure 1. Reactive power compensation in a network with a nonlinear fast-varying load.

An example of a load that generates both harmonics and interharmonics is a nonlinear fast-variable load with an amplitude-modulated current waveform, in which the amplitude and initial phase change randomly [27–33].

$$i_o(t) = (\xi(t) + 1) \sum_{h=1}^{\infty} I_{mh} \sin(h\omega t + \varphi_h) = (\xi(t) + 1) i(t), \tag{13}$$

where $\xi(t)$ is the centered stationary random process with zero expected value and a given correlation function, I_{mh} are the constant amplitudes of the harmonic current components, φ_h are the mutually independent random variables of the initial phase of the harmonic components, evenly distributed in the interval $(-\pi, \pi)$, $i(t) = \sum_{h=1}^{\infty} I_{mh} \sin(h\omega t + \varphi_h)$ load current containing fundamental and higher harmonics.

The autocorrelation function of the random nonlinear load current process has the following form:

$$K_{i_o}(\tau) = \left(K_{\xi}(\tau) + 1\right) \sum_{h=1}^{\infty} D_h \cos(h\omega\tau), \tag{14}$$

where $K_{\xi}(\tau)$ is the given autocorrelation function of the modulating random process $\xi(t)$; D_h is the variance of the h-th harmonic is equal to $D_h = I_{mh}^2/2$.

The modulating random process can in most cases be characterized by one of three types of autocorrelation functions:

- exponential

$$K_{\xi}(\tau) = D_{\xi} e^{-\alpha|\tau|}, \tag{15}$$

- exponential-cosine

$$K_{\xi}(\tau) = D_{\xi} e^{-\alpha|\tau|} \cos \omega_0 \tau, \tag{16}$$

- exponential-cosine-sine

$$K_{\xi}(\tau) = D_{\xi} e^{-\alpha|\tau|} \left(\cos \omega_0 \tau + \frac{\alpha}{\omega_0} \sin \omega_0 |\tau| \right), \tag{17}$$

where D_{ξ} is the variance of current oscillations; α is the damping factor of the autocorrelation function; ω_0 is the angular frequency of the autocorrelation function.

The choice of a specific type of autocorrelation function depends on the problem being solved. Thus, when solving several problems related to the phenomenon of heating of wires and current-carrying parts of electrical devices when a non-sinusoidal current flows, the autocorrelation function in most cases can be represented using any of the three types (15)–(17). In such a case, a probabilistic assessment of the heating processes of current-carrying parts can be performed on the basis of the spectral current density, determined as the Fourier transform of the autocorrelation function. When modeling random processes to study heating phenomena, it is possible to replace one type of autocorrelation function with another, because the determining factor in this case is the energy of the random process, which is related to the area under the autocorrelation function curve. For example, when replacing an exponential-cosine autocorrelation function of the form (16) with an exponential one of the form (15), its damping coefficient will be determined by the expression found in the equality of the areas under the curves of autocorrelation functions (15) and (16) [26].

$$\alpha_e = \alpha + \frac{\omega_0^2}{\alpha}, \tag{18}$$

where α and ω_0 are the parameters of the original exponential-cosine autocorrelation function.

When solving problems directly related to the assessment of electromagnetic compatibility parameters during the operation of fast-changing, nonlinear loads, it is also possible to use any of the three autocorrelation functions (15)–(17) to simulate the random process of changing the load current.

As an example, Figure 2a shows one of the implementations of the current curve of phase A of the arc furnace for melting EAF-100 steel, which is an amplitude-modulated oscillation in accordance with the expression (13). The modulating random process $\xi(t)$ is given by an exponential autocorrelation function (15), where the variance of the current oscillations is $D_{\xi} = 9331.2\ \mathrm{A}^2$. Figure 2b shows the corresponding frequency spectrum of the phase A network current curve of the EAF-100 furnace, obtained using a fast Fourier transform directly from the simulated current curve. Figure 3a shows one of the implementations of the current curve of phase A of the EAF-100 furnace, which is an amplitude-modulated oscillation, where the random modulation process $\xi(t)$ is determined by an exponential-cosine autocorrelation function of the form (16) with the parameters: $D_{\xi} = 9331.2\ \mathrm{A}^2$, $\alpha = 1.47\ \mathrm{s}^{-1}$, and $\omega_0 = 2.69\ \mathrm{s}^{-1}$. Figure 3b shows the corresponding amplitude spectrum. Figure 4a shows the implementation of the phase A current curve of the EAF-100 furnace, which is an amplitude-modulated oscillation, where the random modulation process $\xi(t)$ is determined by the exponential-cosine-sine autocorrelation function of the form (17), with the same parameters as in the case of the exponential-cosine autocorrelation function. Figure 4b shows the corresponding amplitude spectrum.

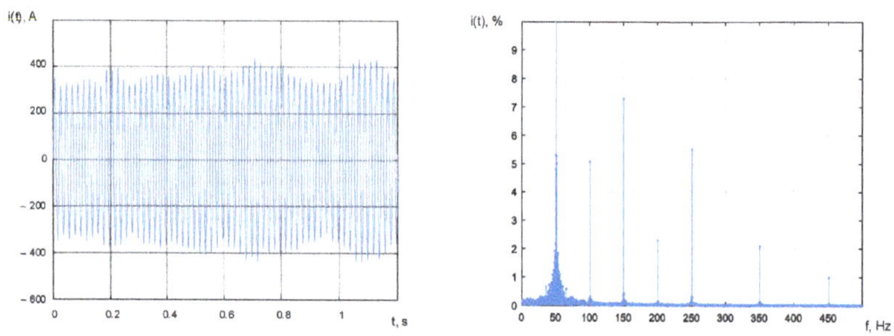

(a) (b)

Figure 2. Modeled phase A current curve of the EAF-100 furnace (**a**) and the corresponding amplitude spectrum (**b**) in the case of the exponential autocorrelation function of the modulating random process.

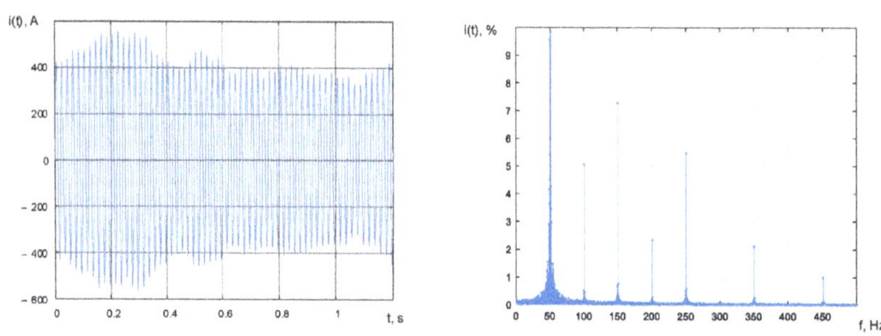

(a) (b)

Figure 3. Modeled phase A current curve of the EAF-100 furnace (**a**) and the corresponding amplitude spectrum (**b**) in the case of the exponential-cosine autocorrelation function of the modulating random process.

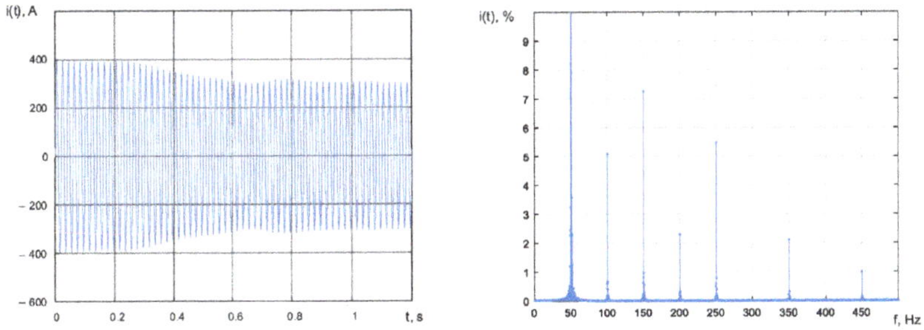

(a) (b)

Figure 4. Modeled phase A current curve of the EAF-100 furnace (**a**) and the corresponding amplitude spectrum (**b**) in the case of the exponential-cosine-sine autocorrelation function of the modulating random process.

As a result of numerous studies, it was found that for the given parameters of the random process, the expected value of the current curve distortion coefficient $E[K_I]$ is

- $E[K_I] = 15.7\%$, when using the exponential autocorrelation function;
- $E[K_I] = 17.3\%$, when using the exponential-cosine autocorrelation function;
- $E[K_I] = 15.6\%$, when using the exponential-cosine-sine autocorrelation function.

Therefore, the error in estimating the current curve distortion factor when using different types of autocorrelation functions to simulate the same random process of changing the arc furnace load current did not exceed 10%. This result allows us to conclude that when conducting various tests simulating the random process of changing the load current of an arc furnace, it is possible to use any type of autocorrelation function (15)–(17), depending on the scope of the problems being solved.

When examining the problem of electromagnetic compatibility during the operation of arc furnaces for steel melting by modeling random processes of changes in load currents, one type of autocorrelation function was replaced by another based on the equality of areas under the autocorrelation function curves.

In general, such an approach is in principle unacceptable. This is due to the fact that the factor determining the value of the distortion coefficient of the current waveform is not the energy of the random process of changing load current, but its speed of change, which is characterized by the parameters of the autocorrelation function. The test results have shown that as the damping coefficient α of the angular frequency of the autocorrelation function ω_0 increases, the distortion coefficient of the current curve also increases. In such a case, replacing the exponential-cosine autocorrelation function in the form (15) with an exponential function in the form (14) and determining the equivalent damping coefficient α_e using the expression (18) may lead to an overestimation of the value of the distortion coefficient of the current curve.

5. Calculation of Compensating Device Parameters for the Random Nature of Electrical Loads

Let us consider the electrical system shown in Figure 1 for the case without reactive power compensation. The RMS value of the voltage at the point of connection of the variable load is defined as the difference between the network voltage and the voltage drop across the network impedance:

$$U_o = U_s - I_s X_s \sin\varphi, \tag{19}$$

where φ is the load phase shift angle, and I_s is the RMS value of the network current, which in the absence of compensation is equal to the load current determined similarly to the instantaneous waveform (13):

$$I_s = I_o = (\xi(t) + 1)I, \tag{20}$$

Then, the RMS value of the voltage on the load is

$$U_o = U_s - (\xi(t) + 1)IX_s \sin\varphi = U_s - IX_s \sin\varphi - \xi(t)IX_s \sin\varphi. \tag{21}$$

In expression (21), the difference between the first two components is the effective value of the voltage at the point of load connection, without taking into account the random process modulation of the current waveform, while the network reactance can be determined as

$$X_s = \frac{U^2}{S''_k}, \tag{22}$$

where S''_k is the short-circuit power of the network.

Then, taking into account relation (6), we obtain

$$U_o = U - \xi(t)I\frac{U^2}{S''_k}\sin\varphi = \left(1 - \xi(t)\frac{Q_o}{S''_k}\right)U, \tag{23}$$

where Q_o is the reactive power of the load.

Moving to instantaneous voltage values, an expression can be written for the voltage at the load connection point, taking into account the modulation $\xi(t)$ of the load current curve:

$$u_o(t) = \left(1 - \xi(t)\frac{Q_o}{S''_k}\right)u(t) = (1 - a\xi(t))u(t), \tag{24}$$

where $a = Q_o/S''_k$.

Let us consider the issue of reactive power compensation for the case when a compensating device in the form of a capacitor bank is connected in parallel with a nonlinear fast-changing load (Figure 1).

As in the case of reactive power compensation in a network with higher harmonics, it is necessary to meet condition (7), in which the network current is equal to

$$i_s(t) = i_o(t) + i_k(t), \tag{25}$$

where $i_k(t)$ is the current flowing through a capacitor bank with capacity C,

$$i_k(t) = C\frac{du(t)}{dt} = C\frac{d(1 - a\xi(t))u(t)}{dt} = C\left[(1 - a\xi(t))u'(t) - a\xi'(t)u(t)\right], \tag{26}$$

wherein

$$\xi'(t) = \frac{d\xi(t)}{dt}. \tag{27}$$

Taking into account expressions (13), (25) and (26) we obtain

$$i_s(t) = (\xi(t) + 1)i(t) + C\left[(1 - a\xi(t))u'(t) - a\xi'(t)u(t)\right] \tag{28}$$

The square of the RMS value of the current in the supply network is

$$I_s^2 = \frac{1}{T}\int_0^T [\xi(t) + 1]^2 [i(t)]^2 dt + \frac{1}{T}\int_0^T 2C\,[\xi(t) + 1]\,i(t)\left\{[1 - a\xi(t)]\,u'(t) - a\xi'(t)u(t)\right\}dt + \frac{1}{T}\int_0^T C^2\left\{[1 - a\xi(t)]\,u'(t) - a\xi'(t)u(t)\right\}^2 dt \tag{29}$$

and the derivative of the square of the current with respect to the capacitance C takes the following form:

$$\frac{dI_s^2}{dC} = \frac{2}{T}\int_0^T [\xi(t) + 1]\,i(t)\left\{[1 - a\xi(t)]\,u'(t) - a\xi'(t)u(t)\right\}dt + \frac{2C}{T}\int_0^T \left\{[1 - a\xi(t)]\,u'(t) - a\xi'(t)u(t)\right\}^2 dt\,. \tag{30}$$

By equating the derivative $\frac{dI_s^2}{dC}$ to zero, we can determine the capacity C and then the power of the capacitor bank:

$$Q_k = U_N^2 \omega C = U_N^2 \omega \frac{\int_0^T (\xi(t)+1)(i(t))[(1-a\xi(t))u'(t) - a\xi'(t)u(t)]dt}{\int_0^T [(1-a\xi(t))u'(t) - a\xi'(t)u(t)]^2 dt}, \tag{31}$$

where U_N is the rated voltage of the capacitor bank.

The expected value of the capacitor bank power can be determined using the random process linearization method:

$$EQ_k = U_N^2 \omega \frac{\int_0^T E[(\xi(t)+1)(u'(t) - a\xi(t)u'(t) - a\xi'(t)u(t))]i(t)dt}{\int_0^T E[(u'(t) - a\xi(t)u'(t) - a\xi'(t)u(t))]^2 dt}. \tag{32}$$

Taking into consideration that the expected values of the random processes $\xi(t)$ and $\xi'(t)$ are equal to zero ($E[\xi(t)] = 0$ and $E[\xi'(t)] = 0$) and that the coefficient of their mutual autocorrelation $K_{\xi\xi'} = 0$, we get

$$EQ_k = \frac{U_N^2 \omega (1 - aD_\xi) \int_0^T i(t)u'(t)dt}{(1 + a^2 D_\xi) \int_0^T u'(t)^2 dt + \frac{a^2 D_{\xi'}}{\omega^2} \int_0^T u(t)^2 dt}, \tag{33}$$

where D_ξ is the variance of the centered stationary random process $\xi(t)$, and $D_{\xi'}$ is the variance of the derivative of this process.

Substituting into this expression the expansion of current and voltage into the Fourier series (3) and (4), we obtain:

$$EQ_k = \frac{U_N^2 (1 - aD_\xi) \sum_{h=1}^{\infty} hU_h I_h \sin\varphi_h}{(1 + a^2 D_\xi) \sum_{h=1}^{\infty} h^2 U_h^2 + \frac{a^2 D_{\xi'}}{\omega^2} \sum_{h=1}^{\infty} U_h^2}. \tag{34}$$

The requirement for the differentiation of the random process $\xi(t)$ is the continuity of the derivative of its autocorrelation function around the point $\tau = 0$. This requirement is met by the exponential-cosine-sine autocorrelation function in the following form:

$$K_\xi(\tau) = D_\xi e^{-\alpha|\tau|} \left(\cos\omega_0\tau + \frac{\alpha}{\omega_0} \sin\omega_0|\tau| \right), \tag{35}$$

and hence the variance of the derivative of the process $\xi(t)$:

$$D_{\xi'} = -\frac{d^2 K_\xi(\tau)}{d\tau^2}\bigg|_{\tau=0} = D_\xi\left(\alpha^2 + \omega_0^2\right) e^{-\alpha|\tau|} \left(\cos\omega_0\tau + \frac{\alpha}{\omega_0} \sin\omega_0|\tau| \right)\bigg|_{\tau=0} = D_\xi\left(\alpha^2 + \omega_0^2\right) \tag{36}$$

Substituting expression (36) into Equation (34), we obtain

$$EQ_k = \frac{U_N^2 (1 - aD_\xi) \sum_{h=1}^{\infty} hU_h I_h \sin\varphi_h}{(1 + a^2 D_\xi) \sum_{h=1}^{\infty} h^2 U_h^2 + \frac{a^2 D_\xi (\alpha^2 + \omega_0^2)}{\omega^2} \sum_{h=1}^{\infty} U_h^2}. \tag{37}$$

6. Calculation Results

Let us consider an example of calculating the power of the capacitor bank needed to compensate for reactive power during the operation of a steel arc furnace EAF-100 with the following parameters: rated power of the transformer S_{NT} = 45 MV·A, rated voltage of the primary winding of the transformer U_N = 35 kV, rated current of the furnace I_N = 400 A. Let us consider three cases.

6.1. Calculation of Capacitor Bank Power Based on the Fundamental Harmonic of Current and Voltage

In this case, we use the furnace ratings as the RMS values of the fundamental harmonics of current and voltage. The value of the phase angle was determined experimentally and was $\varphi_1 = 26°$. Then, the capacity of the capacitor bank can be defined as

$$Q_k = U_1 I_1 \sin\varphi_1 = 35 \cdot 0.4 \cdot \sin 26° = 6.137 \text{ Mvar}. \tag{38}$$

6.2. Calculation of the Power of Capacitor Banks Taking into Account Higher Harmonics

Table 1 contains the relative values of higher harmonics of current and voltage and the corresponding phase angles obtained experimentally on the basis of the measurements made during the operation of the EAF-100 arc furnace in question at one of the metallurgical enterprises (Mariupol, Ukraine). The measurements were performed using the Power Quality analyzer Fluke 435.

Table 1. Parameters of higher harmonics of current and voltage measured during the operation of the arc furnace EAF-100.

Parameter	Harmonic Number of Order h							
	2	3	4	5	6	7	8	9
I_h/I_1, %	5.1	7.2	2.3	5.5	0	2.1	0	1.0
U_h/U_1, %	2.5	3.6	1.2	3.0	0	1.0	0	0.5
φ_h, °	87	−89	52	−55	0	98	0	−90

To calculate the power of the compensating device, taking into account higher harmonics, the source of which is the EAF-100 arc furnace, we use the expression (12). As a result of the calculations, we get

$$Q_k = 5.734 \text{ Mvar}. \tag{39}$$

6.3. Calculation of the Power of Capacitor Banks Taking into Account Higher Harmonics and Interharmonics

In this case, it is necessary to estimate the expected value of the capacitor bank capacity according to Equation (35). Let us assume that in order to take into account interharmonics, the EAF-100 current curve is modulated by a random process with an exponential-cosine autocorrelation function with the following parameters: oscillation current variance D_ξ = 9331.2 A^2, damping factor of the autocorrelation function α = 1.47 s^{-1}, angular frequency ω_0 s^{-1}. The parameters of the autocorrelation function of the modulating random process were obtained experimentally using measurement data for the melting phase of the EAF-100 arc furnace.

Then, as a result of calculations according to relation (31), we obtain

$$EQ_k = 5.71 \text{ Mvar}. \tag{40}$$

For most practical cases $0 < a < 0.2$; $0 < D_\xi < (0.5)^2 = 0.25$; $\alpha << \omega$, $\omega_0 << \omega$, so expression (38) can be simplified to

$$EQ_k = \left(1 - aD_{\xi}\right)U_N^2 \frac{\sum_{h=1}^{\infty} hU_h I_h \sin\varphi_h}{\sum_{h=1}^{\infty} h^2 U_h^2}. \tag{41}$$

The comparison of Equations (12), (38) and (41) shows directly that in order to minimize energy losses in the power supply network, the load variability should be taken into account in the process of selecting the power of reactive power compensation devices. In the case considered in this study, when the compensator is a capacitor bank, its power will be lower when taking into account the rapidly changing nature of the load.

7. Conclusions

The research conducted shows that in power networks supplying rapidly changing nonlinear loads causing an occurrence of interharmonics in current and voltage waveforms, a power correction of reactive power compensating devices is required.

When examining non-sinusoidal conditions in electrical networks with rapidly changing nonlinear loads, the non-sinusoidal current curve can be represented as an amplitude modulated waveform with randomly varying amplitude and initial phase.

The use of a centered stationary random process with a given autocorrelation function as a modulating signal allows us to obtain analytical expressions for calculating the probabilistic current characteristics of fast-varying loads and parameters of reactive power compensating devices.

When solving problems related to both the phenomenon of the heating of wires and parts of electrical devices carrying current during the flow of non-sinusoidal current, as well as the assessment of electromagnetic compatibility parameters during the operation of rapidly changing nonlinear loads, the autocorrelation function of a modulating random process can be represented by any of the three types according to expressions (15)–(17).

In the study of processes related to the heating of wires and current-carrying parts of electrical devices and energy losses, it is possible to replace one type of autocorrelation function modulating a random process with another; however, when examining the problem of electromagnetic compatibility, such an exchange turns out to be unacceptable. When solving problems related to reactive power compensation in electrical networks with rapidly changing nonlinear loads, the autocorrelation function of the modulating random process must have a continuous derivative around the point $\tau = 0$ to ensure the condition of differentiability of the random process. This requirement corresponds to the autocorrelation function in expression (17).

If a capacitor bank is used as a compensation device, its power, determined by the condition of minimal energy losses in the power supply network, should be lower than without, taking into consideration the rapidly changing nature of the load and interharmonics.

In the calculation example presented herein, taking into account higher harmonics and interharmonics in the voltage and current waveforms of unsettled receivers, the power of the capacitor bank was reduced by approximately 7%, which allows for reduced investment costs. An additional effect is the reduction of energy losses in the power supply network, estimated at 3–5%.

This study is the first step in solving the problem of reactive power compensation in electrical networks containing interharmonics. The analysis concerns the use of capacitor banks as compensating devices for reactive power in rapidly changing loads, such as arc furnaces and welding equipment.

The authors plan further research in two directions: The first direction is the use of passive filter-compensating devices, active filters, and STATCOM compensators to compensate the reactive power of fast-varying loads and to improve power quality.

The second direction results from the fact that the sources of interharmonics are also various types of power electronic converters, especially frequency converters used in speed-controlled drive systems. The completely different nature of interharmonic generation in such loads requires a different approach to the methods for calculating and compensating reactive power.

Author Contributions: Conceptualization, Y.S., R.P. and T.B.; methodology, Y.S. and R.P.; validation, Y.S., R.P. and T.B.; investigation, Y.S., R.P. and T.B.; simulation, Y.S. and T.B.; writing—original draft preparation, Y.S. and R.P.; writing—review and editing, Y.S. and R.P.; visualization, Y.S. and T.B.; supervision, Y.S. All authors have read and agreed to the published version of the manuscript.

Funding: This research received no external funding.

Data Availability Statement: Data are contained within the article.

Conflicts of Interest: The authors declare no conflict of interest.

References

1. Budeanu, C.I. *Puissances Reactives et Fictives*; Instytut Romain de l'Energie: Bucharest, Romania, 1927; pp. 1–360.
2. Fryze, S. Moc rzeczywista, urojona i pozorna w obwodach elektrycznych o przebiegach odkształconych prądu i napięcia (Part I). *Prz. Elektrotech.* **1931**, *XIII*, 194–203. Available online: https://www.dbc.wroc.pl/dlibra/publication/15119/edition/13678/content?ref=struct (accessed on 1 January 2008).
3. Fryze, S. Moc rzeczywista, urojona i pozorna w obwodach elektrycznych o przebiegach odkształconych prądu i napięcia (Part II). *Prz. Elektrotech.* **1931**, *XIII*, 226–234. Available online: https://www.dbc.wroc.pl/dlibra/publication/15120/edition/13679/content (accessed on 1 January 2008).
4. Shepherd, W.; Zakikhani, P. Suggested definition of reactive power for nonsinusoidal systems. *Proc. Inst. Elect. Eng.* **1972**, *119*, 1361–1362. [CrossRef]
5. Kusters, N.L.; Moore, W.J.M. On the definition of reactive power under non-sinusoidal conditions. *IEEE Trans. Power Appar. Syst.* **1980**, *PAS-99*, 1845–1854. [CrossRef]
6. Akagi, H.; Kanazawa, Y.; Nabae, A. Instantaneous Reactive Power Compensators Comprising Switching Devices without Energy Storage Components. *IEEE Trans. Ind. Appl.* **1984**, *IA-20*, 625–630. [CrossRef]
7. Czarnecki, L.S. Considerations on the Reactive Power in Nonsinusoidal Situations. *IEEE Trans. Instrum. Meas.* **1985**, *IM-34*, 399–404. [CrossRef]
8. Czarnecki, L.S. Physical interpretation of reactive power in terms of the CPC power theory. *Electr. Power Qual. Util. J.* **2007**, *XIII*, 89–95. Available online: www.epqu.agh.edu.pl/archives/journal/v13i1/v13i1_13.pdf (accessed on 1 April 2007).
9. Czarnecki, L.S. Currents' Physical Components (CPC) concept: A fundamental power theory. In Proceedings of the International School on Nonsinusoidal Currents and Compensation, Lagow, Poland, 10–13 June 2008; pp. 1–11. [CrossRef]
10. Canturk, S.; Balci, M.E.; Hocaoglu, M.H. On the Definition of Apparent Power. *Electr. Power Qual. Util. J.* **2015**, *18*, 1–9. Available online: www.epqu.agh.edu.pl/archives/journal/v18i2/v18i2_01.pdf (accessed on 1 July 2015).
11. Pawelek, R.; Gburczyk, P.; Wasiak, I. Analysis of current distortion of the unsteady non-linear loads. In Proceedings of the IEEE 13th International Conference on Harmonics and Quality of Power, Wollongong, NSW, Australia, 28 September–1 October 2008; pp. 1–6. [CrossRef]
12. *EN 61000-4-7*; Electromagnetic Compatibility (EMC). Part 4–7: Testing and Measurement Techniques–General Guide on Harmonics and Interharmonics Measurements and Instrumentation, for Power Supply Systems and Equipment Connected Thereto. CENELEC: Brussels, Belgium, 2002.
13. Bohorquez, V.B. Fast Varying Loads. In Proceedings of the 9th International Conference on Electrical Power Quality and Utilisation, Barcelona, Spain, 9–11 October 2007; pp. 1–6. [CrossRef]
14. Merlin Sajini, M.L.; Suja, S.; Merlin Gilbert Raj, S. Impact analysis of time-varying voltage-dependent load models on hybrid DG planning in a radial distribution system using analytical approach. *IET Renew. Power Gener.* **2021**, *15*, 153–172. [CrossRef]
15. Iqteit, N.A.; Arsoy, A.B.; Çakır, B. The random varying loads and their impacts on the performance of smart grids. *Electr. Power Syst. Res.* **2022**, *209*, 107960. [CrossRef]
16. Rohouma, W.; Balog, R.S.; Peerzada, A.A.; Begovic, M.M. Reactive Power Compensation of Time-Varying Load Using Capacitor-less D-STATCOM. In Proceedings of the 10th International Conference on Power Electronics-ECCE Asia, Busan, Republic of Korea, 27–30 May 2019; pp. 2296–2301. [CrossRef]
17. Dudek, G. A Comprehensive Study of Random Forest for Short-Term Load Forecasting. *Energies* **2022**, *15*, 7547. [CrossRef]
18. Li, X.; Li, H.; Li, S.; Jiang, Z.; Ma, X. Review on Reactive Power and Voltage Optimization of Active Distribution Network with Renewable Distributed Generation and Time-Varying Loads. *Math. Probl. Eng.* **2021**, *2021*, 1196369. [CrossRef]
19. Maevskiy, O.A. *Energy Indicators of Valve Converters (In Russian: Энергетические показатели вентильных преобразователей)*; USSR: Moscow, Russia, 1978; pp. 1–320.
20. Sayenko, Y.L. Reactive Power in Power Supply Systems with Non-Linear Loads (In Russian: Реактивная мощность в системах электроснабжения с нелинейными нагрузками). In *Zeszyty Naukowe Politechniki Śląskiej, Elektryka*; Wydawnictwo Politechniki Śląskiej: Gliwice, Poland, 1991; Volume 123, pp. 1–118. Available online: https://delibra.bg.polsl.pl/dlibra/publication/43377/edition/39423/content (accessed on 1 January 2010).

21. Mujumdar, B.; Joshi, J.S. Optimal reactive power compensation under non sinusoidal conditions using current minimization method. In Proceedings of the International Conference on Power and Energy Systems, Chennai, India, 22–24 December 2011; pp. 1–6. [CrossRef]
22. Jeon, S.J. Passive-component-based reactive power compensation in a non-sinusoidal multi-line system. *Electr. Eng.* **2020**, *102*, 1567–1577. [CrossRef]
23. Seong-Jeub, J. Considerations on a reactive power concept in a multiline system. *IEEE Trans. Power Deliv.* **2006**, *21*, 551–559. [CrossRef]
24. Nicolae, P.-M.; Nicolae, I.-D.; Nicolae, M.-Ş. Powers and Power Factor in Non-Sinusoidal and Non-Symmetrical Regimes in Three-Phase Systems. *Energies* **2022**, *15*, 5130. [CrossRef]
25. Balci, M.E.; Hocaoglu, M.H. Comparison of power definitions for reactive power compensation in nonsinusoidal conditions. In Proceedings of the 11th International Conference on Harmonics and Quality of Power, Lake Placid, NY, USA, 12–15 September 2004; pp. 519–524. [CrossRef]
26. Shepard, W.; Zakikhani, P. Power factor correction in nonsinusoidal systems by the use of capacitance. *J. Phys. D Appl. Phys.* **1973**, *6*, 1850–1861. [CrossRef]
27. Sayenko, Y.L.; Baranenko, T.K.; Sayenko, I.Y. The spectral correlation theory application of random processes for estimation of power losses involving higher harmonic and interharmonic. *Electrif. Transp.* **2017**, *13*, 129–133. [CrossRef]
28. Sayenko, Y.; Baranenko, T.; Kalyuzhniy, D. Compensation of reactive power in electrical supply systems of large industrial enterprises. *Prz. Elektrotech.* **2015**, *91*, 77–80. [CrossRef]
29. Chen, M.; Wen Lu, C. Statistic characteristic estimations of harmonic and flicker on electric arc furnace feeders. In Proceedings of the 9th Probabilistic Methods Applied to Power Systems Conference (PMAPS), Stockholm, Sweden, 11–15 June 2006; pp. 1–5. [CrossRef]
30. Collantes-Bellido, R.; Gomez, T. Identification and modelling of a three phase arc furnace for voltage disturbance simulation. *IEEE Trans. Power Deliv.* **1997**, *12*, 1812–1817. [CrossRef]
31. Cano Plata, E.A.; Ustariz Farfan, A.J.; Soto Marin, O.J. Electric arc furnace model in distribution systems. *IEEE Trans. Ind. Appl.* **2015**, *51*, 4313–4320. [CrossRef]
32. Samet, H.; Masoudipour, I.; Parniani, M. New reactive power calculation method for electric arc furnaces. *Measurement* **2016**, *81*, 251–263. [CrossRef]
33. Lozynskyi, O.Y.; Paranchuk, Y.S.; Paranchuk, R.Y.; Matico, F.D. Development of methods and means of computer simulation for studying arc furnace electric modes. *Electr. Eng. Electromech.* **2018**, *3*, 28–36. [CrossRef]

energies

MDPI

Article

Voltage Rise Mitigation in Medium-Voltage Networks with Long Underground Cables and Low Power Demand

Deni Ćetković [1,*], Josip Žutolija [2] and Vitomir Komen [1]

[1] Faculty of Engineering, University of Rijeka, 51000 Rijeka, Croatia; vkomen@riteh.hr
[2] HEP—Distribution System Operator, Elektroprimorje, 51000 Rijeka, Croatia; josip.zutolija@hep.hr
* Correspondence: dcetkovic@student.uniri.hr

Abstract: Medium-voltage (MV) distribution networks that are spread through larger territory and threatened by extreme weather conditions are sometimes formed by very long underground cable lines. In such circumstances, a significant amount of capacitive reactive power flow can be generated. If, concurrently, there is low power demand in the network, it can result in significant reverse reactive power flows and voltage rise issues. This paper proposes a general approach for analyzing and mitigating voltage rise issues and demonstrates it using an example of a real distribution network that operates under the described conditions. Previous studies that dealt with this problem did not include the allocation of multiple shunt reactors in a larger distribution network, modeling a high number of lines that create reverse reactive power flows, and modeling the main distribution transformers, which are the locations where voltage rise predominantly occurs. In this paper, we demonstrate that precise allocation and placement of multiple shunt reactors in a fully modeled, larger distribution system, including transformer models, can reduce reverse reactive power flows, thereby improving voltage in the distribution system. If hourly control of the power factor from the distributed generation unit is also implemented, the voltage can be further improved.

Keywords: voltage rise; long cable lines; shunt reactors; distributed generation power factor control; voltage profile improvement

1. Introduction

Medium-voltage overhead distribution networks that are spread through a large territory are vulnerable to severe weather conditions. A study [1] showed that in some places, weather-related outage events account for 78.09% of the total, with lightning, contributing 63.26%, and wind, contributing 12.27%, as the two main weather factors. Additionally, wildfires account for 2.14%, icing for 0.16%, and geological events for 0.11%.

In conditions where overhead line damage is frequent, medium-voltage networks are sometimes formed by long underground cable lines, ensuring enhanced reliability in the supply of electrical energy [2]. Electric power distribution networks were primarily developed for power flows directed towards consumers [3]. Due to the inductive characteristics of consumption, reactive power flows are commonly directed towards consumers, with variations in the inductive component roughly corresponding to fluctuations in electrical demand, especially when the medium-voltage network is predominantly implemented using overhead lines.

Underground medium-voltage cable lines generate reverse reactive power flows [4] toward the distribution network feeding point, where the quantity of reactive power flows increases with the length of the lines. In distribution networks with very low power demand and very long cable lines, significant reverse reactive power flows may occur, leading to voltage rises in the distribution network.

Standard indicators of voltage quality are defined by the EN 50160 standard [5]. The permissible voltage deviation is typically specified as ±10% of the nominal voltage. Voltage fluctuation limits are set to ensure that short-term flicker does not exceed 1.0 and long-term flicker remains below 0.8, as outlined in the same standard. Additionally, the standard specifies that voltage asymmetry between phases (voltage unbalance) should be $\leq$2%, ensuring balanced distribution across three-phase systems. Limits for non-sinusoidal voltages dictate that total harmonic distortion (THD) should not exceed 8%, maintaining the integrity of voltage waveforms and minimizing interference with electrical equipment.

The problems with voltage rise can result in the need for power curtailment of DG units [6]. The voltage rise issue is typically investigated in the context of the connection of distributed generation sources [7]. Papers related to voltage rises typically address the reduction in voltage rises resulting from reverse power flows caused by the connection of distributed generation units. In Ref. [8], the author proposed a phase shifting strategy to mitigate local voltage rises and reverse power flows caused by high photovoltaic penetration. In Ref. [9], the power curtailment of distributed generation is proposed to mitigate voltage rise. In Ref. [10], the authors considered the voltage rise issue during the distributed generation allocation process. In Ref. [11], the authors proposed the dynamic operation of a voltage regulator for voltage regulation in a distribution system with a significant integration of distributed generation. In Ref. [12], the authors proposed the usage of distributed energy storage for mitigation of the voltage rise impact caused by rooftop solar PV systems. In Ref. [13], the authors proposed a decentralized control strategy using droop-based control to adjust electric vehicle (EV) charging rates in real time based on nodal voltages, voltage unbalance, and state of charge (SOC) to mitigate voltage rise from rooftop solar PV systems. The paper [14] introduces a novel reactive power control technique tailored for low-voltage distribution networks with high photovoltaic (PV) system penetration, addressing voltage rise challenges by dynamically adjusting reactive power based on both PV active power injection and real-time network voltage conditions. It demonstrates superior voltage regulation and reduced losses compared to conventional fixed power factor and voltage-based methods, highlighting its adaptability to varying scenarios despite a potentially slightly slower response time. The paper [15] addresses voltage regulation challenges arising from increased distributed PV installations in distribution systems. It introduces a PID closed-loop-based volt-var (VV) function and a mode-switching function for PV smart inverters to enhance voltage regulation effectiveness. The paper [16] introduces a control strategy for grid-connected inverters interfacing with battery storage systems (BSS) to enhance PV penetration in low-voltage (LV) grids. The strategy optimizes inverter set points for active and reactive power based on grid voltage, considering battery specifications and state of charge (SoC). The paper [17] explores leveraging electric minibus taxi (EmBT) charging alongside residential solar photovoltaic (PV) generation to mitigate voltage rises in low-voltage (LV) networks with high penetration of distributed energy resources (DERs).

In contrast to the literature addressing voltage rises due to distributed generation units, the literature that investigates voltage rises caused by large capacitive currents in underground cables and variations in power demand is much less common. In Ref. [18], a reactive power control method using a variable shunt reactor and the reactive power control of a wind power plant was proposed to compensate for the capacitive reactive power of the cable. In Ref. [19], the authors addressed voltage issues resulting from substantial load variations by proposing the installation of shunt capacitor banks for voltage drop issues and shunt reactors (ShRs) for voltage rise issues. In Ref. [20], the authors deployed switchable shunt reactors in combination with DGs that have the capability of absorbing reactive power for voltage rise mitigation in a low-voltage distribution network with high penetration of photovoltaic DGs. The authors concluded that DGs have limited capacity to serve as a source of reactive power and are unable to provide sufficient reactive power for full voltage rise mitigation. In Ref. [21], the authors demonstrated that the shunt reactor can mitigate the power frequency voltage rise caused by the capacitance effect of long lines under no-load or light-load conditions in high-voltage systems. The paper [22] explores the voltage rise mitigation caused by minimal loading and excessive reactive power generated by line capacitance in magnetically controlled shunt reactors.

This paper focuses on analyzing voltage rise issues that are not primarily caused by DGs, but rather by very low power demand and a significant number of long underground cable lines that create a high amount of capacitive reactive power. The analysis was conducted on a real distribution network spread over two voltage levels. The network is located in Primorsko-Goranska County, Croatia. This distribution network is specific due to the very low power demand resulting from a significant population decrease, as well as a large number of long underground cable lines installed to withstand extreme weather conditions caused by the ice storm in 2014. A method for reducing voltage rise was introduced, involving the allocation of multiple fixed shunt reactors within the distribution network and the implementation of hourly power factor control for the existing distributed generation. The distribution system model contains two voltage levels of the distribution network with two distribution transformers that are connected between the 35 kV and 20 kV distribution networks. Modeling the distribution transformers and including them in the analysis of voltage rise issues is significant, as the majority of voltage rises occur in them. The analysis includes not only the 20 kV network and 35/20 kV transformers but also the feeder lines of the 35 kV network, which are expected to contribute to voltage rise issues due to their length and characteristics. The main 110/35 kV transformer is not included in the analysis since it has an on-load tap changer so the voltage is fixed to a set value of 1.05 p.u.

The main innovations introduced in this paper are summarized in four points:

(1) This paper proposes a novel general approach for analyzing and mitigating voltage rise issues where the entire system is accurately modeled and various elements that contribute to voltage rises are unified into a single method for the mitigation of voltage rises on multiple voltage levels of a distribution network.

(2) One of the novel aspects of our approach lies in the inclusion of main distribution transformers in the analysis of the voltage rise issue. The voltage rise issue is analyzed with the modeling of distribution transformers. Earlier research addressing this issue overlooked the incorporation of main distribution transformers, which are pivotal sites where voltage rise predominantly occurs.

(3) Previous studies on this topic typically incorporated just one shunt reactor and a limited number of distribution lines. In contrast, our paper introduces a method for addressing voltage rise by strategically allocating multiple shunt reactors and implementing hourly power factor control for a distributed generation unit within extensive distribution networks featuring a high volume of lines generating reactive power flows. The allocated multiple shunt reactors simultaneously influence the reduction of voltage rise both in the lines and in the transformers.

(4) The proposed method also unifies and accurately models the components of distribution lines responsible for creating reverse flows of capacitive reactive power, particularly the parallel branches that inject capacitive reactive power. The voltage rise issue is analyzed with the modeling of lines on two voltage levels of the distribution network.

Despite the relatively small size of the higher-level network, we demonstrated the feasibility of mitigating voltage rises in distribution lines of both voltage levels of the distribution network, as well as in the transformers that serve as connections between these voltage levels, by using shunt reactors and hourly power factor control of DGs.

2. Materials and Methods

The main objective of optimization is to mitigate voltage rise issues caused by low power demand and significant reverse power flows generated by underground cables in the distribution system. This is achieved by the allocation of up to eight fixed shunt reactors and the hourly power factor control of the existing distributed generation.

2.1. Objective Function for Voltage Rise Mitigation

The mitigation of voltage rises will be achieved by defining the objective function for voltage profile improvement across all distribution network nodes and time intervals, which can be mathematically expressed as follows:

$$Min\ f = \sum_{t=1}^{Nt} \sum_{i=1}^{Nn} \left(U_n - U_i{}^t\right)^2 \quad (1)$$

In Equation (1), N_n and N_t represent the total number of nodes and time intervals, while $U_i{}^t$ represents the ith node voltage at interval t. U_n represents the desired system voltage in the distribution network. The nominal voltage is as follows:

$$U_n = 21000 < 0^\circ\ V \quad (2)$$

2.2. Constraints

2.2.1. Positions of Shunt Reactors

The algorithm is permitted to suggest up to eight new shunt reactors, which can be connected to nodes on the secondary side of the 35/20 kV transformers; the nodes satisfy the following constraints, respectively:

$$(3 \leq \text{ShRs\_position} \leq 42) \vee (44 \leq \text{ShRs\_position} \leq 71) \quad (3)$$

2.2.2. Installed Power of Shunt Reactors

The algorithm is permitted to suggest the following installed reactive power of one ShR unit:

$$Q_{ShR} = n \cdot 100\ [\,kVAr] \quad (4)$$

In Equation (4), n represents the factor that must satisfy the constraint (5):

$$n_{min} \leq n \leq n_{max} \quad (5)$$

In this paper, n_{min} was set to a value of 0, and n_{max} was set to a value of 11.

2.2.3. Permitted Power Factor of Existing DG

Various types of DG generators have different operational constraints for the capability of reactive power generation and absorption [23]. Since the DG in the presented network is hydropower, the synchronous generator is considered. Therefore, the permitted power factor of the existing DG used in this paper ranges from 0.85 lagging to 1.0 p.u. and from 0.95 leading to 1.0 p.u. in steps of 0.005.

2.2.4. Current Magnitudes

The magnitude of the current that flows through all lines and transformers must satisfy the following constraint:

$$I_i{}^t \leq I_n \quad (6)$$

In Equation (6), $I_i{}^t$ stands for the current of the ith branch in the time interval t, and I_n is the nominal current of each branch.

2.3. Power Demand Modeling and Distribution System Model Used for Voltage Rise Issue Analysis

The analysis was conducted using a model based on a segment of the actual distribution network supplied from the transformer station 110/35 kV Delnice. From this substation, two other substations are taken for analysis. The substations are supplied through two separate feeders.

The distribution system model is shown in Figure 1. The substation TS 35/20 kV Kupjak is supplied through an underground cable, while the substation TS 35/20 kV Gerovo is supplied through a 35 kV overhead line. The transformer station 110/35 kV

Delnice has a transformer installed with an on-load tap changer, which sets the secondary side voltage at 36.75 kV, corresponding to 1.05 p.u. Therefore, the voltage in the 20 kV distribution networks is also initially set to 21 kV, corresponding to 1.05 p.u. The voltage is set to 1.05 p.u. because a lower set voltage could result in excessively low voltages in some other parts of the distribution networks that are also supplied from TS 110/35 kV Delnice. Therefore, the voltage sought to be achieved by objective function (1) during the mitigation of excessive voltage rises is 1.05 p.u.

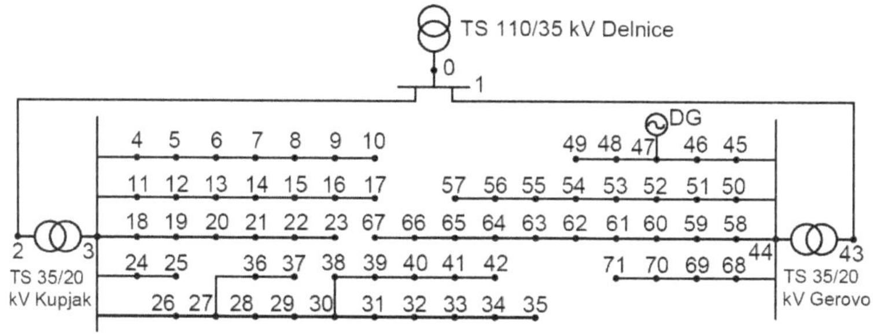

Figure 1. Distribution system model used for analysis.

In addition to the two transformers, the model of the network includes 69 distribution lines with a total length of 252 km, the majority of which are underground cable lines.

Power demand is modeled in nodes 4–42 and 45–71, based on maximum active and reactive power in each node and real measured data for overall daily power demand in the presented network. The maximum power in each node is shown in Table 1.

Table 1. Maximum active and reactive power in nodes that create power demand.

Node	Maximum Active Power (kW)	Maximum Reactive Power (kVAr)	Node	Max. Active Power (kW)	Max. Reactive Power (kVAr)	Node	Max. Active power (kW)	Max. Reactive Power (kVAr)
4	160	52.59	24	350	115.04	46	450	147.91
5	370	121.61	25	250	82.17	47	1090	358.27
6	630	207.07	26	150	49.30	48	1030	338.54
7	1250	410.86	27	210	69.02	49	850	279.38
8	650	213.64	28	630	207.07	50	150	49.30
9	350	115.04	29	850	279.38	51	150	49.30
10	460	151.19	30	650	213.64	52	150	49.30
11	350	115.04	31	150	49.30	53	50	16.43
12	630	207.07	32	150	49.30	54	300	98.61
13	630	207.07	33	250	82.17	55	100	32.87
14	1250	410.86	34	350	115.04	56	400	131.47
15	650	213.64	35	610	200.50	57	50	16.43
16	200	65.74	36	760	249.80	58	200	65.74
17	150	49.30	37	500	164.34	59	410	134.76
18	1030	338.54	38	300	98.61	60	810	266.23
19	150	49.30	39	300	98.61	61	790	259.66
20	200	65.74	40	100	32.87	62	320	105.18
21	450	147.91	41	150	49.30	63	510	167.63
22	460	151.19	42	150	49.30			
23	500	164.34	45	880	289.24			

The normalized power demand curves are shown in Figure 2. To obtain these curves, the active power demand was measured in TS Gerovo and TS Kupjak in the period from 22 January 2023 to 21 July 2023. This six-month period is well suited for analysis as it includes both high and low consumption periods, making it a sufficiently robust dataset for representing the method. The data was obtained from the SCADA system, which provides the average active power consumption for each hour for all days within the specified period.

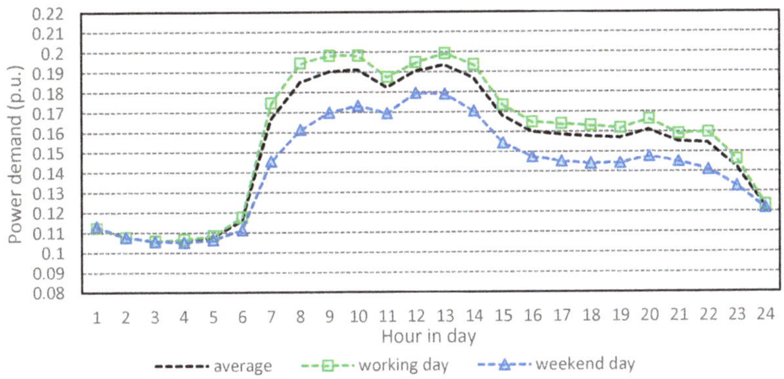

Figure 2. Normalized power demand curves.

In this paper, the average curve shown in Figure 2 was used. The working day and weekend day curves are similar. Using an average curve is sufficient to present the method effectively. However, due to fluctuations in power consumption throughout the week, it is expected that voltage levels during weekends might be slightly higher than anticipated. Consequently, if the algorithm were to base its operation on the weekend day curve, it would likely decide on a slightly higher delivery of reactive power, either through increased deployment of ShRs or increased reactive power that the DG absorbs.

For even more accurate results, further segmentation could be considered by differentiating between weekdays and weekends. This would enable a more precise representation of power demand fluctuations throughout the week.

For each hour, the average power demand for the whole distribution system was calculated using the following equation:

$$PD^t = \frac{P_{measured\_total}{}^t}{\sum_{i=1}^{Ni} P_{i\_max}} \tag{7}$$

In Equation (7), $P_{measured\_total}{}^t$ represents the total measured power demand of the whole distribution system during time interval t (including measurements from both substations). P_{i_max} represents the maximum allowed power demand in node i, while N_i represents the number of nodes.

In this paper, the demand changes in every node on an hourly basis, according to the curve. The active and reactive power demand at each node for time interval t is calculated as follows:

$$P_i{}^t = PD^t \cdot P_{i\_max} \tag{8}$$

$$Q_i{}^t = \tan\varphi \cdot P_i{}^t \tag{9}$$

In Equation (9), $\tan\varphi$ is determined based on the chosen value for $cos\varphi$, which, in this paper, was 0.95. Therefore, the maximum reactive power in Table 1 represents the maximum reactive power of that value of $cos\varphi$.

2.4. Distribution Line Modeling, Data, and Voltage Rise Calculation

In this paper, the "backward/forward sweep" algorithm described in [24] was used for modeling distribution lines and load flow calculations. Matlab, Version R2015A, a widely used software platform known for its capabilities in numerical analysis, data processing, and visualization, was employed for implementing both the load flow calculation and the optimization algorithm and for processing the experimental results.

The distribution line model is shown in Figure 3. The longitudinal branch contains the distribution line series resistance R_{DLi} and the series inductive reactance X_{DLi}. Since the distribution system model contains distribution lines of two voltage levels, all series line impedances were reduced to the 21 kV voltage level. Real and reactive power P_i, Q_i, P_{i-1}, and Q_{i-1} are calculated using the load flow algorithm from P_{Pi}, Q_{Pi}, $P_{P,i-1}$, and $Q_{P,i-1}$, which represent the active and reactive power demand in starting node $i-1$ and end node i.

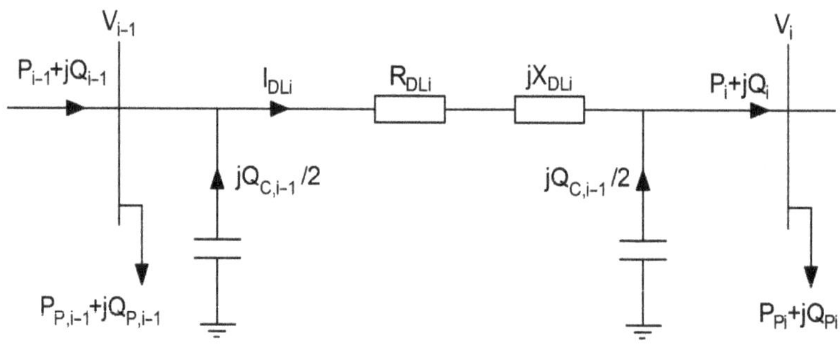

Figure 3. Distribution line model.

The model also contains parallel branches, which are introduced to model the capacitive currents that underground distribution lines create. For each line, one part of that reactive power is added to the distribution line starting node and the other half to the distribution line end node. The reactive power is calculated using Equation (10).

$$\frac{\overline{Q_{Ci}}}{2} = U_i^2 \cdot \omega \cdot \frac{C_i}{2} \cdot (cos\varphi_i + jsin\varphi_i)^2 \tag{10}$$

In Equation (10), C_i represents the capacitance of the ith distribution line, U_i represents system voltage, and φ_i represents the phase angle of the voltage.

The current that flows through the distribution transformer and creates voltage rises or drops is calculated using Equation (11).

$$\overline{I_{DLi}} = \frac{P_i \cdot cos\varphi_i + (Q_i + Q_{Ci}/2) \cdot sin\varphi_i}{\sqrt{3}\,U_i} + j\frac{P_i \cdot sin\varphi_i - (Q_i + Q_{Ci}/2) \cdot cos\varphi_i}{\sqrt{3}\,U_i} \tag{11}$$

In Equation (10), Q_{Ci} represent the reactive power of the parallel branch of ith distribution line.

The voltage rise on the distribution line is calculated using Equation (12):

$$\overline{V_i} - \overline{V_{i-1}} = -(R_{DLi} + jX_{DLi}) \cdot \overline{I_{DLi}} \tag{12}$$

In Equation (12), $\overline{V_{i-1}}$ represent the voltage in the node at the beginning of the ith distribution line and $\overline{V_i}$ the voltage at the end of the distribution line. R_{DLi} and and X_{DLi} represent the serial resistance and reactance of the ith distribution line, respectively. Distribution line lengths are shown in Table 2, and other data are shown in Table 3.

Table 2. Distribution lines' length.

Line No.	Length (km)	Line No.	Length (km)	Line No.	Length (km)	Line No.	Length (km)
2	7.92	20	2.37	37	6.18	55	1.03
4	1.35	21	2.34	38	1.89	56	3.86
5	2.4	22	6.45	39	4.09	57	4.13
6	1.33	23	4.83	40	10.42	58	1.51
7	0.59	24	0.29	41	3.38	59	5.53
8	1.18	25	8.63	42	11.42	60	1.17
9	1.34	26	3.68	43	28.28	61	0.22
10	2.96	27	2.2	45	5.82	62	0.81
11	5.6	28	1.35	46	1.9	63	1.38
12	2.55	29	6.62	47	6.91	64	3.84
13	1.15	30	1.55	48	6.15	65	5.54
14	1.75	31	1.51	49	2.13	66	0.7
15	4.4	32	3.12	50	5.15	67	1.82
16	0.56	33	4.84	51	7.28	68	0.25
17	1.59	34	2.74	52	1.99	69	3.817
18	1.56	35	4.68	53	2.29	70	2.75
19	2.27	36	4.78	54	2.95	71	3.53

Table 3. Distribution lines' data.

Line No.	Type	R (Ω/km)	X (Ω/km)	B (μS/km)
2	35 kVcable	0.176	0.12	63
43	35 kV overhead	0.253	0.35	3.3
4, 5, 7–19, 22–30	20 kV cable	0.164	0.122	85.5
32, 33, 35–42,	20 kV cable	0.164	0.122	85.5
45–54, 56, 57, 59,	20 kV cable	0.164	0.122	85.5
61, 62, 65–71	20 kV cable	0.164	0.122	85.5
6, 21, 58, 60	20 kV overhead	0.43	0.345	3
20, 31, 34, 55, 63, 64	20 kV overhead	0.86	0.365	3

2.5. Distribution Transformers' Modeling, Data, Voltage Rise Calculation, and Importance in Analysis

Two distribution transformers are integrated into the model because most voltage rises are expected to occur at the transformers. The transformer model is shown in Figure 4, and transformer data are shown in Table 4.

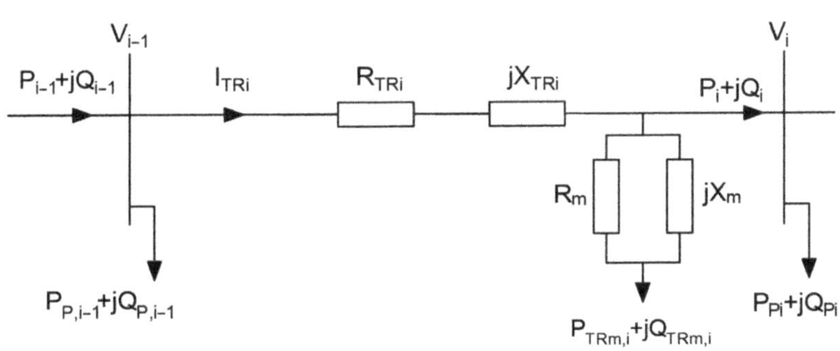

Figure 4. A distribution transformer model.

Table 4. Distribution transformer data.

Transformer No.	From Node	To Node	Installed Power (MVA)	P_{TRm} (W)	i_0 (%)	u_k (%)
1	2	3	8	47,000	0.8	7
2	43	44	4	24,800	0.8	6

The series resistance R_{TRi} and the series inductive reactance of the distribution transformer X_{TRi} have the largest impact on voltage rises and voltage drops that occur. The series resistance and reactance could be calculated as follows:

$$R_{TRi} = P_{Cu} \cdot \frac{{U_n}^2}{{S_n}^2} \tag{13}$$

$$Z_{TRi} = \frac{u_k}{100} \cdot \frac{{U_n}^2}{{S_n}^2} \tag{14}$$

$$X_{TRi} = \sqrt{{Z_t}^2 - {R_t}^2} \tag{15}$$

The aforementioned impedance values of the transformers were calculated for the nominal voltage value of 21 kV because this voltage is the voltage on the secondary side of the two transformers in the actual network. Additionally, all the line impedances were recalculated for the same voltage. In Equations (13)–(15), S_n represents the transformer installed power, P_{Cu} denotes transformer copper loss, and u_k is the transformer short-circuit voltage.

The current that flows through the distribution transformer and creates voltage rises or drops is calculated using Equation (16):

$$\overline{I_{TRi}} = \frac{(P_i + P_{TRm,i}) \cdot cos\varphi_i + (Q_i + Q_{TRm,i}) \cdot sin\varphi_i}{\sqrt{3}\, U_i} + j\frac{(P_i + P_{TRm,i}) \cdot sin\varphi_i - (Q_i + Q_{TRm,i}) \cdot cos\varphi_i}{\sqrt{3}\, U_i} \tag{16}$$

$P_{TRm,i}$ and $Q_{TRm,i}$ represent active and reactive power in the parallel branch of the transformer, where $P_{TRm,i}$ is obtained from the transformer open-circuit test and $Q_{TRm,i}$ is calculated as follows:

$$Q_{TRm,i} = \frac{{U_n}^2}{{X_{mi}}^2} \tag{17}$$

$$X_{m,i} = \frac{U_n}{\sqrt{3} \cdot \frac{i_{0\%}}{100} \cdot I_n} \tag{18}$$

In Equation (18), $i_{0\%}$ stands for the transformer open-circuit current, and I_n denotes the transformer nominal current.

The voltage rise in the power transformer is calculated using Equation (19):

$$\overline{V_i} - \overline{V_{i-1}} = -(R_{TRi} + jX_{TRi}) \cdot \overline{I_{TRi}} \tag{19}$$

Based on Equations (13) and (15) and the distribution transformer data shown in Table 4, the serial impedance of the distribution transformer TS 35/20 kV Kupjak can be calculated to be 0.324 + j3.845 [Ω] and the impedance of the distribution transformer TS 35/20 kV Gerovo to be 0.684 + j6.58 [Ω]. If we compare these impedances with the impedances of the distribution lines given in Table 3, we can conclude that the transformer reactances are larger. We can also conclude that the reverse power flow through distribution transformers is larger because it contains the sum of the reverse power flows from all distribution lines after the transformer. Therefore, from Equation (19), used for the calculation of the voltage rise in the transformer, we can conclude that the voltage rise in the transformer will have a significant impact on the overall voltage rise in the distribution system at all nodes located after the distribution transformer.

From Equation (19), we can see that voltage rises could be decreased by decreasing the reverse reactive part of the current $\overline{I_{TRi}}$ by using shunt reactors and the power factor control of DGs.

2.6. Distributed Generation Unit Data

In this paper, one existing DG with a constant power output is considered with a connection at node 47. The installed power of this DG is 2500 kW.

2.7. Concise Theory and Specific Advantages of Genetic Algorithms

The genetic algorithm (GA) was chosen for the optimization. GA is a computational method inspired by natural selection and genetics, initially proposed by John Holland in the 1960s and further developed by researchers such as David Goldberg. GA simulates the process of natural evolution to efficiently solve optimization and search problems.

At the core of GA is a population of potential solutions, represented as chromosomes encoding candidate solutions. The algorithm iteratively evolves this population over generations, guided by principles akin to biological evolution.

The iterative process of the genetic algorithm begins with an initial population of chromosomes:

$$P^{(0)} = \left\{ {x_1}^{(0)}, {x_2}^{(0)}, \ldots, {x_N}^{(0)} \right\} \tag{20}$$

Each chromosome represents a potential solution to the problem at hand. These chromosomes undergo evaluation, where their fitness is assessed based on an objective function defined by the problem domain. This fitness evaluation determines how well each chromosome solves the problem and guides the selection process. During selection, chromosomes with higher fitness are more likely to be chosen as parents for the next generation. Various selection methods, such as roulette wheel selection or tournament selection, are employed to determine which chromosomes proceed to the next stage. Selected parent chromosomes undergo crossover, where genetic material (genes) is exchanged to create offspring. Crossover points are randomly selected to maintain genetic diversity within the population. Additionally, GA introduces occasional mutations, where random changes are applied to offspring chromosomes. This mutation process promotes the exploration of new solutions and prevents premature convergence to suboptimal solutions. After generating offspring through crossover and mutation, the algorithm replaces less fit individuals in the current population with these new offspring. Elitism strategies may be employed to ensure that the best-performing individuals survive across generations, maintaining the population's overall fitness. The iterative process continues until a termination criterion is met, such as reaching a maximum number of generations or achieving convergence where further improvement is unlikely.

GA was chosen because each solution of optimization in this research has 40 optimization variables, and GA is particularly effective in tackling intricate combinatorial problems with a large number of optimization variables [25]. Therefore, one of the main advantages is that genetic algorithms excel at exploring solution spaces and finding global optimal solutions, which is crucial in complex optimization problems where the search space is large or uneven. Another benefit is that by combining global exploration through selection and crossover with local exploration through mutation, genetic algorithms strike a balance between intensive and extensive search, often yielding superior results compared to metaheuristics focused solely on one aspect of search. Genetic algorithms can be easily adapted to meet specific optimization problem requirements, such as different types of constraints or multi-objective optimization, ensuring their wide applicability.

2.8. Implementation of an Optimization Algorithm for Allocation of Shunt Reactors and Hourly DG Power Factor Recommendation

The optimization variables for one solution of the optimization algorithm are shown in Table 5. The process of optimization is shown in Figure 5, and the optimization process is similar to the GA optimization process described in [26].

Table 5. Genetic algorithm optimization variables.

Proposed DG Power Factor for All Hours	Position of 8 Shunt Reactors	Sizes of 8 Shunt Reactors
Gene 1–Gene 24	Gene 25–Gene 32	Gene 33–Gene 40

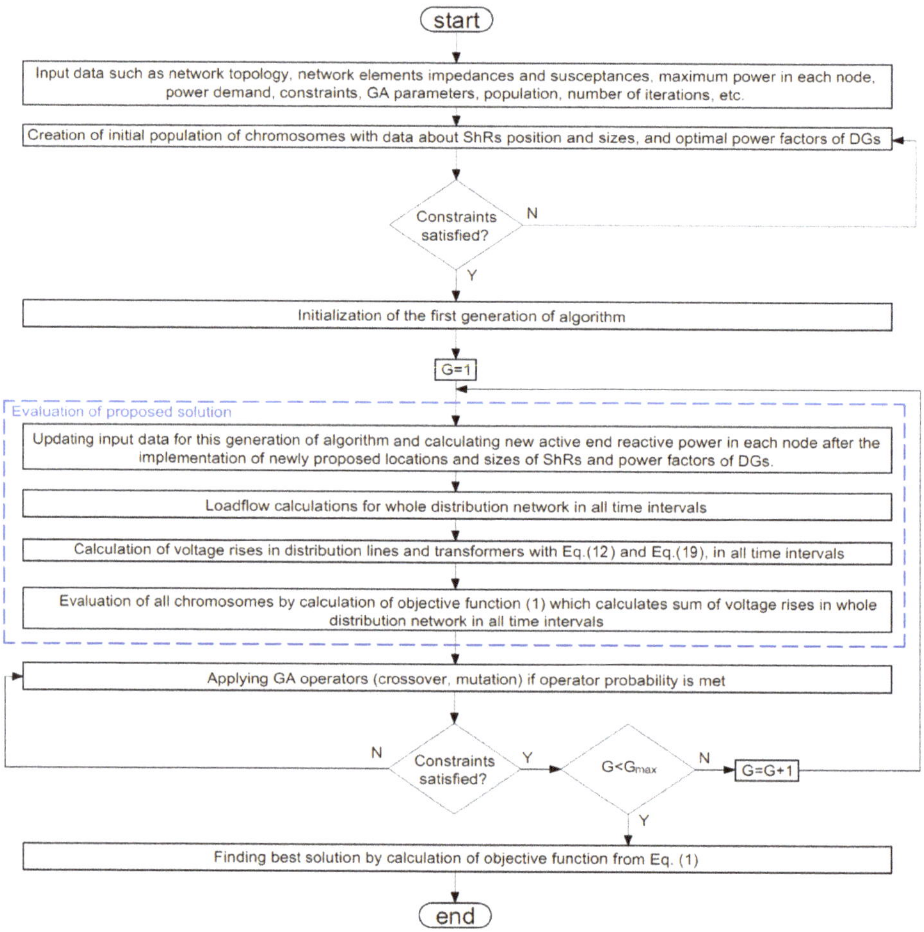

Figure 5. Genetic algorithm flowchart.

The algorithm starts with inputting data about distribution network topology, network element impedances and susceptances, maximum power at each node, power demand, constraints, and genetic algorithm parameters such as population size, number of generations, and operator factors. The value of the chosen crossover factor is 0.9, and the chosen mutation factor is 0.1.

The first step is the creation of an initial population of chromosomes with data about proposed positions and sizes for shunt reactors and optimal power factors of DGs, if applicable. All proposed solutions for all chromosomes must adhere to the constraints described in Sections 2.2.1–2.2.4 of this paper. After that, the first generation of the algorithm is initiated. For the proposed solution for the ShRs and power factors of DGs, the new data about active and reactive power in each node is calculated. Then, load flow calculations are made using a backward–forward sweep algorithm for the whole distribution network at all time intervals. After that, voltage rises are calculated for all distribution lines and transformers using Equations (12) and (19). Then, all chromosomes are evaluated using the calculation of the objective function given in Equation (1), which calculates the overall sum of voltage rises in the whole distribution network in all time intervals. After that, genetic algorithms apply crossover and mutation operators if the probability of each operator is met. All proposed solutions for all chromosomes after applying crossover and mutation factors must adhere to the constraints described in Sections 2.2.1–2.2.4 of this paper.

The process repeats until the last generation of the algorithm is executed. After that, the algorithm finds the best solution by calculating the objective function given in Equation (1).

2.9. Comparision with Other Established Methods for Voltage Rise Mitigation

This study proposes a novel approach for addressing voltage rise issues in distribution networks through a comprehensive methodology that integrates various contributing factors into a unified solution. By accurately modeling the entire system, including medium-voltage lines, distribution transformers, and the effects of low power demand and long underground cable lines, the method aims to mitigate voltage rises effectively across multiple voltage levels. Existing methods typically focus on mitigating voltage rises at single voltage levels, whereas our approach uniquely addresses these challenges across multiple voltage levels of the distribution network.

Unlike traditional approaches that may focus on singular aspects such as shunt reactors or power factor control alone [19], this method combines the allocation of fixed shunt reactors with hourly power factor control of distributed generators (DGs). This integrated approach ensures a more balanced reduction in voltage rises by effectively managing reactive power flows at critical nodes. The study emphasizes the role of distribution transformers in voltage rise occurrences, showing that significant voltage rises often occur at these points due to reverse reactive power flows. This insight underscores the importance of modeling transformers to effectively tackle voltage rise challenges, a perspective that distinguishes this study from the existing literature.

2.10. Cost Analysis of Shunt Reactor Deployment

The existing studies on the installation of shunt reactors [18–20] have excluded a comprehensive cost–benefit analysis. Shunt reactors are typically installed in power grids, facing issues with excessive reverse flows of reactive power. Installing compensating shunt reactors is a necessary solution to mitigate these reverse flows. An alternative would be to deactivate the cables, generating these reverse flows and connecting existing consumers via new overhead lines. However, such an approach is often not economically viable compared to the installation of shunt reactors.

According to [27], an MV aluminum conductor with a cross-sectional area of 185 mm^2, which is the most prevalent in the network in our study, generates a capacitive reactive power of 34.2 kVAr/km. This means that to reduce the capacitive reactive power by 1 MVAr, approximately 29.24 km of the mentioned cable would need to be removed and replaced with overhead lines. According to [28], the construction cost of overhead lines would amount to at least USD 62,150 per kilometer. Therefore, the construction of a 29.24 km overhead line needed to remove a 29.24 km cable line could cost around USD 1.8 million.

According to [29], the cost of installing shunt reactors is USD 54,000/MVAr. Therefore, installing 1 MVAr of shunt reactors at an approximate cost of USD 54,000 would present a much more cost-effective solution compared to alternatives, such as replacing a 29.24 km cable with overhead lines, which could cost around USD 1.8 million.

3. Results

3.1. Cases Studied

The simulations were made to demonstrate the effectiveness of the proposed method for mitigating voltage rises by allocating shunt reactions and using the hourly recommended power factors of the DG unit. The simulation was made for a single objective function for voltage profile improvement based on the fitness function given in Equation (1).

To compare the results, three simulation cases were considered. Before applying the method, load flow calculations were made for the initial network state without connected shunt reactors and with the existing DG unit working at unity power factor. Then, in Case 1, the algorithm recommended the hourly power factor of the existing DG without connecting new shunt reactors. In Case 2, the algorithm allocated and sized up to eight shunt reactors without hourly power factor control of the existing DG. Finally, in Case 4, the algorithm used both new shunt reactors and hourly power factor control of the DG.

3.2. Results and Comments

The results for the locations and sizes of new shunt reactors for various cases are shown in Table 6. The results for recommended DG power factors for previously installed DG units are shown in Table 7.

Table 6. Allocation of shunt reactors.

	Case 1		Case 2		Case 3	
New ShR	Size (kVAr)	Location	Size (kVAr)	Location	Size (kVAr)	Location
1	-	-	1000	49	700	48
2	-	-	900	40	600	30
3	-	-	800	3	600	42
4	-	-	100	37	800	49
5	-	-	200	25	700	4
6	-	-	300	57	700	48
7	-	-	800	49	1000	47
8	-	-	1100	24	1100	18
Overall	-	-	5200	-	6200	-

Table 7. Recommended DG power factors.

Hour	Power Factor (Case 1)	Power Factor (Case 2)	Power Factor (Case 3)
1–4	0.95_{cap}	1	0.96, 0.97, 0.975, 0.975
5–8	0.95_{cap}	1	0.97, 0.96, 0.895, 0.865
9–12	0.95_{cap}	1	0.86, 0.85, 0.875, 0.865
13–16	0.95_{cap}	1	0.855, 0.865, 0.895, 0.915
17–20	0.95_{cap}	1	0.91, 0.91, 0.91, 0.905
21–24	0.95_{cap}	1	0.915, 0.915, 0.93, 0.955

In Case 1, voltage mitigation is achieved only by controlling the power factor of one DG installed in part of the distribution network powered by TS 35/20 kV Gerovo in nodes 45 to 71. The algorithm in all parts of the day proposed the maximum leading power factor of 0.95. With this power factor, the existing DG behaves as a consumer of excessive reactive power. However, since the DG produces 2.5 MW of real power, with a leading power factor of 0.95, the DG unit can consume only around 0.84 MVAr of reactive power. Figure 6 shows the overall reactive power in branch 44 of the power transformer in TS 35/20 kV Gerovo. The large reverse reactive power flow is partially reduced, but is still high.

Therefore, the voltage rise is slightly mitigated, as shown in Figure 7, which shows the voltage in the whole distribution network at 3 AM. In nodes 45 to 71, which are powered from TS 35/20 kV Gerovo, the maximum voltage rise is reduced, but only from 1.13 p.u. to 1.11 p.u. The voltage at the end of MV line 43 is around 1.07 p.u., which shows that even the voltage rise issue in that MV line is not fixed. The voltage in nodes 2 to 42 is not affected because changing the power factor of the DG unit in node 47 does not have an impact on that part of the distribution network.

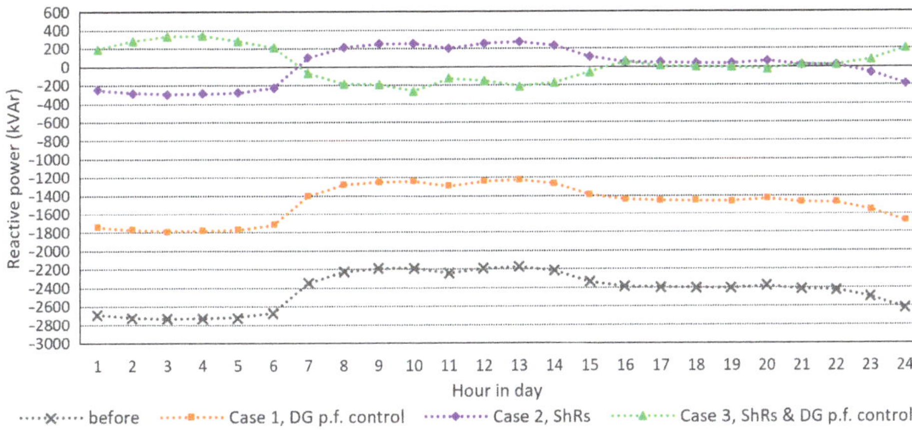

Figure 6. Reactive power in branch 44 of the TS Gerovo transformer.

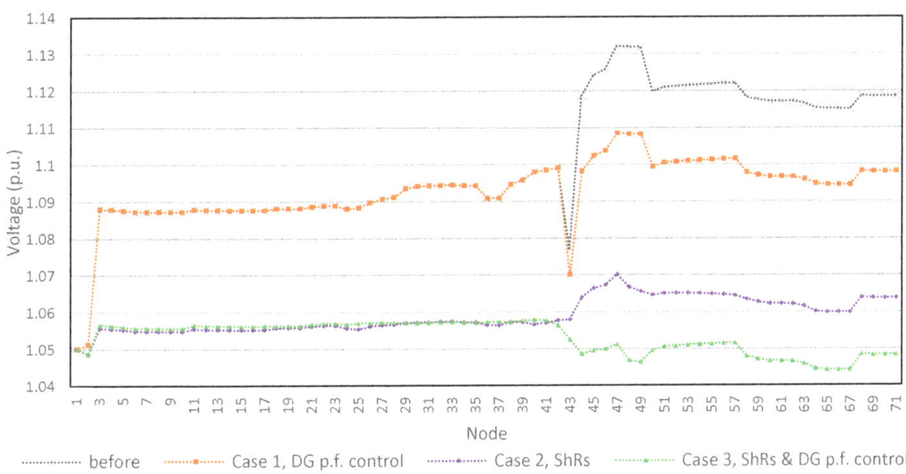

Figure 7. Voltage profile for nodes 1–71 at 3 AM.

In Case 2, the algorithm proposed the locations and sizes of new shunt reactors, but without power factor control of the existing DG. The algorithm recommended the allocation of shunt reactors with a total installed power of 5.2 MVAr. Figure 7 shows that the voltage magnitudes at all nodes are significantly reduced. The maximum voltage rise is 1.07 p.u. and the voltage at all nodes is near the desired magnitude of 1.05 p.u. Figure 8 shows the voltage in node 44 on the secondary side of the transformer in TS 35/20 kV Gerovo. The voltage in this case is within the range of 1.04 p.u. to slightly above 1.06 p.u., which is much better than in Case 1.

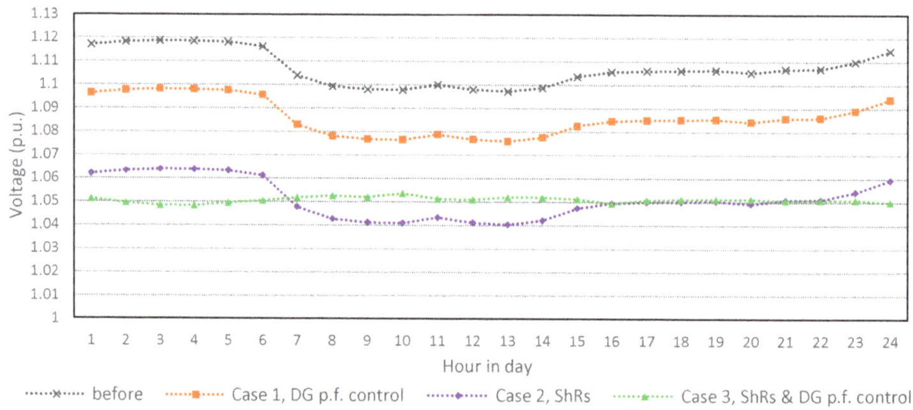

Figure 8. Voltage at node 44 on the secondary side of TS 35/20 kV Gerovo at 3 AM.

In Case 3, the algorithm proposed the locations and sizes of new shunt reactors and also the optimal power factor of the existing DG, as shown on Figure 9. The algorithm recommended the allocation of shunt reactors with a total installed power of 6.2 MVAr. The algorithm proposed a lagging power factor of the existing DG in the range of 0.85 to 0.975 p.u., as shown in Figure 8.

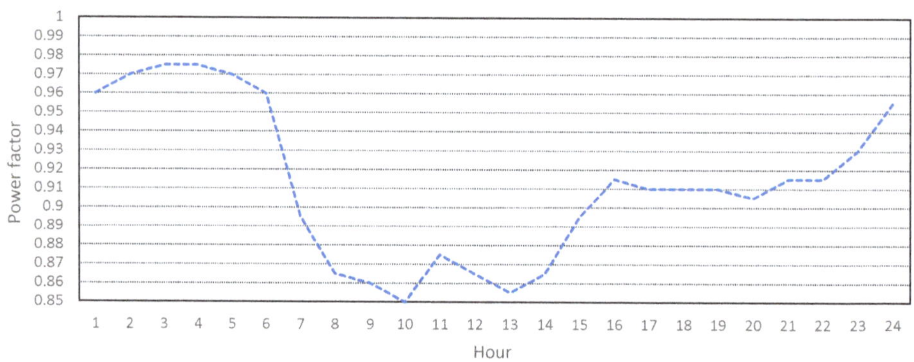

Figure 9. Proposed DG power factor for Case 3.

The combination of new shunt reactors and DG power factor control achieved the best results in voltage rise mitigation. Figure 7 shows that the voltage in all nodes in Case 3 is very close to the desired 1.05 p.u. at 3 AM. Figure 8 shows that the voltage in node 44, on the secondary side of the transformer in TS 35/20 kV Gerovo, is very close to the desired value of 1.05 p.u at all times of the day.

To summarize, Figure 6 shows the reactive power flow in branch 44 at the TS 35/20 kV Gerovo transformer. In Case 1, voltage rise mitigation adjusted the DG power factor, resulting in a modest reduction in the reverse reactive power flow. Cases 2 and 3 achieved significant reductions in reactive power flows through the deployment of SHRs, with Case 3 also optimizing DG power factor control. Figure 7 displays the voltage profile of the whole network at 3 A.M. In Case 1, adjusting DG power factors slightly reduced the voltage rises to around 1.11 p.u. Case 2, using SHRs, saw voltages peak at 1.07 p.u. Case 3, with SHRs and optimized DG factors, maintained stable voltages of around 1.05 p.u. across all nodes. Figure 8 shows the voltage at node 44 on the secondary side of TS 35/20 kV Gerovo at 3 AM. Case 1 limits voltage to 1.1 p.u., Case 2 ranges between 1.04 to slightly above

1.06 p.u., and Case 3 is near the desired 1.05 p.u. Figure 9 illustrates the proposed DG power factor for Case 3, ranging from 0.85 to 0.975 p.u., as recommended by the algorithm.

3.3. The Importance of Modeling Distribution Transformers in Voltage Rise Analysis

Figure 10 shows part of the distribution network, which includes one 35 kV overhead line between nodes 1 and 43, one 35/20 kV distribution transformer between nodes 43 and 44, and five 20 kV distribution lines from node 44 to node 49.

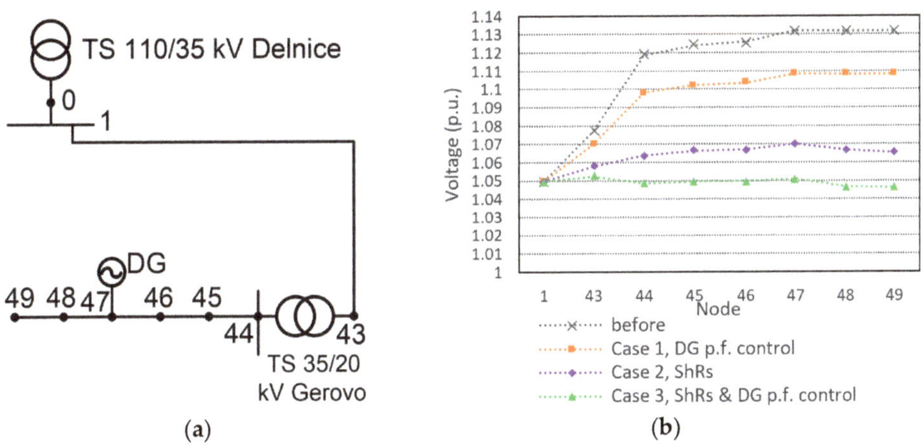

Figure 10. Part of the distribution network (**a**) and the voltage profile in that part of the network (**b**).

Without the implementation of the presented method, it is notable (Figure 10b) that the overall voltage rise from the start to the end node of this part of the distribution network is approximately 0.08 p.u. The voltage rise in the 35 kV line between nodes 1 and 43 is approximately 0.03 p.u. The voltage rise in the 35/20 kV distribution transformer between nodes 43 and 44 is 0.04 p.u. The voltage rise in the 20 kV part of the distribution network between nodes 44 and 49 is 0.01 p.u. It is clear that the voltage rise in the distribution transformer represents approximately half of the overall voltage rise. It is also important to note that the voltage rise in the distribution transformer is caused not only by reverse power flows from the distribution network between nodes 44 and 49 but also by reverse power flows from all other distribution lines that are modeled after that distribution transformer. The results for Cases 1 to 3 show that with the implementation of the presented method, the voltage rises decreased. In Case 3, the algorithm even recommended a small voltage drop in the distribution transformer in order to keep the voltage in the remaining part of the network close to the desired one. Therefore, modeling distribution transformers is very important in voltage rise analysis.

4. Conclusions

This paper proposed a novel general approach for analyzing and mitigating voltage rise issues in distribution networks where the entire system is accurately modeled and various elements that contribute to voltage rises are unified into a single method for the mitigation of voltage rises on multiple voltage levels of a distribution network. The proposed method may prove beneficial in the analysis and mitigation of voltage rises caused by low power demand and very long underground cable lines, which create a significant amount of reverse reactive power flow. The primary objective of this paper was to reduce voltage rises. The proposed method has been demonstrated in the analysis of voltage rise issues within a real distribution network operating under these conditions, with a focus on exploring potential solutions.

The distribution system model used in this paper contains MV lines at two voltage levels and distribution transformers that are connections between these two voltage levels. Through simulations, it was established that by implementing hourly power factor control of the DG unit, voltage rise issues could be mitigated. The maximum voltage rise was reduced, but only from 1.13 p.u. to 1.11 p.u., during a critical part of the day when the demand is lowest. Another case showed that precise allocation and sizing of fixed shunt reactors can also decrease reactive power flows and reduce voltage rises. The maximum voltage rise during a critical part of the day was reduced from 1.13 p.u. to 1.07 p.u.

The best results in the mitigation of voltage rises are achieved by combining the allocation of new shunt reactors and the control of the power factor of the existing DG. In this case, the voltage in all nodes in the part of the network with the installed DG is held very close to the set value of 1.05 p.u. The fixed shunt reactor consumes a large amount of reactive power, while changing the DG power factor precisely injects the remaining reactive power of the DG unit, through which the transformer voltage is controlled during different time intervals.

The scientific innovations of this paper are summarized as follows:

(1) In this paper, we proposed the use of a new method for an improved approach to evaluating voltage rise, where the entire system is more accurately modeled and various elements that contribute to voltage rise are unified into a single method for the mitigation of voltage rises. In addition to the usual modeling of distribution lines, the method precisely models transformers due to their significance for voltage rises. It also models distribution lines at two voltage levels, not just at one. The method also unifies and accurately models the components of distribution lines responsible for creating reverse flows of capacitive reactive power, particularly the parallel branches that inject capacitive reactive power.

(2) Unlike traditional methods that often address voltage rise issues in isolation or at a single voltage level, our method provides an integrated approach that considers multiple voltage levels and unifies various mitigation elements into a coherent strategy. This holistic modeling significantly enhances the accuracy and effectiveness of voltage rise mitigation.

(3) In the analysis of voltage rises caused by low power demand and long underground cable lines that generate reverse reactive power flows, it is important to model distribution transformers because the transformers are the location where voltage rise predominantly occurs. The research showed that in the selected network example, which is based on a real distribution network, in the moments of the highest voltage increase, half of the voltage rise occurs in the transformer.

(4) It is possible to mitigate voltage rises with the use of shunt reactors or DG power factor control. The research results based on a real example showed that the voltage at 3 a.m. could be decreased from 1.13 to 1.11 p.u. with DG power factor control and from 1.13 to 1.07 p.u. with shunt reactors. However, the best results are achieved by combining the allocation of new shunt reactors and hourly control of the power factor of DGs. In this case, the highest voltage is decreased very close to the desired value of 1.05 p.u. The best results in mitigation of voltage rises are achieved by using a fixed shunt reactor that consumes a large amount of reactive power and changing the DG power factor to precisely inject the remaining reactive power of the DG unit.

Future research should include testing the obtained results using other optimization techniques and focus on influencing voltage rise prevention during the development planning of distribution networks. Strategies such as reducing the cross-sectional area of cable conductors to minimize transverse capacitance and planning overhead lines can effectively decrease reverse capacitive currents, thereby mitigating voltage rises without the need for additional investments in ShRs or DG power factor control. Moreover, evaluating the impact of integrating new DGs on voltage conditions and optimizing their integration into existing networks are important areas for further investigation. Exploring network reconfiguration possibilities also offers promising avenues for reducing voltage rises in distribution systems. Additionally, it is necessary to investigate how excess reverse reactive power affects higher-level transmission networks. Given that during the analysis of this study, the authors noted frequent instances of elevated input voltages from the transmission network, voltage rise represents a problem that extends across multiple voltage levels and should thus be addressed accordingly. This study provides a solution for analyzing the voltage rise issue on multiple voltage levels of a network.

Author Contributions: Conceptualization, D.Ć., J.Ž. and V.K.; software, D.Ć.; methodology, D.Ć.; formal analysis, D.Ć. and J.Ž.; data curation, D.Ć.; writing—original draft, D.Ć.; writing—review and editing, D.Ć.; visualization, D.Ć.; supervision, V.K. All authors have read and agreed to the published version of the manuscript.

Funding: This research received no external funding.

Data Availability Statement: The data are contained within the article.

Conflicts of Interest: The authors declare no conflict of interest.

References

1. Yang, S.; Zhou, W.; Zhu, S. Failure probability estimation of overhead transmission lines considering the spatial and temporal variation in severe weather. *J. Mod. Power Syst. Clean Energy* **2019**, *7*, 131–138. [CrossRef]
2. Souto, L.; Santoso, S. *Overhead Versus Underground: Designing Power Lines for Resilient, Cost-Effective Distribution Networks under Windstorms*; Resilience Week (RWS): Salt Lake City, UT, USA, 2020; pp. 113–118. [CrossRef]
3. Marot, A.; Kelly, A.; Naglic, M.; Barbesant, V.; Cremer, J.; Stefanov, A.; Viebahn, J. Perspectives on Future Power System Control Centers for Energy Transition. *J. Mod. Power Syst. Clean Energy* **2022**, *10*, 328–344. [CrossRef]
4. Karki, U.; Gunasekaran, D.; Fang, Z.P. Reactive compensation of overhead AC transmission lines using underground power cables. In Proceedings of the 2015 IEEE Power & Energy Society General Meeting, Denver, CO, USA, 26–30 July 2015; pp. 1–5. [CrossRef]
5. Ferreira, P.D.F.; Carvalho, P.M.S.; Ferreira, L.A.F.M.; Ilic, M.D. Distributed Energy Resources Integration Challenges in Low-Voltage Networks: Voltage Control Limitations and Risk of Cascading. *IEEE Trans. Sustain. Energy* **2013**, *4*, 82–88. [CrossRef]
6. Markiewicz, H.; Klajn, A. *Voltage Disturbances Standard EN50160—Voltage Characteristics of Electricity Supplied by Public Electricity Networks*; Wroclaw University of Technology: Wrocław, Poland, 2004.
7. Tonkoski, R.; Turcotte, D.; EL-Fouly, T.H.M. Impact of High PV Penetration on Voltage Profiles in Residential Neighborhoods. *IEEE Trans. Sustain. Energy* **2012**, *3*, 518–527. [CrossRef]
8. Raval, D.Y.; Pandya, S.N. Phase Shifting Strategy for Mitigation of Local Voltage Rise in Highly PV Penetrated Distribution Network. In Proceedings of the 2021 9th IEEE International Conference on Power Systems (ICPS), Kharagpur, India, 16–18 December 2021; pp. 1–6. [CrossRef]
9. Wang, J.; Zhou, N.; Ran, Y.; Wang, Q. Reducing the active power curtailment of distributed generation based on soft open points in unbalanced distribution networks. *IET Renew. Power Gener.* **2020**, *14*, 3252–3258. [CrossRef]
10. Singh, R.K.; Goswami, S.K. Optimum allocation of distributed generations based on nodal pricing for profit, loss reduction, and voltage improvement including voltage rise issue. *Int. J. Electr. Power Energy Syst.* **2010**, *32*, 637–644. [CrossRef]
11. Mufaris, A.L.M.; Baba, J.; Yoshizawa, S.; Hayashi, Y. Determination of dynamic line drop compensation parameters of voltage regulators for voltage rise mitigation. In Proceedings of the 2015 International Conference on Clean Electrical Power, Taormina, Italy, 16–18 June 2015; pp. 319–325. [CrossRef]
12. Alam, M.J.E.; Muttaqi, K.M.; Sutanto, D. Distributed energy storage for mitigation of voltage-rise impact caused by rooftop solar PV. In Proceedings of the 2012 IEEE Power and Energy Society General Meeting, San Diego, CA, USA, 22–26 July 2012; pp. 1–8. [CrossRef]
13. Then, J.; Agalgaonkar, A.P.; Muttaqi, K.M. Coordinated Charging of Spatially Distributed Electric Vehicles for Mitigating Voltage Rise and Voltage Unbalance in Modern Distribution Networks. *IEEE Trans. Ind. Appl.* **2023**, *59*, 5149–5157. [CrossRef]
14. Jasni, J.; Mohd Radzi, M.A.; Azis, N. *Novel Reactive Power Control Technique for Mitigating Voltage Rise in Distribution Network with High Pv Penetration*; SSRN: Rochester, NY, USA, 2023; Available online: https://ssrn.com/abstract=4661084 (accessed on 22 June 2024).
15. Xie, S.; Kaneko, A.; Hayashi, Y. A Decentralized Voltage Regulation Scheme Using Improved Volt-Var Function of PV Smart Inverter. *IEEJ Trans. Electr. Electron. Eng.* **2024**. [CrossRef]
16. Omar, M.A.; Hamdan, A. Control strategy of battery inverter for voltage profile improvement in low voltage networks with high PV penetration level. *Int. J. Power Energy Convers.* **2024**, *15*, 25–41. [CrossRef]
17. Waswa, L.; Chihota, M.J.; Bekker, B. Coordinated deployment of electric minibus taxis to enhance solar photovoltaic hosting capacity of residential networks. In Proceedings of the 27th International Conference on Electricity Distribution (CIRED 2023), Rome, Italy, 12–15 June 2023; pp. 3939–3943. [CrossRef]
18. Song, S.-H.; Kim, S.-B. Coordinated Reactive Power Control with a Variable Shunt Reactor and an Inverter-Based Wind Power Plant. *Energies* **2022**, *15*, 4739. [CrossRef]

19. Sithole, S.; Mbuli, N.; Pretorius, J. Voltage regulation in the douglas area using shunt capacitor banks and controllable shunt reactors. In Proceedings of the 2013 13th International Conference on Environment and Electrical Engineering (EEEIC), Wroclaw, Poland, 1–3 November 2013; pp. 85–90. [CrossRef]
20. Teshome, D.F.; Xu, W.; Bagheri, P.; Nassif, A.; Zhou, Y. A Reactive Power Control Scheme for DER-Caused Voltage Rise Mitigation in Secondary Systems. *IEEE Trans. Sustain. Energy* **2019**, *10*, 1684–1695. [CrossRef]
21. Hu, J.; Yuan, P.; Li, X.; Liu, Y. Analysis on the Necessity of High-Voltage Shunt Reactors in Power Grid. In Proceedings of the 2020 10th International Conference on Power and Energy Systems (ICPES), Chengdu, China, 25–27 December 2020; pp. 83–87. [CrossRef]
22. Bryantsev, A.; Dorofeev, V.; Zilberman, M.; Smirnov, A.; Smolovik, S. *Magnetically Controlled Shunt Reactor Application for AC HV and EHV Transmission Lines*; CIGRE: St. Petersburg, Russia, 2006; pp. 1–8.
23. Pereira, B.R.; da Costa, G.R.M.; Contreras, J.; Mantovani, J.R.S. Optimal Distributed Generation and Reactive Power Allocation in Electrical Distribution Systems. *IEEE Trans. Sustain. Energy* **2016**, *7*, 975–984. [CrossRef]
24. Prenc, R.; Škrlec, D.; Komen, V. A novel load flow algorithm for radial distribution networks with dispersed generation. *Technol. Gazzete* **2013**, *20*, 969–977.
25. Prenc, R.; Škrlec, D.; Komen, V. Distributed generation allocation based on average daily load and power production curves. *Int. J. Electr. Power Energy Syst.* **2013**, *53*, 612–622. [CrossRef]
26. Ćetković, D.; Komen, V. Optimal Distributed Generation and Capacitor Bank Allocation and Sizing at Two Voltage Levels. *IEEE Syst. J.* **2023**, *17*, 5831–5841. [CrossRef]
27. Vlahinić, S.; Franković, D.; Komen, V.; Antonić, A. Reactive Power Compensation with PV Inverters for System Loss Reduction. *Energies* **2019**, *12*, 4062. [CrossRef]
28. Miranda, L.; Mueller-Stoffels, M.; Whitney, E. An Alaska case study: Electrical transmission. *J. Renew. Sustain. Energy* **2017**, *9*, 061702. [CrossRef]
29. Grup Arge, Energy and Control Systems 2023/1 Price List. Available online: https://www.gruparge.com/wp-content/uploads/2023/02/grup-arge-price-list-2023.pdf (accessed on 22 June 2024).

energies

MDPI

Article

Investigation and Analysis of the Power Quality in an Academic Institution's Electrical Distribution System

Khaled M. Alawasa * and Abdullah H. Al-Badi

Department of Electrical and Computer Engineering, Sultan Qaboos University, P.O. Box 33, Muscat 123, Oman; albadi@squ.edu.om
* Correspondence: k.alawasa@squ.edu.om

Abstract: The presence of non-linear, time-variant loads in power networks introduces harmonics and other power quality issues in voltage and current waveforms. With the growing prevalence of power electronics, managing harmonic distortion has become a significant challenge in modern power distribution systems. This study aims to assess and analyze power quality at various locations within an academic institution (Sultan Qaboos University). The study encompasses the outputs of PV converters, uninterruptible power supplies (UPS), chillers with variable-speed motors, 11 kV/415 V transformers at both the College of Engineering and the Center of Information Systems, as well as two main substations supplying power to the university (33 kV/11 kV). Data collected—such as harmonic content in current and voltage waveforms and Total Harmonic Distortion (THD)—were meticulously analyzed to identify the factors affecting waveform quality. Measurements were conducted using advanced power quality instruments (Fluke 435-II), and the data were analyzed using MATLAB software. The study reveals that most results adhered to both the Oman grid code and international standards. The findings and methodology presented in this paper provide a critical benchmark for guiding standards compliance as Sultan Qaboos University, and other academic institution, undertakes the modernization of aging equipment and the expansion of new high-tech facilities.

Keywords: academic institution; harmonics distortion; non-linear loads; power electronics; power quality

1. Introduction

End users and electric utilities companies are highly concerned about power quality owing to the use of powerful electronic devices with microprocessor-based controls, continuous growth in the application of variable speed drives and capacitors banks, and the equipment being more interconnected in the system so the impact of a problem becomes more severe. These devices could include arc furnaces, computer power supplies, HV DC systems, and renewable energy systems, etc. These nonlinear loads could draw harmonic and neutral currents, cause system imbalances, reduce system efficiency and power factors, interfere with nearby communication systems, and disturb other consumers. The current distortion causes a voltage distortion due to the presence of system impedance. All loads connected to the same point of common coupling to which the original, disturbing load is connected will also be affected.

There are several methods that can be utilized to reduce harmonics in power systems, such as using active and passive filters, reducing the harmonic currents produced by the load by adding a line reactor or connecting specialized transformers, and modifying the frequency response of the system by connecting filters, inductors, or capacitors. However, the problem with passive filters is their large size, resonance, and fixed compensation. These drawbacks can be overcome by using active filters, which is a well-developed technology. Many active filter configurations exist to compensate for harmonic currents, reactive power, neutral currents, harmonics, and unbalanced currents [1].

If grid-connected PVs utilize inverters, the harmonic distortion at the point of common coupling (PCC) should not affect the quality of supply to customers connected to the point. The harmonic content of the waveform can be evaluated by Total Harmonic Distortion. The harmonic voltage distortion limits of nominal fundamental frequency voltage and the harmonic current distortion limits current at the PCC are given by IEEE Standard 519-2022 [2]. Furthermore, there are IEC standards, such as the IEC 61727 about PV systems [3] and IEC 61000-3-2,12 [4,5], which apply to low-voltage systems and define maximum limits for individual harmonic currents and THD limits, depending on the short-circuit ratio of the grid and the nominal power of the generating unit.

Several researchers investigated harmonics distortion due to the integration of the PV-inverter in electrical systems. An investigation on the power quality characteristics of residential PV inverters installed on small-sized rooftop PV stations was presented in [6]. Reference [7] investigated the power quality of single-phase grid-connected PV in a low-voltage system. The study reveals that the THDI fell below 10% when the inverter exceeded 50% of its rating. The 7th and 9th harmonic orders were the main contributors to THDI across all conditions. Reference [8] discussed the presence of current and voltage harmonics caused by the connection of PV systems and various load types on the power grid, suggesting that system performance could be improved by compensating for harmonics at the point of common coupling. An investigation and characterization of the harmonic interactions and distortion level on modern residential loads were reported in [9]. The measurement showed that the distortion in the current waveform reached as high as 45%. A case study on the harmonic profile in an electrical distribution system was presented in Reference [10], which included measurements of both three-phase and single-phase harmonics. The findings indicated a significant current unbalance, reaching up to 56% of the phase current, along with a high level of total harmonic distortion (THD) in the current, peaking at 30.36%, and a Total Current Demand Distortion (TDD) level of up to 10%.

Reference [11] examined the use of a unified power quality conditioner to mitigate the power quality problems existing in the grid and the harmonics penetrated by the non-linear loads. The conditioner is supported by a photovoltaic and battery energy storage system. An optimized unified power-quality conditioner (UPQC), which aims to integrate series and shunt active power filters with a minimum volt–ampere (VA) loading of the UPQC, was presented in [12]. The validity of the proposed scheme was proved through an experimental investigation in the laboratory. In reference [13], a transformerless hybrid series active filter, using a sliding-mode control algorithm and a notch harmonic detection technique, was implemented to compensate for source current harmonics. The realized active power filter enhances the power quality while cleaning the point of common coupling (PCC) from possible voltage distortions, sags, and swells. Reference [14] assessed the level of harmonic pollution in a healthcare facility. The investigation focused on assessing the level of THD in this facility. In this paper, a detailed power quality assessment was conducted at Sultan Qaboos University (SQU) focused on identifying harmonic penetration across the campus infrastructure. Measurements were taken at the end-load level, building distribution panels, and main substation buses to quantify distortion propagation through the system. The data were analyzed using MATLAB software. The results highlight the impacts of diverse and aggregate linear loading, with noticeable differentiation between more and less heavily loaded sections of the distribution network. Recommendations are provided for ongoing monitoring and consideration of targeted mitigation efforts to maintain power quality as campus loads modernize and electricity demands increase. The presented methodology and dataset provide a benchmark to guide standards compliance as SQU upgrades aging equipment and expands new high-tech facilities.

2. Harmonic Analysis and Decomposition of Distorted Waveform

Every periodic waveform can be analyzed as a combination of a fundamental frequency sinusoid and multiple sinusoids at harmonic frequencies. The presence or absence of coefficients for these sinusoidal terms depends on the nature of the waveform. In some cases, there may also be a direct current (d.c.) component that complements these purely sinusoidal terms. This concept can be mathematically described using the following equation [7,8].

$$f(t) = \frac{a_o}{2} + \sum_{n=1,2,3\ldots}^{\infty} (a_n \cos n\omega t + b_n \sin n\omega t) \tag{1}$$

where

$f(t)$ is a generic periodic waveform;
a_o is the d.c. component;
a_n, b_n are the coefficients of the series;
n is an integer number between 1 and infinity.
The coefficients of the series can be calculated as follows:

$$a_o = \frac{1}{(T)}\int_0^T \sqrt{2}V_p \sin\omega t \, d(\omega t)$$
$$a_n = \frac{1}{(T/2)}\int_0^T v_s \cos n\omega t \, d(\omega t) = \frac{1}{(\pi)}\int_0^{2\pi} V_p \sin\omega t \, \cos n\omega t \, d(\omega t)$$
$$b_n = \frac{1}{(T/2)}\int_0^T v_s \sin n\omega t \, d(\omega t) = \frac{1}{(\pi)}\int_0^{2\pi} V_p \sin\omega t \, \sin n\omega t \, d(\omega t)$$

T = 2π is the period. Harmonic pollution on a power line can be quantified by a measure known as Total Harmonic Distortion or THD [1]. THD is the ratio of the root mean square values of harmonic components to the root mean square value of the fundamental component.

The Total Harmonic Current Distortion factor (THD_I) can be calculated by:

$$THD\%|_I = 100\% \times \sqrt{\sum_{h\neq 1}\left(\frac{I_h}{I_1}\right)^2} \tag{2}$$

where

I_1 is the fundamental root mean square current component;
I_h is the harmonics root mean square current component.
The Total Harmonic Voltage Distortion factor (THD_V) can be calculated by:

$$THD\%|_V = 100\% \times \sqrt{\sum_{h\neq 1}\left(\frac{V_h}{V_1}\right)^2} \tag{3}$$

where

V_1 is the fundamental root mean square voltage component;
V_h is the harmonics root mean square voltage component.
Another useful index for the current can be used, which is the Total Demand Distortion (*TDD*). Equation (4) describes this index. In this index, the sum of the harmonics is compared to the demand current, not to the fundamental current, as in (*THD*).

$$TDD\%| = 100\% \times \sqrt{\sum_{h\neq 1}\left(\frac{I_h}{I_L}\right)^2} \tag{4}$$

where I_L is the demand current during the test period, *TDD(I)* = Total Current Demand Distortion.

3. System Information and Measurement Setup

The power grid of SQU has an internal 11.0 kV network ring supplied by three substations (A, B, and C) that step down the voltage from 33 kV to 11 kV. Each building and load center has a step-down transformer (11/0.415 kV), as depicted in the single line diagram (SLD) in Figure 1. The harmonics measurement and analysis were undertaken at different locations within the SQU campus, at the load levels, such as UPS and VFDs, at the 415 V busbar of the step-down transformer (11/0.415 kV) and the main substation at the 11 kV bus. Such diversity helps map and quantify the harmonics penetration and spread at the main PCC within the grid and also within the university networks. At the measurement points, three voltage phases, current, THD, and harmonics content, were monitored as the typically required measurement qualities for a power quality investigation at this site.

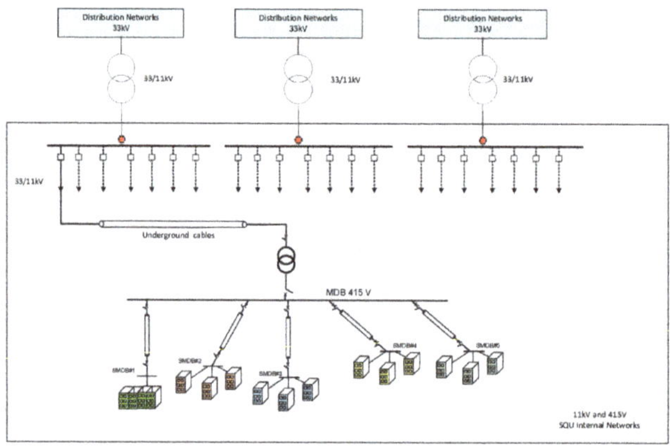

Figure 1. Representative SLD for the SQU 11 kV and 415 V networks.

3.1. PQ Monitoring Results and Analysis for the PV-Inverter System

Figure 2 shows the SLD for the PV system under study and the location of the measurement point, the PQ meter. This is a Car Park PV project located at the SQU campus, and the system consists of 333 solar PV panels with a rated power of 255 Wp for each panel. The panels are grouped and connected into four on-grid inverters. The size for each inverter is 20 kW with 4 strings of photovoltaics [15].

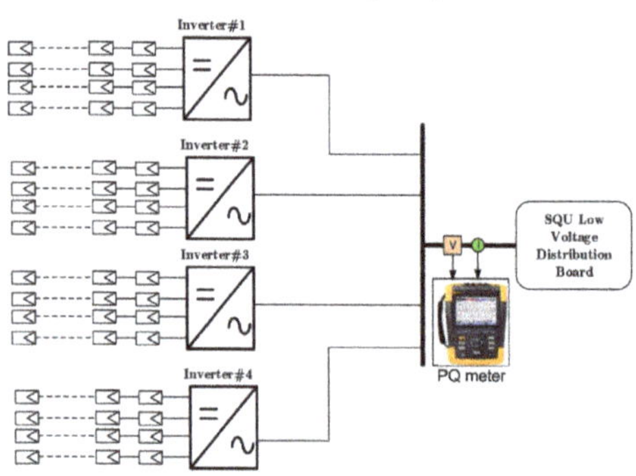

Figure 2. SLD of the PV-inverter system.

Figure 3 shows the profile of the current, voltage, THDI, and THDV for one week. The THDV variation was between 0.9% and 1.7%, while the THDI had a wider range of values, and thus, TDD analysis was required in this case. Figure 4 shows the zoom for two days. It is observed that TDD showed a steady variation for the generation period with a maximum value of 3.315%. THDI decreased when the PV system generated power but exhibited significant spikes during the transition periods at sunrise and sunset when energy production from the PV started and stopped. Figure 5 shows the variation in THDI with the output current. Analysis revealed that THDI dropped to 10% when the generated power reached 17% and further decreased to 5% at 37% of the generated power. Figure 6 illustrates the variation in THDI at different times of the day, indicating a higher THDI at the beginning and at the end of power generation from the PV inverter (i.e., at sunrise and sunset). The THDI was higher in the afternoon compared to that in the morning and toward sunset compared to sunrise, which is likely due to the grid status and background voltage harmonics. Figure 7 shows the variation in voltage THD at the grid connection point for the same day. It is clearly observed that the voltage THD was higher in the afternoon and at sunset compared to sunrise. As is clearly observed, the THD-I value in the afternoon was higher than that in the morning for the same percentage of generated current. This increase was mainly due to an increase in the grid distribution. Since the inverter controller used grid voltage for its synchronization and feedback control, voltage harmonics impaired control performance, leading to increased current harmonics.

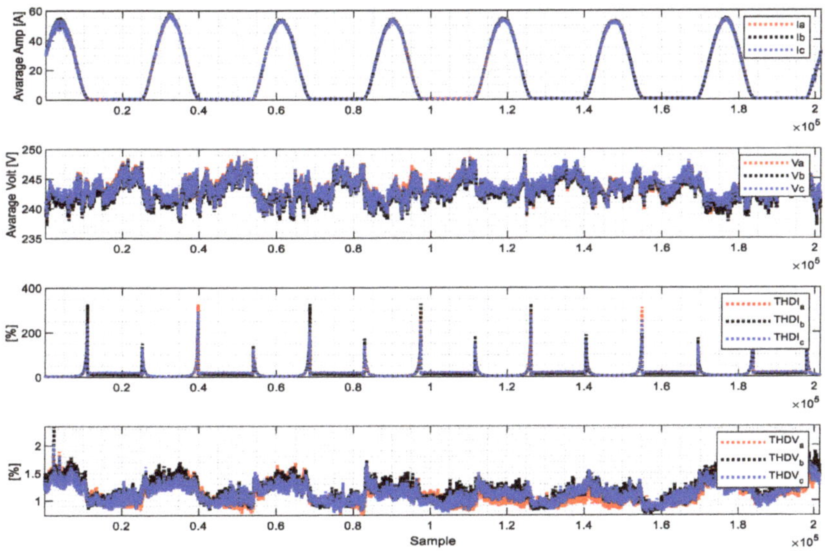

Figure 3. Current, voltage, THDI, and THV of the PV-inverter.

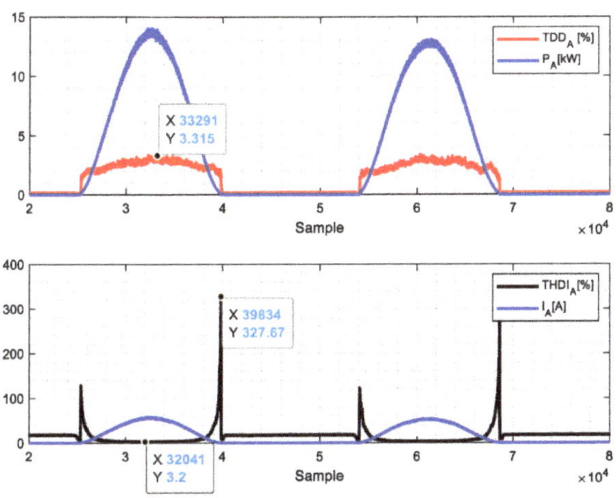

Figure 4. Variation in THDI and TDD over a two-day period varies in current and active power.

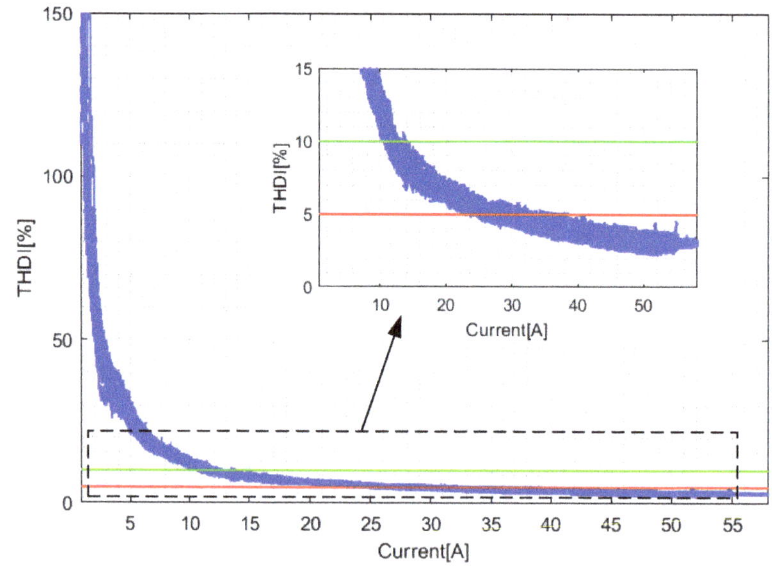

Figure 5. Variation in THDI versus the output current.

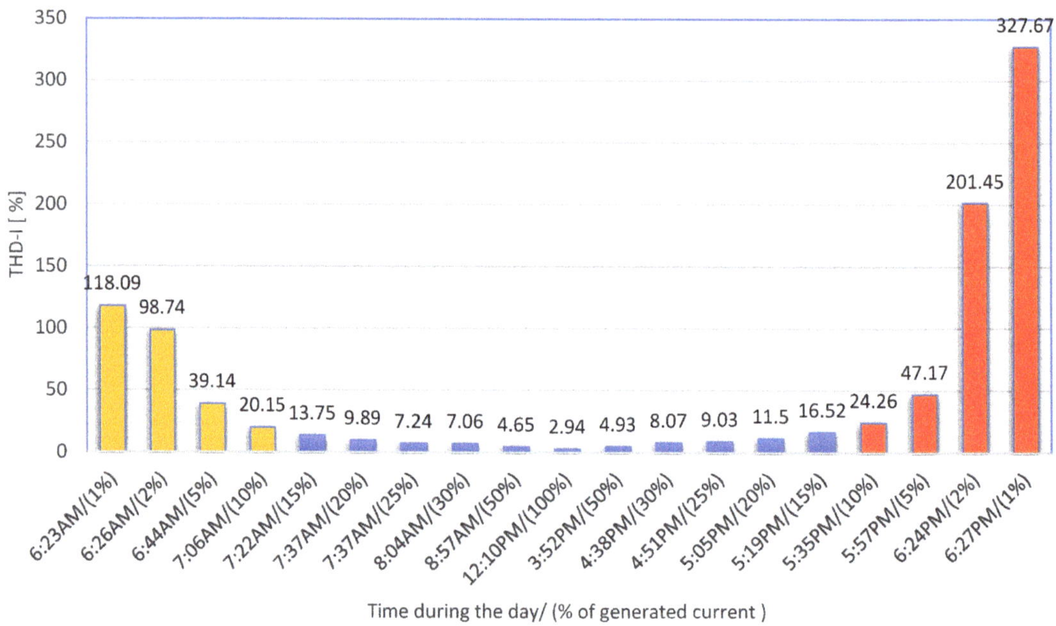

Figure 6. Variation in the THD-I during the day with respect to the percentage of generated current from the PV system.

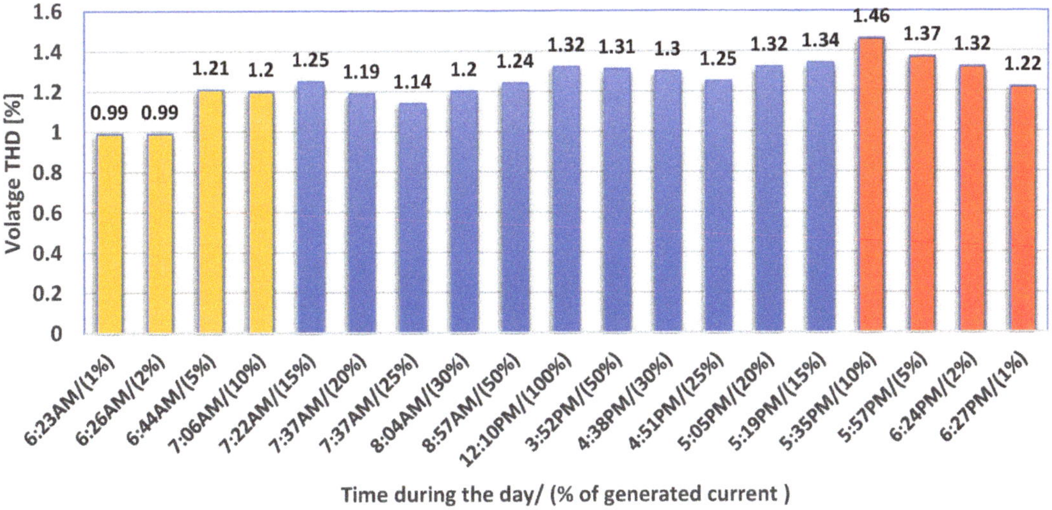

Figure 7. Variation in the THD-V at the PCC point at a percentage of generated current from PV system.

3.2. Monitoring Results and Analysis CIS and UPS

Figure 8 shows an SLD for the electrical system in the Computer Information Center (CIS) at the university. This system consists of an uninterruptible power supply (UPS), lighting loads, air conditioning units (ACs), and other single-phase loads. The system is supported by a standby diesel generator (for emergencies). This site is supplied by an internal 11 kV distribution system within the university through a distribution transformer (11/0.415 kV). The measurements were conducted at two measurement points (PCC#1 and PCC#2), as depicted in Figure 8.

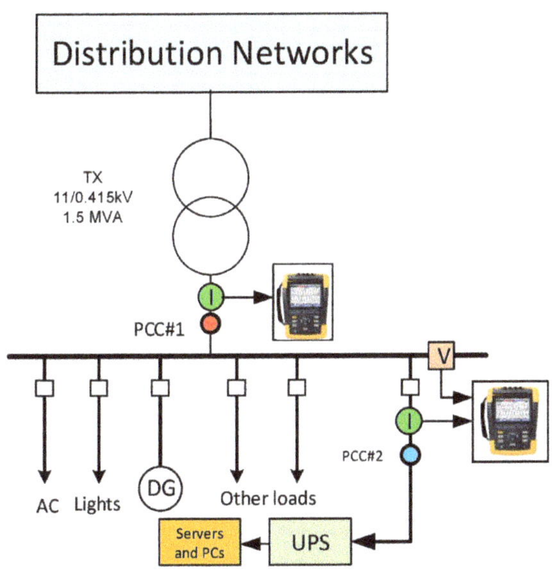

Figure 8. Single line diagrams for the CIS system under study.

3.2.1. Voltage Profile and Quality at the CIS Bus (PCC#1)

Plots in Figure 9 show the variation in three voltage phases. (The plots show three values: minimum, average, and max.). The voltage varied between 243.5 and 228.71 V, which is within the distribution code of Oman, ±6% of the normal value (240 V) [16]. Figure 10 shows the THD and harmonics voltage contents. The maximum records THDV for the three phases, A, B, and C, are 1.69%, 2.1%, and 1.49%, respectively, with an average value for all phases of 0.88%, which is within the acceptable value based on IEEE519 [15]. The dominant harmonics in all the phases were the 5th and 7th. The THD voltage showed varying values among the phases, which can be attributed to the unbalanced loading between them.

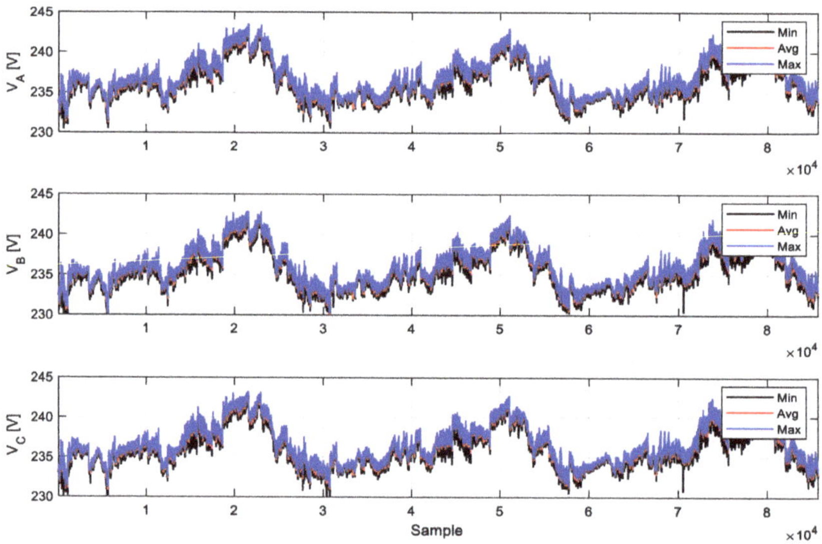

Figure 9. Voltage profile at the CIS main bus.

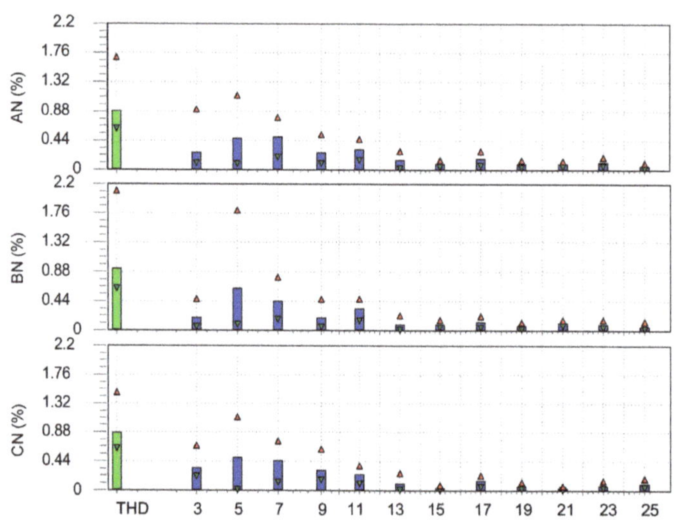

Figure 10. THD and Harmonics Spectrum for the phase voltages (Red triangle: maximum value, green triangle: minimum value, blue/green bar: average value).

3.2.2. Current Profile and Quality of the UPS System (PCC#2)

Plots in Figure 11 represent the root mean square average current patterns for the UPS for almost three days. As observed, the current had the same span variation (with min 24 A to max 30 A). This is a typical profile for such a system. The three current phases had some level of unbalance, especially for phase B, but were still within the acceptable limit.

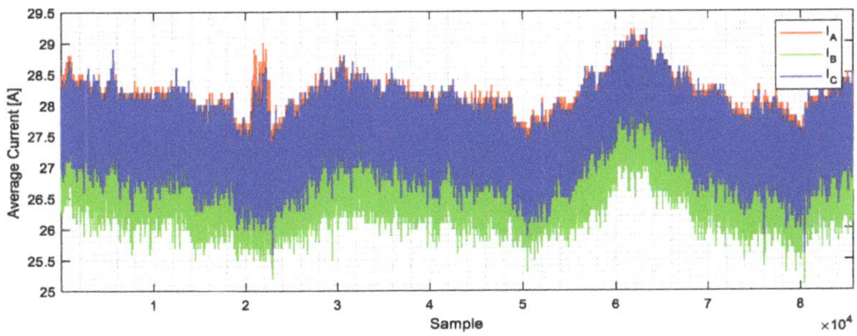

Figure 11. The three phases' root mean square average current patterns at PCC#2 (at the UPS connection point).

Figure 12 shows the variation in TDD. It is clearly understood that the UPS system exhibited merely a constant harmonics pattern for its complete daily cycles. Figure 13 shows the average TDD, THD, and Harmonics Spectrum for phase currents. The current was highly distorted, and it had odd as well as even harmonics. The dominant harmonics were different between the phases: for phases A and C, it was the 7th, while for phase B, both the 3rd and 7th were dominant. The TDD reached 11.95%, 14.83%, and 13.03% for phases A, B, and C, respectively. Phase B showed higher harmonics current distortion, which explains the increase in voltage distortion, as observed in Figure 13. Figure 14 shows the active, reactive, and power factors consumed by the UPS. The active power varied between 17 and 21 kW, with an average power factor of 0.98. The UPS-injected reverse reactive power varied between −1.8 and −2.8 kvar. This negative var was due to the filter capacitance existing on the front side of the UPS. The maximum registered current was 29 A (as observed in Figure 8). The short circuit level at PCC#1 can be calculated by $I_{sc(pu)} = V_{pu}/Z_{pu}$. Z_{pu} is the transformer impedance of 4.5%. Hence, $I_{sc} = 1.0/0.045 = 22.22$ p.u. (i.e., 2318.7 A). The ratio of I_{sc}/I_L was ratio = 2318.7/29 = 80, as per the IEEE519. The TDD limit was 12%, revealing that the TDD was over the limit at PCC#2.

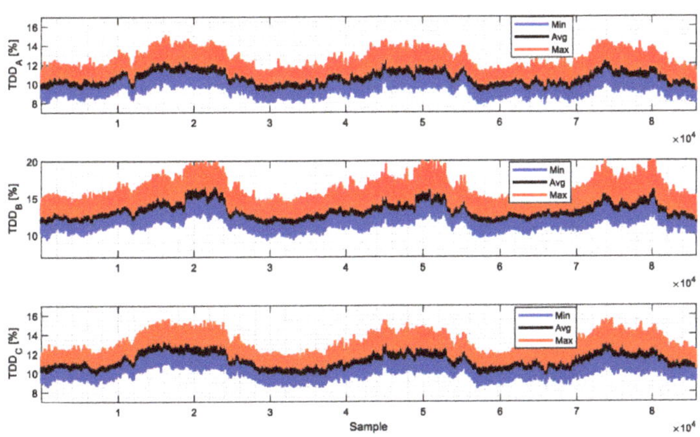

Figure 12. Variation in TDD against load current.

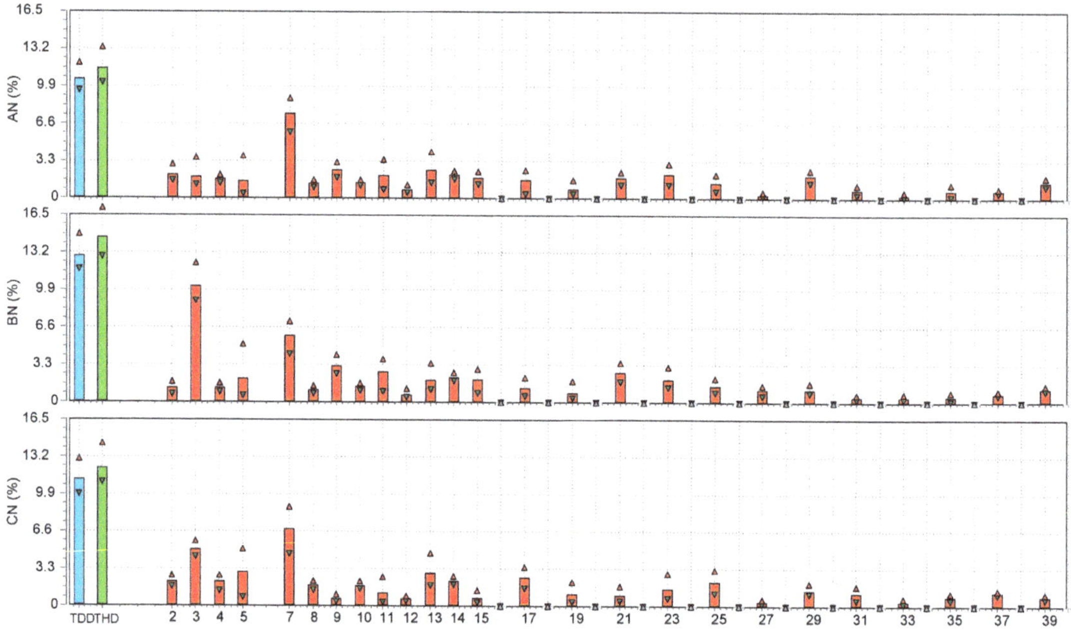

Figure 13. TDD, THD, and Harmonics Spectrum for the phase currents (Red triangle: maximum value, green triangle: minimum value, blue/red/green bar: average value).

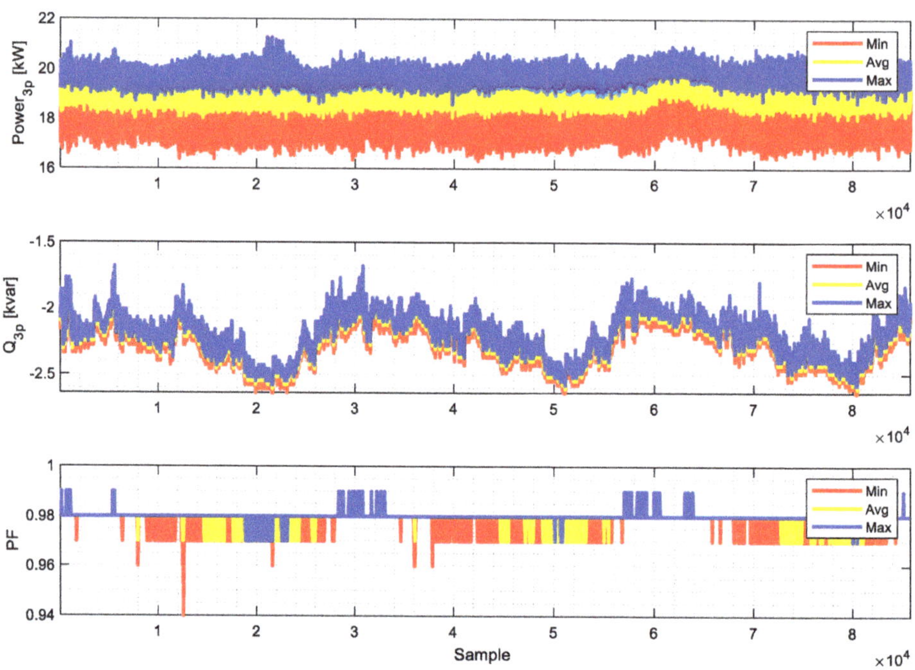

Figure 14. Active, reactive, and power factor for the UPS.

3.2.3. Current Profile and Quality for the Main Incomer (PCC#1)

Figure 15 represents the root mean square current patterns during the ten-day test period. The plots show the variation in three values of the phase currents: minimum, average, and maximum, and Figure 16 shows the associated THDI profile. Figure 17 shows TDD, THD, and Harmonics Spectrum for phase currents. At this point, the current was slightly distorted. The level of harmonics at the PCC point was reduced due to harmonics cancelation and increased with linear load compared with no linear load. The TDD reached an average of 2.3% for phases A, B, and C, respectively. The dominant harmonics were the 3rd, 5th, and 7th. According to IEEE519, the TDD level at PCC#1 was within the standard. It is worth mentioning that the distortion created at PCC#1 by the UPS system did not significantly impact the overall TDD at PCC#1. This is attributed to the size of the UPS compared to the overall system size.

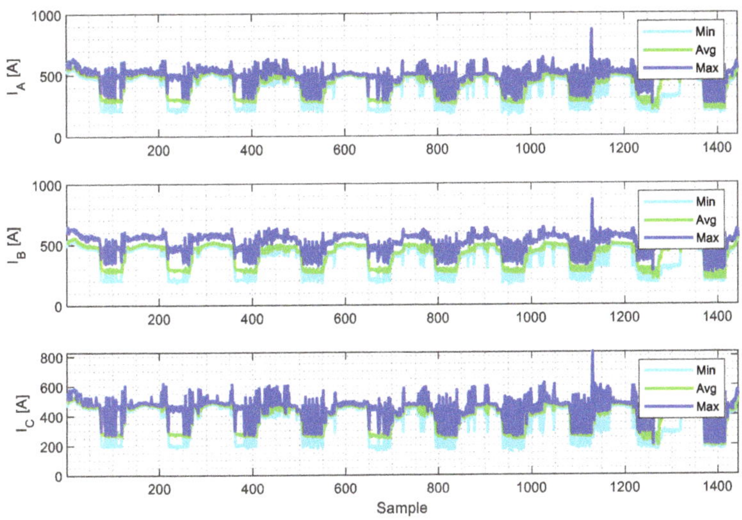

Figure 15. The three phases' current profiles during the test period.

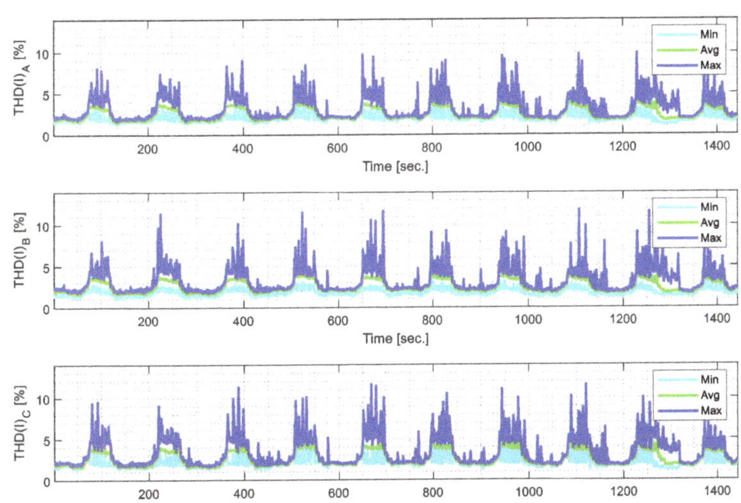

Figure 16. The three phases' THD current profiles during the test period.

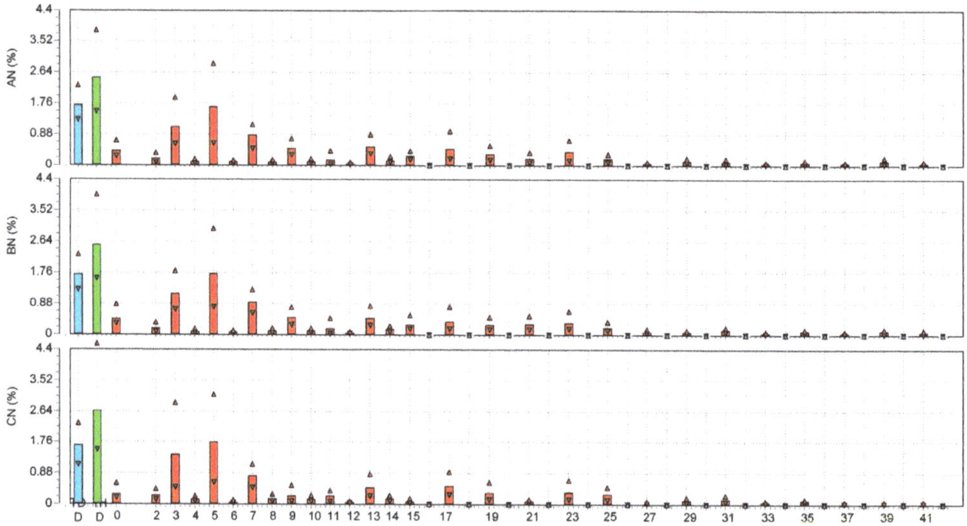

Figure 17. TDD, THD, and Harmonics Spectrum for the phase currents. (Red triangle: maximum value, green triangle: minimum value, blue/green/red bar: average value).

Figure 18 shows the patterns for the selected phase (phase A, other phases showed similar profile), its current, THDI, TDD, and the dominant harmonics. The current indicated two loading levels: a day loading level where the current reached an average of 530 amp and a night loading level with an average value of 250 amp. The harmonic levels increased at night and decreased during the day, as observed in Figure 19. The THD ranged between 2 and 4%, and the dominant harmonic was the fifth. Such a profile can be attributed to the increase in linear loads during the day (mainly AC motors). However, for both operational scenarios, the TDD did not exceed 2.2%.

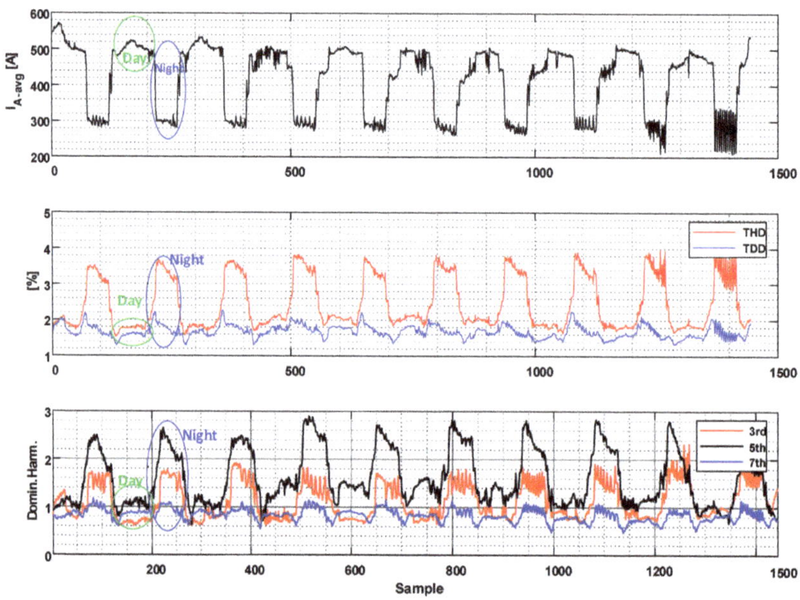

Figure 18. Phase current, TDD, THD, and Harmonics Spectrum for phase A.

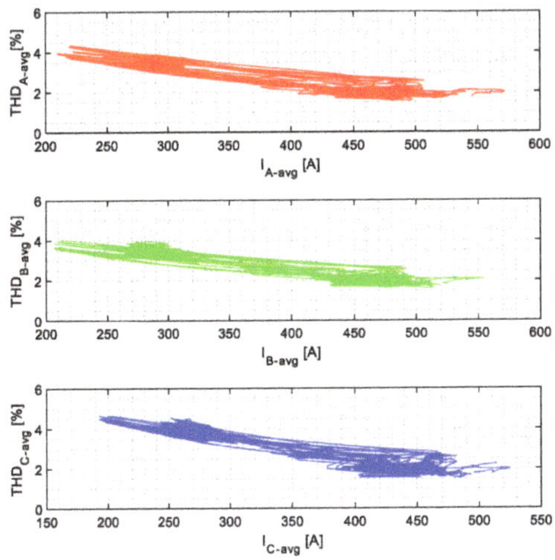

Figure 19. Variation in THD against load current.

3.3. *Current Profile and Quality for a Motor Drive (Variable Speed Drive, VDF)*

Most of the cooling systems on campus are based on a chiller system, where chilled water is circulated among the buildings using motor pumps. These motor pumps were recently upgraded with variable-speed drives (VFDs) to enhance control and efficiency. This subsection shows the measurement and analysis for two types of VFD.

3.3.1. Current Profile and Quality for VFD-I

Figure 20 shows the average three phases' currents, THDIs, and TDDs for the VFD used in the HVAC system (for circulating the cooled water). The measurements were taken for almost 24 h. The currents varied between 7.5 A and 9.0 A. Figure 21 shows a screenshot of the current waveforms, which represent 6-pulses. The current of the phase C was highly distorted. The average value of THDI was 130%, and the average value of TDD was 63%. The dominant harmonics currents were 5th, 7th, 11th, and 13th, as depicted in Figure 22.

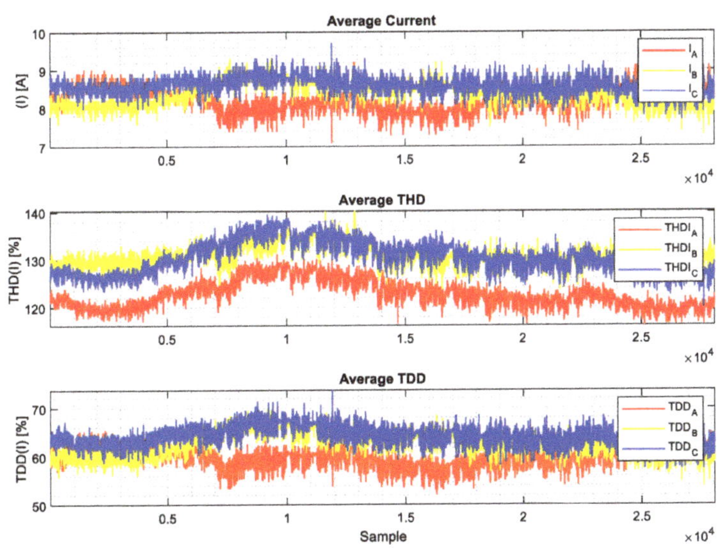

Figure 20. Patterns for the three phases' current, TDD, THD for VFD-1.

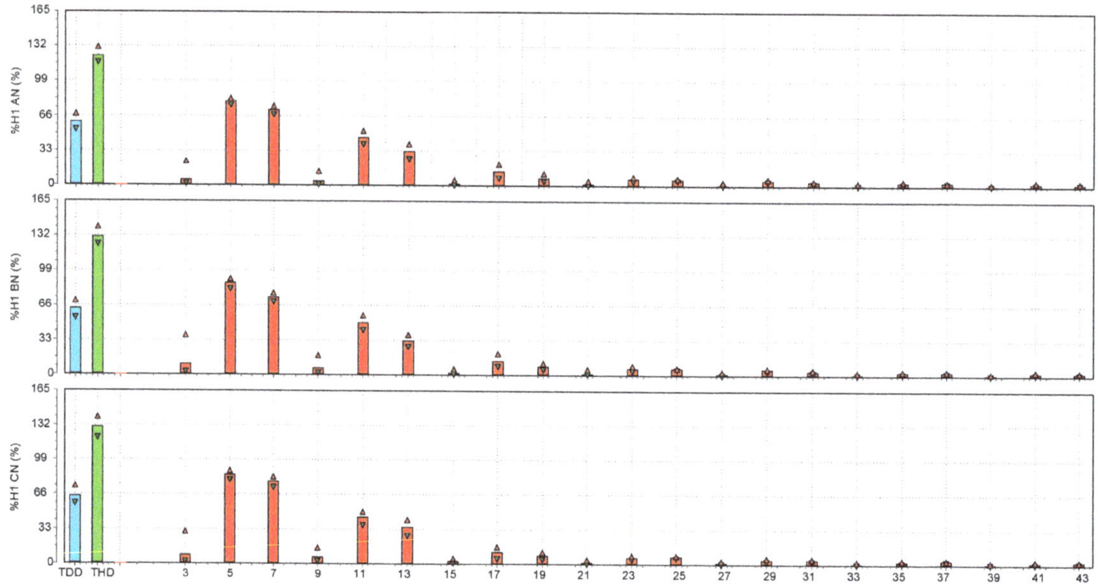

Figure 21. TDD, THD, and Harmonics Spectrum for the phase currents of VFD-I. (Red triangle: maximum value, green triangle: minimum value, blue/green/red bar: average value).

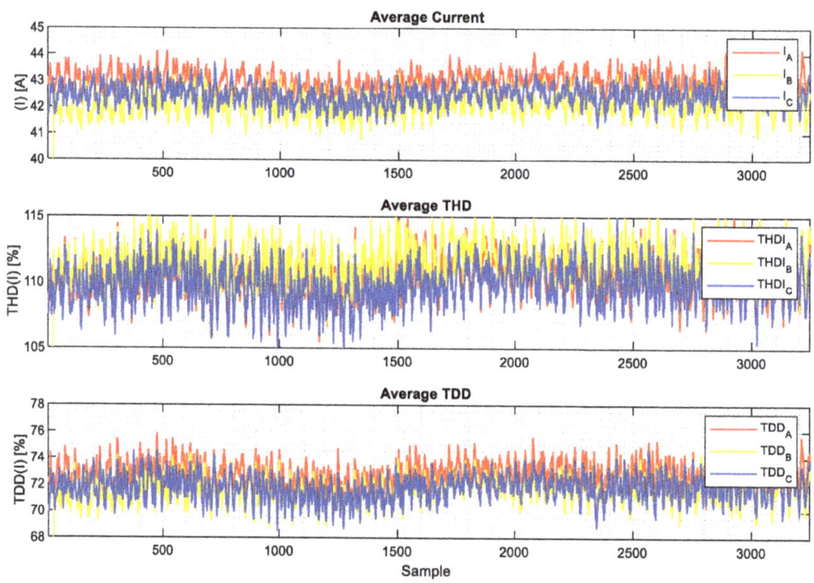

Figure 22. Patterns for the three phases' current, TDD, THD for VFD-II.

3.3.2. Current Profile and Quality for VFD-II

Figure 22 shows the average three phases' current, THDI, and TDD for the VFD used in the HVAC system (for circulating the cooled water). The currents changed around a value of 43.0 A. Figure 23 shows a screenshot of the current waveforms, which represent 6 pulses operation. The current in phase C was highly distorted. The average value of

THDI was 110%, and the average value of TDD was 72%. The dominant harmonics currents were the 5th, 7th, 11th, and 13th, as shown in Figure 24.

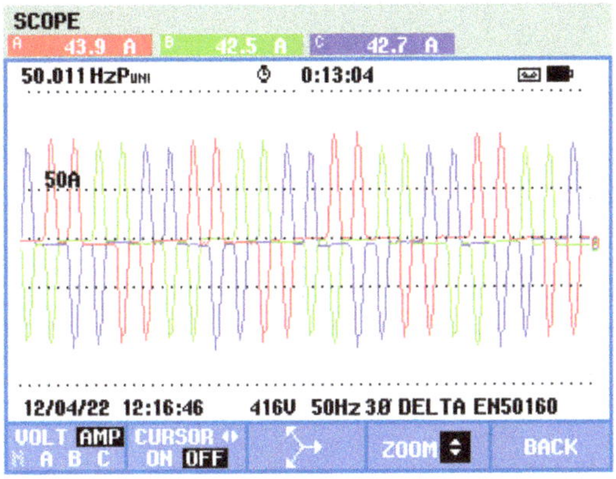

Figure 23. Screenshot of the three phases' waveforms for VFD-II.

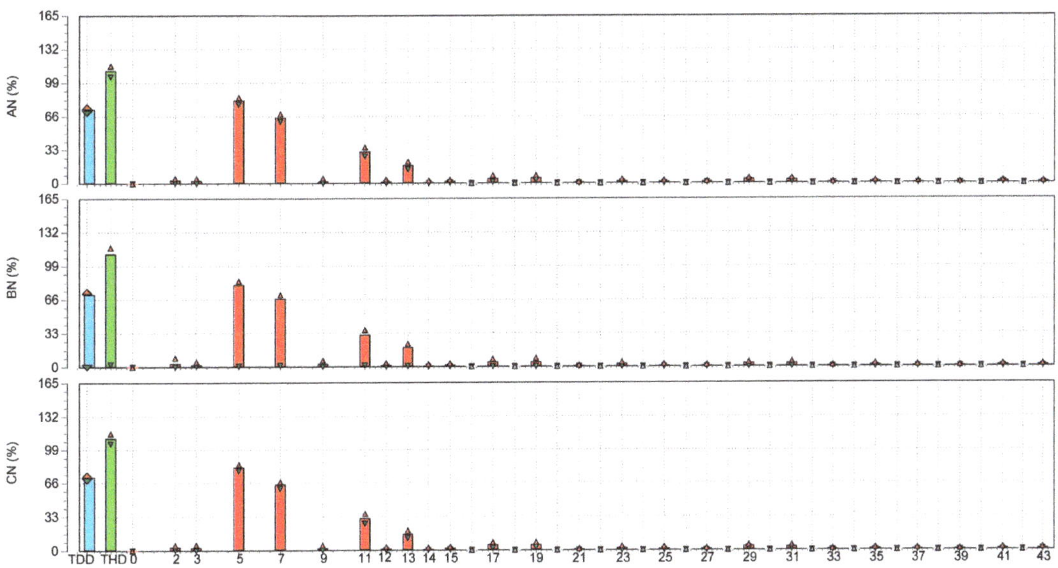

Figure 24. TDD, THD, and Harmonics Spectrum for phase currents of VFD-2. (Red triangle: maximum value, green triangle: minimum value, blue/green/red bar: average value).

3.4. Current Profile and Quality at Engineering Building

A sample from the colleges on the SQU campus was monitored, which was the College of Engineering. Figure 25 shows a single line diagram of the electrical system used in the engineering building. This system consists of an HVAC based on variable frequency drives (VFDs), light loads, and other loads. Figure 26 shows the current, THDI, and TDD measured at the 0.415 kV bus on the LV side of the building transformer. The current showed two levels: a day level that reached 430 A, and a night level dropped to 155 A.

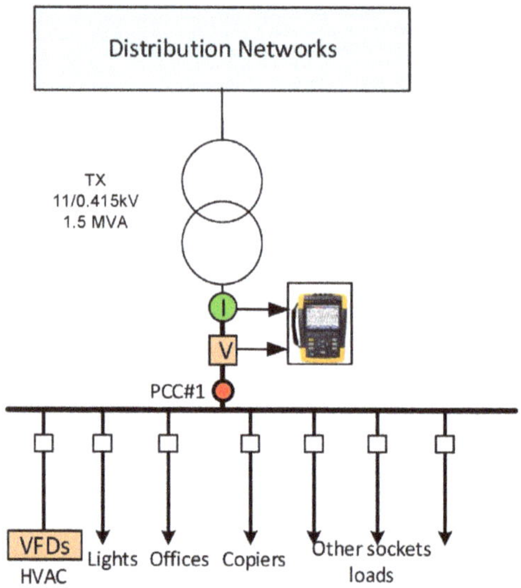

Figure 25. Single line diagrams for the engineering building system.

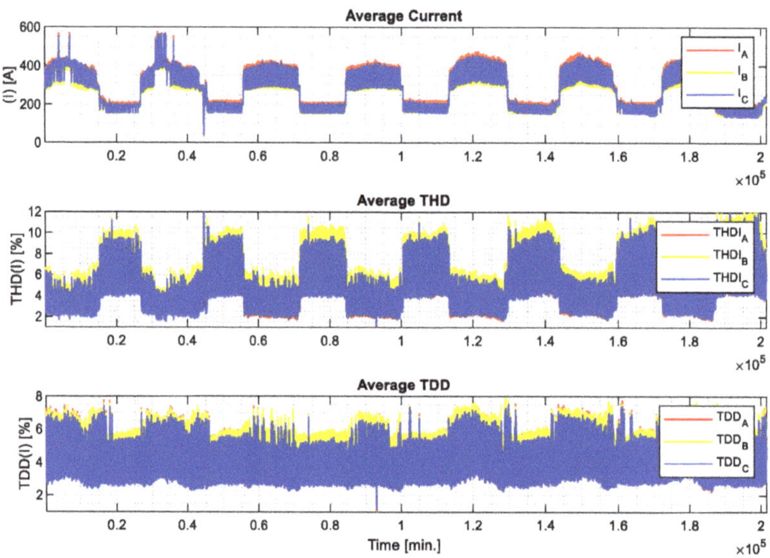

Figure 26. The three phases' current, TDD, THD.

3.5. *Current Profile and Quality at Main Substation C*

Figure 27 shows a representative SLD for substation C 33/11 kV. The measurements were taken on the 11 kV side of the transformer, as pointed out in the figure. The measurement was limited to current due to the accessibility limitation to the voltage circuit and we observed that there were no issues with voltage harmonics, as the downstream measurement showed that the voltage distortion was below the limit.

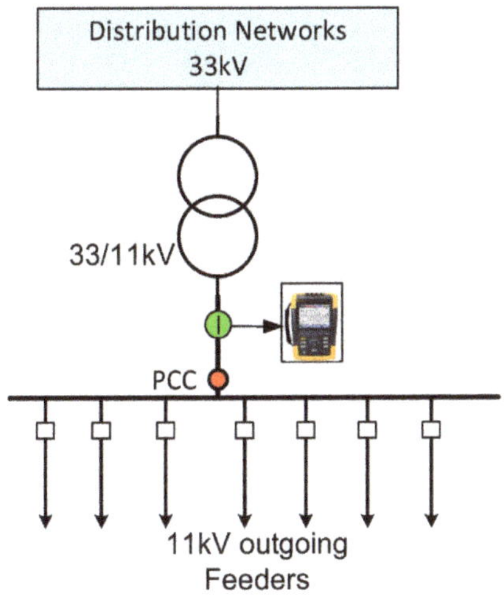

Figure 27. Representative SLD for Substation C.

Figure 28 shows the measured phase currents and the THDI at the measured point, indicated in Figure 27, which was the 11 kV incomer side of the main substation C that feeds the university outgoing feeders. From Figure 28, the variation and relationship between the load current level and the THD-I was not consistent over different periods. As depicted in point 1 of Figure 28, the relationship was directly proportional during certain periods. However, in other periods, it was inversely proportional, as shown in point 2. This kind of pattern created two group levels of the THD-I chat, which are shown in Figure 29, which shows THD with respect to the current that flowed in the phases.

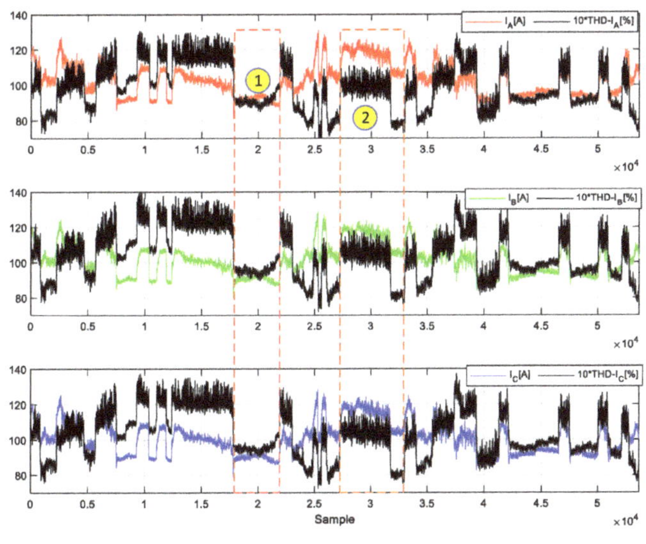

Figure 28. The three phases' currents and their THDs patterns at Substation C.

It is interesting to note that, for the same load current, there were multiple values for the THD-I, as highlighted by the rectangular area in Figure 29. The THD-I varied from 7% to 14%, depending on the load's operational mode. When linear loads were switched on, the THD was lower, whereas the addition of nonlinear loads increased the

THD. For the transition to a higher current, as seen in Zone 2 of Figure 29, the THD-I point in Figure 29 shifted from Zone A to Zone B. For the same Zones A and B, the THD-I showed a downward trend. This can be explained by the transition to a higher current.

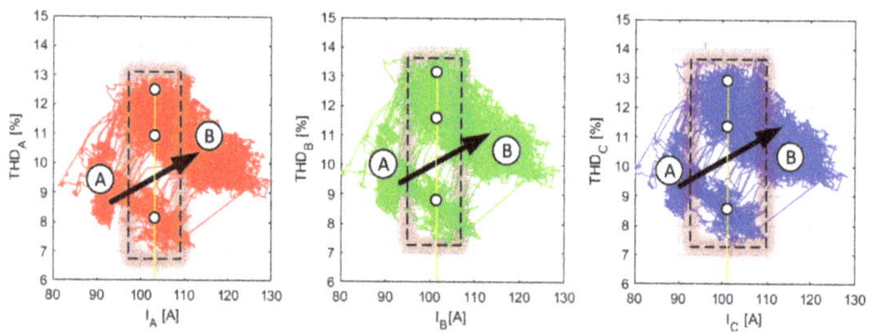

Figure 29. Variation in Total Harmonic Distortion (THD) for the different phases with respect to the current that flowed in the phases at Substation C.

The variation in THD exhibited a unique profile that was highly dependent on the timing of operations. During certain periods, as the current increased, the THD decreased (Period A in the graph), indicating that the added loads were linear. Conversely, in other periods, as the current increased, the THD increased, indicating that the added loads were nonlinear. This behavior was consistent for both low- and high-current operations. However, when transitioning from low- to high-current groups, there was a noticeable increase in THD.

Figure 30 illustrates the TDD, THD, and Harmonics Spectrum for the phase currents of Substation C. The Total Demand Distortion for different phases was similar, whereas the THD for phase A was slightly lower compared with the other phases. The average THD was 10.45%, while the average TDD was 8.35%. The dominant harmonics currents were 5th, 7th, 11th, and 13th, as depicted in the figure.

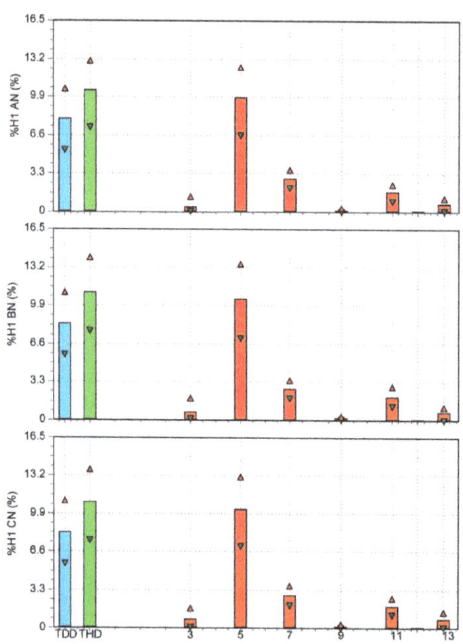

Figure 30. TDD, THD, and Harmonics Spectrum for the phase currents of Sub C. (Red triangle: maximum value, green triangle: minimum value, red/blue/green bar: average value).

3.6. Current Profile and Quality at the Main Substation A

SQU has built a new substation, substation A, to meet the new demand associated with an increase in the system capacity and new buildings for the hospital. The new building has been equipped with modern electrical devices, and the majority are power electronics-based devices (LEDs, LCDs, medical equipment, etc.). However, the cooling load was still the majority load, which represented around 70% of these buildings. The measurements were repeated at the main substation A, and an almost similar trend in results was obtained to that of substation C. The load current consumed from this substation was more (almost 300 A) compared with substation C (around 100 A). Figure 31 shows the measured phase currents, and the THD-I measured at the 11 kV bus of the main substation station. Compared with substation C, the levels of THD and TDD were lower at this location. The reduction in THD and TDD was around 60%, and phase A had the lowest distortion for both substations. The loads on the two phases were slightly lower compared with the load on the third phase, as shown in Figure 31. These two phases will have slightly more THDI compared with the third phase. The variation in THD-I across different phases relative to the load current is depicted in Figure 32. Generally, at this site, as the load increased, THD-I decreased. This is because the additional load consisted mostly of linear loads, resulting in a lower THD. There was more variation in phases B and C compared with that of phase A since the load on phase A was compared more with the other phases.

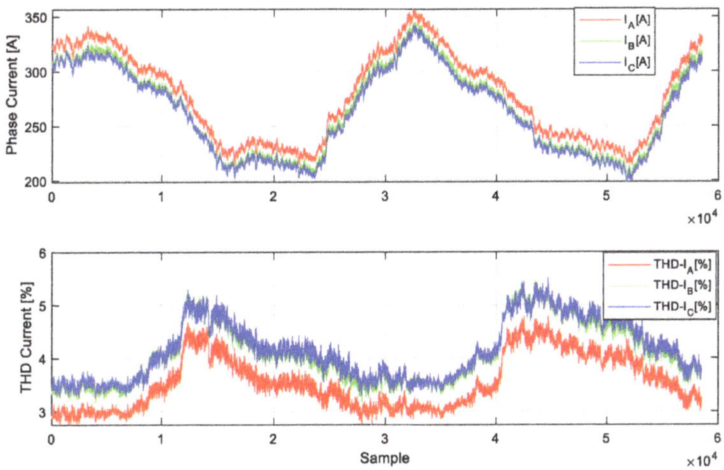

Figure 31. The three phases' currents and the THD for the different phase currents at Substation A.

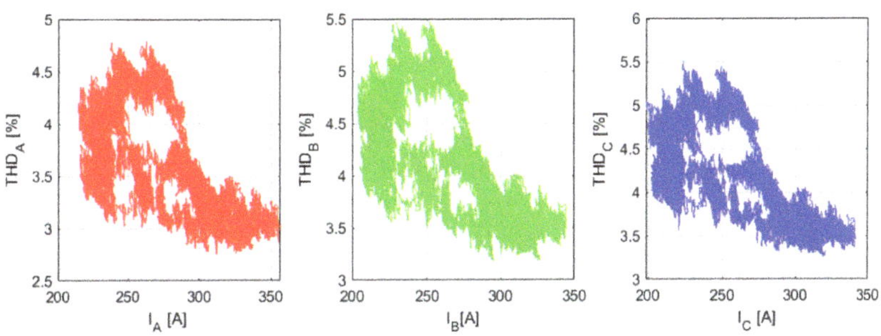

Figure 32. Variation in Total Harmonic Distortion (THD) for the different phases versus the phases current at Substation A.

Figure 33 illustrates the TDD, THD, and Harmonics Spectrum for the phase currents of Substation A. The Total Demand Distortion for different phases was identical, whereas the THD for phase A was slightly lower compared with the other phases.

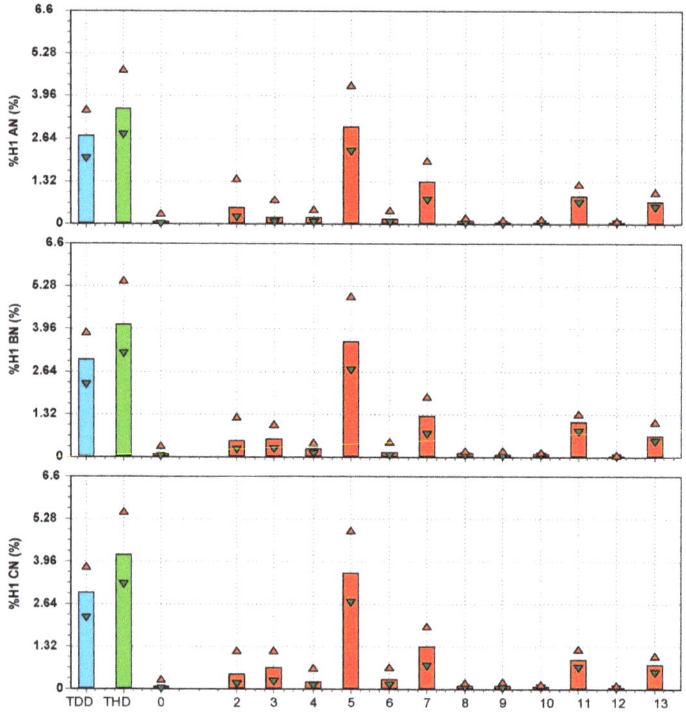

Figure 33. TDD, THD, and Harmonics Spectrum for the phase currents of Substation A. (Red triangle: maximum value, green triangle: minimum value, blue/green/red bar: average value).

The Total Demand Distortion (TDD) for the different phases was identical, whereas the THD for phase A was slightly lower compared with the other phases. The average THD was 3.96%, while the average TDD was 2.6%. The dominant harmonics currents were the 5th, 7th, 11th, and 13th, as shown in Figure 33.

4. Summary and Conclusions

This paper presented a comprehensive power quality assessment of and investigation into the electrical distribution system at Sultan Qaboos University (SQU). Harmonics measurements and analysis were conducted at various locations across the SQU campus, including at the load level (UPS, PV inverters, VFDs), at the 415 V busbar of the distribution transformers level, and at the MV level, 11 kV bus of the main substations. This diverse measurement approach helped to map and quantify the penetration and spread of harmonics at the main point of common coupling (PCC) with the grid and within the university networks. The key findings are

- The voltage harmonics THDV at all load-side connected points (415 V) exhibited low voltage distortion with a THDV of between 0.9 and 1.7%, within the distribution code and standards;
- Table 1 summarizes current harmonics THDI and TDD measurement results performed at different locations on the SQU campus, including the 11 kV bus at two substations;

- The PV inverter system exhibited noticeable current distortion. THDI had higher variations during sunrise/sunset transitions but TDD remained within acceptable limits per the standards;
- The UPS system showed highly distorted current waveforms with dominant odd and even harmonics. TDD reached 11.95–14.83% and THDI over 100% for some phases. However, at the UPS input (PCC#1), the harmonics attenuated significantly due to cancellations and linear load mixing, reducing TDD to 2.3%;
- The TDD values at the UPS connection point exceeded the IEEE519 limit. Thus, it is crucial to prioritize the design and implementation of passive filter solutions. Specifically, enhancing the design and configuration of passive filters can effectively mitigate harmonic distortions. This approach involves selecting appropriate filter types and capacities tailored to the specific harmonic frequencies present in the system. Regular monitoring and assessment of filter performance should also be conducted to ensure sustained compliance with regulatory standards and optimal operational efficiency;
- The VFDs demonstrated extreme current distortion, with THDI over 100% and TDD over 60%, with dominant 5th, 7th, 11th, and 13th harmonics. This is expected for 6-pulse VFDs. A passive filter with proper values is recommended. It is expected that with a passive filter, the THD will reduce to 10%;
- At both the building and substation levels, current harmonics were noticeable but remained within IEEE 519 limits when accounting for the short circuit current. The Total Harmonic Current Distortion (THDI) ranged from 2 to 10%, and the Total Demand Distortion (TDD) ranged from 2 to 8%, depending on the loading;
- SQU plans to move toward a green smart system with integrated large-scale PV, upgraded and automated cooling systems, and the replacement of lights with LEDs within the networks. Therefore, the following recommendations are proposed:
    - Conduct regular power quality audits (health checks) and maintenance of equipment to prevent the escalation of harmonic issues;
    - Suppress the harmonics level associated with future expansion and upgrade of the PV system and VFDs by adding filters to the VFD systems and adopting advanced control techniques, such as active front-end converters;
    - Utilizing the isolation transformers with specific winding configurations (e.g., K-factor transformers) designed to handle non-linear loads;
- The results highlighted the impacts of diverse and aggregate linear loading, with noticeable differentiation between more and less heavily loaded sections of the distribution network. Recommendations are provided for ongoing monitoring and consideration of targeted mitigation efforts to maintain power quality as campus loads modernize and electricity demand increases. The presented methodology and dataset provide a benchmark to guide standards compliance as SQU upgrades aging equipment and expands new high-tech facilities.

Table 1. Measurement results.

Type of System	THD-I	TDD
PV Inverter system	2.94–327.67%	3.315%
UPS	Exceed 100%	11.95–14.83%
VFD-1-HVAC	130%	63%
VFD-2-HVAC	110%	72%
Engineering Building	3–10%	3–7%
Main Substation C	9.9%	7.4%
Main Substation A	3.9%	2.6%

Author Contributions: Methodology, K.M.A.; Software, K.M.A.; Formal analysis, A.H.A.-B.; Investigation, K.M.A.; Data curation, K.M.A.; Writing—original draft, K.M.A.; Writing—review & editing, A.H.A.-B.; Visualization, K.M.A.; Supervision, A.H.A.-B.; Project administration, K.M.A.; Resources A.H.A.-B. All authors have read and agreed to the published version of the manuscript.

Funding: This research received no external funding.

Data Availability Statement: The original contributions presented in the study are included in the article, further inquiries can be directed to the corresponding author.

Acknowledgments: The authors gratefully acknowledge the support of Sultan Qaboos University (SQU) in achieving these research outcomes. Special thanks are extended to the staff of the College of Engineering and the Maintenance &Project Department for their invaluable assistance.

Conflicts of Interest: The authors declare no conflict of interest.

References

1. Singh, B.; Al-Haddad, K.; Chandra, A. A Review of Active Filters for Power Quality Improvement. *IEEE Trans. Ind. Electron.* **1999**, *46*, 960–971. [CrossRef]
2. *IEEE Standard 519-2022*; IEEE Standard for Harmonic Control in Electric Power Systems. IEEE: New York, NY, USA, 2022; pp. 1–31.
3. *IEC 61727*; Photovoltaic (PV) Systems—Characteristics of the Utility Interface. International Electrotechnical Commission (IEC): Geneva, Switzerland, 2004.
4. *IEC 61000-3-2*; Electromagnetic Compatibility (EMC) Part 3-2: Limits for Harmonic Current Emission (Equipment Input Current <16 A per Phase). International Electrotechnical Commission (IEC): Geneva, Switzerland, 2000.
5. *IEC 61000-3-12*; Electromagnetic Compatibility (EMC) Part 3-12: Limits for Harmonic Currents Produced by Equipment Connected to Public Low-Voltage Systems with Input Current <75 A per Phase. International Electrotechnical Commission (IEC): Geneva, Switzerland, 2003.
6. Kontogiannis, K.P.; Vokas, G.A.; Nanou, S.; Papathanassiou, S. Power Quality Field Measurements on PV Inverters. *Int. J. Adv. Res. Electr. Electron. Instrum. Eng.* **2013**, *2*, 11.
7. Alawasa, K.; Al-Odienat, A. Power Quality Investigation of Single Phase Grid-connected Inverter of Photovoltaic System. *J. Eng. Technol. Sci.* **2019**, *51*, 597–614. [CrossRef]
8. Khatri, M.; Kumar, A. Experimental Investigation of Harmonics in a Grid-Tied Solar Photovoltaic System. *Int. J. Renew. Energy Res. IJRER* **2017**, *7*, 901–907.
9. Alawasa, K. Harmonic distortions and interactions of modern residential electric loads. *Int. J. Eng. Technol.* **2018**, *7*, 4174–4181.
10. Alawasa, K.; Al-Mbaideen, A. Power quality assessment and analysis for low voltage distribution networks. *Jordan J. Electr. Eng.* **2018**, *4*, 165–175.
11. Mansor, M.A.; Hasan, K.; Othman, M.M.; Noor, M.S.Z.; Musirin, A.I. Construction and Performance Investigation of Three-Phase Solar PV and Battery Energy Storage System Integrated UPQC. *IEEE Access* **2020**, *8*, 103511–103538. [CrossRef]
12. Yashomani, Y.; Shyama, K.; Das, P. Experimental Investigation of a Single-Phase UPQC With Minimum VA Loading. *IEEE Trans. Power Deliv.* **2007**, *22*, 373–380.
13. Javadi, A.; Hamadi, A.; Woodward, L.; Al-Haddad, K. Experimental Investigation on a Hybrid Series Active Power Compensator to Improve Power Quality of Typical Households. *IEEE Trans. Ind. Electron.* **2016**, *63*, 4849–4859. [CrossRef]
14. Hartungi, R.; Jiang, L. Investigation of Power Quality in Health Care Facility. In Proceedings of the International Conference on Renewable Energies and Power Quality (ICREPQ'10), Granada, Spain, 23–25 March 2010.
15. Alawasa, K.M.; AlAbri, R.S.; Al-Hinai, A.S.; Albadi, M.H.; Al-Badi, A.H. Experimental Study on the Effect of Dust Deposition on a Car Park Photovoltaic System with Different Cleaning Cycles. *Sustainability* **2021**, *13*, 7636. [CrossRef]
16. Oman Distribution Code. Available online: https://dcrp-oman.com/documents/The%20Distribution%20Code.pdf (accessed on 16 June 2023).

energies

Review

Review of Uncertainty Sources in Optical Current Sensors Used in Power Systems

Marcelo M. Costa [1,*], Maria A. G. Martinez [2] and João C. W. A. Costa [3]

[1] Departament of Tests, Centrais Elétricas Brasileiras S.A. (Eletrobras), Belém 66115-000, Brazil
[2] Department of Electronic Engineering, Federal Center for Technological Education Celso Suckow da Fonseca, Rio de Janeiro 20271-204, Brazil; maria.martinez@cefet-rj.br
[3] Faculty of Electrical and Biomedical Engineering, Federal University of Pará, Belém 66075-110, Brazil; jweyl@ufpa.br
* Correspondence: marcelo.melo@eletrobras.com

Abstract: Optical current sensors have been developed and improved over the past few decades, and they have been increasingly employed in power systems, including smart and high-voltage grids. This is due to their many advantages over conventional electromagnetic current sensors, such as reduced size and weight, greater operational safety, and electromagnetic immunity. Like any measuring instrument or system, their quality and reliability are associated with measurement uncertainty, which quantifies their precision. This measurement uncertainty depends on a series of influencing quantities, such as the wavelength of light used in the sensor, the birefringence of the optical material used in the construction of the sensor, and environmental conditions, such as temperature and vibration. This article presents a review of the main influences that affect the quality and performance of optical current sensors and how these influences can be used to estimate measurement uncertainty. The main objective is to serve as a guide or reference for the identification and evaluation of uncertainty sources in optical current sensors used in power systems.

Keywords: optical current sensor; measurement uncertainty; Faraday effect; power systems

1. Introduction

Electromagnetic induction current transformers (CTs) are measuring devices of significant importance in power systems, being used for the most diverse purposes through the measurement of high-voltage electrical currents. Optical current sensors (OCSs) have many advantages over conventional CTs, such as reduced weight and dimensions, facilitating installation, maintenance, and use in remote or difficult-to-access areas; operational safety, as they are constructed using dielectric materials, and potentially catastrophic events will not cause explosions, as is the case for conventional oil-filled transformers, and there will be no critical situations when opening the secondary sides of the current transformers; linearity throughout their dynamic range; relative lower prices; wide bandwidth; absence of magnetic saturation effects; electromagnetic immunity (as they are non-electric sensors); AC and DC measurements; and low power consumption [1–4]. Table 1 summarizes the main differences between CTs and OCSs.

However, despite the extensive list of advantages of optical current sensors over conventional sensors, CTs are still known to be reliable in terms of accuracy and stability. This is not the case for OCSs, for which evaluations of these properties have not yet yielded conclusive or definitive results. In general, OCSs still need to be better characterized metrologically in terms of error and uncertainty so they can obtain a level of reliability similar to or better than CTs. This is one of the reasons why their use has not yet been disseminated, and it is the current challenge for the development of OCSs [5–8]. OCSs have several components, and each of these has behaviors and performance that affect their reliability. If the effects of these components on the response of the OCSs are not known and adequately modeled, the measurements performed will include large errors and uncertainty, significantly compromising the reliability of the sensor. Depending on the characteristics of the sensor and the measurement to be performed, these error or uncertainty sources can become significant or insignificant. Temperature fluctuations alone can cause variations in the response of some OCSs of up to a few percent, as will be shown throughout this article. If these variations have not been properly modeled, the measurement uncertainty of the respective OCS will be quite high, making its use unacceptable, depending on the application.

Table 1. Main differences between CTs and OCSs.

Characteristic	CTs	OCSs
AC and DC measurements	Only AC	AC and DC
Magnetic saturation	Yes	No
Dimensions and weight	Higher	Lower
Bandwidth	Wider	Narrower
Linearity	Lower	Higher
Operational safety	Lower	Higher
Electromagnetic immunity	No	Yes
Installation and maintenance	Harder	Easier
Power consumption	Higher	Lower
Accuracy	Higher	Lower

There is little literature available on discussions about the modeling and quantification of each of the OCS error or uncertainty sources and their combination to obtain sensor measurement uncertainty. Some authors deal with the metrological aspects of optical current sensors but focus only on error and accuracy [9,10]. There are still few approaches regarding measurement uncertainty or precision. The main objective of this paper is to serve as a reference or guide for identifying the uncertainty contributions that affect the performance of optical current sensors based on the Faraday effect. Another objective is to propose methods to estimate uncertainty in such a way that they are minimized. It is not the objective of this paper to estimate the identified uncertainty contributions, nor to estimate the measurement uncertainty of OCSs. Contributions from authors who determined the influence of quantities on the performance of optical current sensors were researched and are presented in this article, and it was also evaluated how the results obtained can be used to estimate the uncertainty contribution arising from these influencing quantities. Subsequently, the natures of some uncertainty contributions were analyzed to suggest an appropriate way to carry out their estimation. The main configurations used in optical current sensors were evaluated, and the contribution of each sensor functional module—light source, optical material, detection scheme, and indication of measured values—and its components to measurement uncertainty was evaluated. This paper is divided as follows: Section 2 presents the operating principle and the main characteristics of optical current sensors that are relevant for estimating their measurement uncertainty; Section 3 presents a review and analyzes the main factors that cause variations in the responses of OCSs, affect their performance, and can be considered sources of uncertainty; Section 4 summarizes the main sources of uncertainty and presents some suggestions for their estimation. Finally, Section 5 presents some conclusions and future perspectives.

2. Optical Current Sensors

Currently, optical current sensors (often referred to as optical current transducers, OCTs, or optical current transformers) are used for the most diverse purposes in electrical power systems, such as control, protection, revenue metering, power quality monitoring, measurement of the arbitrary form of primary currents, calibration, and measurement of current transients and impulses [11–15]. Rose et al. [16] presented the use and performance of an optical current sensor with an in-line Sagnac interferometer using a spun polarization-maintaining fiber as the sensing element in measurement and protection applications in high-voltage alternating current systems and in control, measurement, and protection applications in power systems in high-voltage direct current (HVDC) systems. In the area of protection of power systems, Azad et al. [17] present a data-based primary fault detection and identification algorithm for HVDC grids where current measurements are performed by OCSs. The algorithm benefits from the wide bandwidth of OCSs. Lei et al. [18] proposed a method for the measurement of transformer inrush currents using an OCS, taking advantage of its immunity to the effects of magnetic saturation. Ivanov et al. [19] provided a comparison between CT and OCS measurements of a 250 MVA auto-transformer re-energizing inrush current obtained during a field test at a 500 kV substation. Li et al. [20] proposed a transfer standard for the calibration of high-voltage DC currents using an OCS that employed spun optical fibers as the sensing element, which achieved an accuracy of 0.1%. Zhao, Shi, and Sun [21] developed and evaluated an OCS for use in power grids up to 1100 kV DC. Zhou et al. [22] proposed a new operating mechanism for a DC optical current transformer with the aim of overcoming situations of improper operation of a power grid protection system.

Applications of OCSs in high-voltage substations usually require accuracies down to 0.2%, often over a temperature range of tens of degrees Celsius. However, research has shown that ensuring this accuracy is challenging. Several influencing factors, such as mechanical disturbances, temperature variation effects on the optical elements, vibration, residual linear birefringence inside sensing coils, and others, deteriorate the performance of OCSs [23,24].

Regarding smart grids, they have been disseminated in recent years, and their monitoring is increasingly crucial for their reliable and safe operation, which is carried out through sensors that measure the generation and consumption of energy, characterize operating states, and detect the current topology of the network. In electrical systems, the most important electrical parameters to be measured are voltage and current. Considering that optical current sensors have the advantages of reduced weight and dimensions, facilitating their installation and maintenance, and even allowing the measurement of electrical currents without contact with the primary circuit while still having low prices, they become promising alternatives for use in smart grids [1,25].

The optical sensors discussed in this paper have their operating principle based on the Faraday Effect. The Faraday effect describes the relation between light and a magnetic field. Also called the magnetic–optical effect, and first observed by the English physicist Michael Faraday in 1845, this phenomenon consists of the rotation of the polarization direction of a linearly polarized (LP) light wave when subjected to the action of a magnetic field throughout its propagation in a transparent medium. As can be seen in Figure 1, a beam of light with a certain wavelength, when passing through an optical medium under the influence of a magnetic field, undergoes a change in the direction of its electric field. This change in the direction of the electric field is called Faraday rotation. Through this phenomenon, it is possible to measure an electrical current as a function of the magnetic field produced by it, using sensors built using optical materials suitable for this purpose. The magnetic field B produced by an electrical current I in the air is given by Equation (1), where μ_0 is the magnetic permeability of air (vacuum). The Faraday rotation angle θ_F of the electric field E resulting from propagation along a path L under the action of a magnetic field B is given by Equation (2), where V is the Verdet constant of the optical material and depends on the temperature T and on the wavelength of the light λ.

$$B = {}^{\mu_0 \cdot I}/_{2\pi \cdot d} \quad (1)$$

$$\theta_F = \int_L V(\lambda, T) \cdot B \cdot dL \quad (2)$$

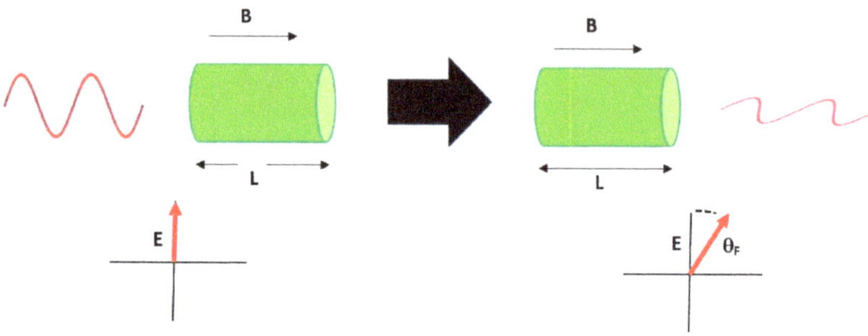

Figure 1. Faraday effect in linearly polarized light.

The measurement of the Faraday rotation angle can be performed using two methods: polarimetric and interferometric detection. In the polarimetric method, the intensity change due to polarization rotation from the induced magnetic field generated by a current is detected. An LP light wave is injected into the optical sensor that will measure the electrical current. The light wave is then analyzed at the output of the sensor using a second polarizer and a photodetector. The analyzer and photodetector combination converts and modulates polarized light into an electrical signal, which in turn corresponds to the electrical current to be measured [26,27]. The bandwidth of the polarimetric detection scheme is only affected by the response speed of the detector, and its upper measurement frequency can be up to 100 MHz, so it can be used for transient current measurements. Polarimetric devices are severely susceptible to the effect of linear and circular birefringence, which can cause distortion of the polarization rotation as well as reducing sensor accuracy and sensitivity. This results in false current readings from these perturbations [28].

In the interferometric method, the linear polarization of light is divided into two orthogonal circularly polarized light waves: one with circular polarization on the left and one on the right. When light waves pass through the sensor, the magnetic field that is created by the current being measured slows down one component and speeds up the other due to the Faraday effect. The change between the two circularly polarized light waves can be used as a detection signal [14,29].

OCSs can have two types of structure regarding the enclosure of the conductor of the electrical current to be measured by the sensing element. Closed-core sensors completely surround the current conductor and, for this reason, they are less susceptible to certain types of influences, such as vibration and stray magnetic fields. Open-core sensors do not completely surround the conductor and are, therefore, more susceptible to vibration and stray magnetic fields but are easy to install and use [30]. Figure 2 shows both types of OCS structures.

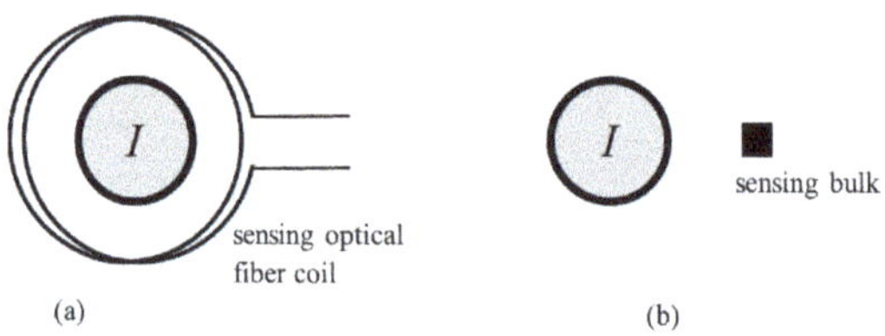

Figure 2. Types of OCS structure in relation to the position of the conductor of the electrical current I to be measured: (**a**) closed-core; (**b**) open-core.

According to their sensing element, OCSs can be classified into two main groups: bulk sensors and fiber sensors. The first group uses a magneto-optical glass structure as a sensor element with a high Verdet constant. Shui et al. [31] used this advantage to build a sensor, the optical material of which had a Verdet constant of approximately 109 rad/(T·m), capable of measuring low currents up to 2.5 A. Bulk sensors have some advantages over fiber sensors. In addition to having a higher Verdet constant, they are smaller and more mechanically rigid in such a way that mechanical gradients, thermal variations, vibrations, and external noise do not significantly alter performance. Bulk sensors are further protected against the influences of temperature variation because they are usually not in contact with the conductor of the electrical current to be measured or are constructed with the encapsulation of a dielectric material. Furthermore, due to their low photoelastic coefficient, their linear birefringence is very small, which allows sensors with high sensitivity to be obtained [32–34]. A disadvantage of bulk sensors occurs when they have an open-core structure. In this situation, they become susceptible to the influence of external magnetic fields. This occurs, for example, in power systems or transmission lines with more than one phase, where the current conductors are generally parallel to each other. The current measurement in one phase can be influenced by the magnetic field generated by the current in the other phases.

In fiber optic current sensors, the sensing element is the fiber itself. The main advantage of these sensors is the possibility of involving the entire conductor of the electrical current to be measured. The sensitivity of these sensors can be adjusted by increasing or decreasing the number of fiber turns around their conductors. This way, the sensors can use several meters of fiber, which will make them more sensitive to environmental variables, such as temperature and vibration, when compared with bulk sensors. However, as fiber optic sensors form a closed path along an electrical current conductor, they become insensitive to external magnetic fields [32].

Figure 3 shows an OCS block diagram. This is useful to identify and classify uncertainty sources. In an OCS, a light wave originating from an optical source with a certain operating wavelength is linearly polarized by a polarizer, becoming LP light. It is transmitted through a magneto-optical material, which is subjected to the magnetic field produced by the electrical current to be measured (the sensor itself). After this propagation, the modified light is received by the detection devices (photodetectors), which in turn measure the rotation angle of the polarization plane caused by the magnetic field, transforming it into an electrical signal. This signal is then transformed into a digital signal through an analog–digital (A/D) board or card. These digital signals are processed by software, which presents the measured electrical current to the user through a panel, communication interface, or other means.

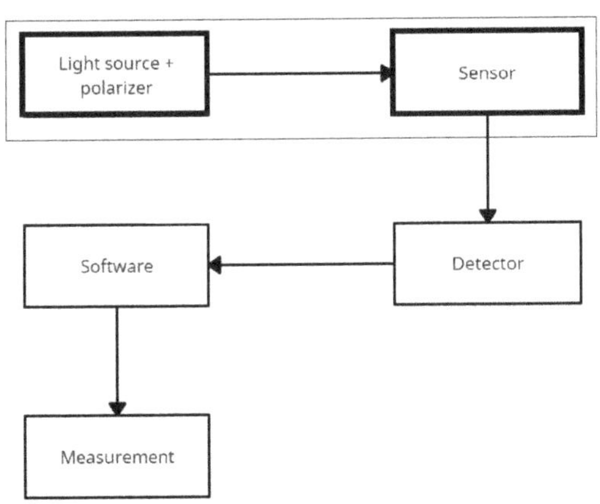

Figure 3. OCS basic block diagram.

3. Uncertainty Sources

This section presents and discusses the most important contributions to OCS measurement uncertainty. Some of the most important uncertainty contributions are those due to the influence of temperature, vibration, birefringence, and aging and stability of optical and signal processing devices in the sensors [16,34,35]. These sources of uncertainty can also be grouped into two classes: those related to the inherent characteristics of the sensors and their components, such as birefringence and short- and long-term stability, and those arising from the environmental conditions and use of the sensors, such as temperature, vibration, and parasitic magnetic fields. Other environmental conditions do not directly contribute to measurement uncertainty, such as relative humidity and solar radiation. In the case of humidity, if it is very high, it can be a limiting factor for the use of optoelectronic devices, such as A/D boards and photodetectors, since these generally have a maximum limit for ambient humidity declared by their manufacturers.

3.1. Temperature

Temperature variation significantly influences the behavior of OCSs, thus contributing to measurement uncertainty. The main mechanisms of how this influence occurs are the following: (a) the Verdet constant of an optical material is a function of temperature, so temperature variations will cause variations in the Verdet constant; (b) the wavelength of the light generated by the light source is a function of temperature, so variations in temperature will cause variations in the generated wavelength, since the Verdet constant of the optical material depends on the wavelength and the rotation angle and, consequently, the response of an OCS will vary (the wavelength of the light generated by the light source also changes with fluctuations in the driving current); (c) temperature variations change the behavior and response of optoelectronic components used in an OCS, such as photodiodes and data acquisition units; (d) temperature induces linear birefringence in the optical material; and (e) temperature variation causes changes in the dimensions of sensor components, which can cause changes in their behavior. When modeling uncertainty contributions arising from temperature, two main situations can be established: one in which temperature fluctuations during measurement modify the performance of the sensor and its components, and another where the sensor is used at a temperature different from that at which it and its components have been calibrated.

To minimize the effect of temperature on measurement uncertainty, the most immediate solution would be to use OCSs in situations where there is reduced temperature variation. However, this solution is not always practical, especially in outdoor applications. Another solution is to implement temperature compensation schemes. Many temperature compensation schemes have been proposed and can be found in the literature [36–41]. Despite this, the residual temperature effects need to be quantified and have an influence on the uncertainty of the measurement evaluated. The temperature-compensated fiber OCS developed by Zhao, Shi, and Sun [21] reached a response variation of about ±0.4% in a temperature range from −40 °C to 70 °C. In a narrower temperature range from 10 °C to 40 °C, this variation was approximately 0.15%. Li et al. [42] evaluated the impact of temperature variations of up to 40 °C on the measurement of the Faraday rotation angle on a bulk sensor, finding values of up to 9% for this variation. Jia et al. [43] proposed a temperature compensation method in a bulk optical current sensor through the insertion of a static bias composed of permanent magnets, where the error due to temperature variation reduced from ±8% to ±1% in a range from −20 °C to 80 °C.

Hoffman [44] presented variations in the Verdet constants of different optical glasses according to wavelength in the visible spectral range. For Schott flint glasses of the SF class, the Verdet constant varies with the inverse square of the wavelength. Figure 4 shows the Verdet constant wavelength dependence of SF-57 flint glass. It was observed that glasses with a higher concentration of PbO presented higher Verdet constants. Chen et al. [45] measured the average wavelength variation of a superluminescent diode (SLD) in a temperature range of 15 °C to 45 °C. When the temperature varied, the variation in the average wavelength had a linear trend. A variation in wavelength of about ±0.6 nm/K was measured in the wavelength region of 1310 nm. Wang et al. [46] measured a ±5 nm variation at the 1310 nm center wavelength of a light source due to temperature fluctuation from 10 °C to 30 °C and a ±12.5 nm variation due to driving current variation from 15 mA to 80 mA. This wavelength variation caused a variation in the Verdet constant of an SF6 glass bulk OCS of ±1.35%. Knowing the values of the variation in the wavelength of the light source as a function of temperature and the variation of the Verdet constant of the sensor's optical material as a function of the wavelength, it is possible to correct the sensor's response to the wavelength at the temperature at which it is being used. The uncertainty of these values and the uncertainty in measuring the sensor's operating temperature must be estimated and considered when estimating the measurement uncertainty. Müller et al. [47] demonstrated a method for self-compensation of the change in the response of an interferometric fiber optic current sensor caused by deviations in the wavelength of the light emitted by the light source caused by temperature variations. This variation in sensor response with respect to wavelength variation is reduced from more than 1% to <0.2% in wavelength ranges of at least 10 nm around 1305 nm.

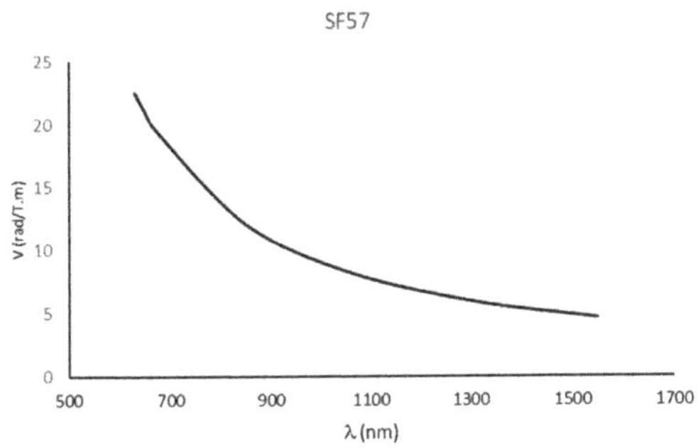

Figure 4. SF-57 glass Verdet constant wavelength dependence.

Williams et al. [48] measured the temperature dependence of the Verdet constants of SiO_2, SF-57, and BK-7. The measured relative Verdet constant variation per K was about $(1.35 \pm 0.08) \times 10^{-4}$ for SF-57, about $(0.69 \pm 0.03) \times 10^{-4}$ for SiO_2, and $(0.71 \pm 0.06) \times 10^{-4}$ for BK-7. When the optical material has a high Verdet constant and its linear thermal expansion coefficient is much lower than its coefficient of relative variation of the Verdet constant as a function of temperature, its linear temperature-induced variation becomes non-significant. For example, for the SF-57 glass, the linear thermal expansion coefficient is 8.3×10^{-6} per K in the range of −30 °C to 70 °C, about 16 times lower than the relative change of the Verdet constant due to temperature [49]. These values of variation in the Verdet constant as a function of temperature can be used to correct the Verdet constant for the temperature at which the sensor is being used, and their uncertainties must be considered when estimating the measurement uncertainty, as well as the uncertainty when measuring the ambient temperature at which the sensor is used.

Temkina et al. [24,50] proposed a method for compensating the response variation of a fiber optic current sensor with an interferometric detection scheme due to the dependence on temperature of the fiber quarter-wave plate. The maximum deviation using the proposed method was ±0.2% in a temperature range of 10 °C to 50 °C. It was also presented that the errors due to dependence of the temperature of the circulator and electro-optical modulator were non-significant, but the errors of the fiber delay line and the spun fiber used as a sensing element due to temperature variation were significant.

3.2. Vibration and Impact

For OCSs with an open core, the conductor of the electrical current to be measured is not surrounded by the optical material of the sensor, meaning that the line integral of the magnetic field along the optical path is not equal to the total current. However, a linear relationship between sensor measurement and electrical current can be determined based on the geometry and distribution of the field. Vibration causes variations in the relative position between the sensor and the conductor. As the magnetic field B is given by Equation (1), variations in the distance between the sensor and the conductor will in turn cause variations in the distribution of the magnetic field across the sensor, introducing changes in the sensor response. Comparing (1) and (2), it is observed that the sensor's response varies with the inverse square of the distance between it and the conductor, so the greater the distance, the lesser the influence of its variation on the sensor's response. Samimi et al. [51] showed that vibration causes large errors in open-core OCSs. These errors become smaller with an increase in the distance between the sensor and conductor, the conductor radius, or the sensor length. Deviations in OCS accuracy were about ±0.5% for a vertical displacement of ±0.5 mm when the sensor head and the conductor were 7.5 cm apart.

Short et al. [27] showed that OCSs with a polarimetric detection method cannot differentiate an acoustic vibration disturbance detected in the sensing element from the electrical current to be measured. They also showed that the OCSs with an in-line Sagnac interferometer are immune to acoustic vibrations in the sensing element; however, they are more sensitive to vibrations that occur farther away from the sensing element.

He et al. [52] presented an analysis of errors caused by linear stress on in-line Sagnac fiber OCSs under impact. A bias error, which is mainly caused by the stress acting on the delay coil leading to a non-reciprocity birefringence, was found.

Another contribution of uncertainty related to vibration refers to the introduction of linear birefringence, which will be addressed later in this article.

3.3. Stability

The detector module of an OCS uses optoelectronic devices, such as photodiodes, photodetectors, and A/D boards. These devices, together with the light source and other components or the sensor as polarizers, couplers, etc., contribute to the measurement uncertainty of an OCS due to their metrological characteristics and their short-term and long-term stability. The metrological data of these devices comes from their technical sheets or manuals. The long-term stability refers to the drift since the last calibration.

Aging of the light source device will change the central wavelength of the emitted light, and as the Verdet constant depends on the wavelength, the sensor response will change. Wang and Guan [34] analyzed some OCSs used in 110 kV voltage substations. The results showed that OCS long-term stability was strongly affected by aging and the performance reduction of the main optical devices of the sensors. Lenner et al. [3] investigated the long-term stability of an interferometric fiber OCS from both component and system perspectives. Some components, such as the SLD used as the light source, the phase modulator, and other passive components, were subject to accelerated aging by extended temperature cycling. Investigation results showed that the long-term drift of the center wavelength of the SLDs from two manufacturers varied from ±0.19 nm to ±0.38 nm.

One way to minimize uncertainty due to drift of the sensor or its components is to perform calibration immediately before and immediately after use. However, this option may be unfeasible for practical or economic reasons.

3.4. Linear Birefringence

Linear birefringence is the difference between the refraction index of two LP lights. A birefringent optical material has different indices of refraction for different directions of light propagation. In this way, light propagates at different speeds, according to its polarization, and a phase difference occurs between them, causing the polarization of light to change from linear to elliptical. In OCSs based on the Faraday effect, birefringence causes a shift in the polarization angle that adds to the rotation angle caused by the Faraday effect, disturbing the behavior of the sensors and decreasing their sensitivity to Faraday rotation and accuracy. Thus, to improve Faraday effect measurement quality in OCSs, the linear birefringence must be reduced [53–56].

The optical fiber birefringence can come from several sources. One of the possibilities is that it is intrinsic to the fiber and generated due to imperfections in the manufacturing process. As multimode fibers disrupt the polarization state of light, single-mode optical fibers are the only option for the sensor head. Unfortunately, single-mode fibers suffer from this inherent birefringence. In general, each type of optical fiber has an intrinsic birefringence characteristic, so several types of optical fibers can be evaluated, and the choice of appropriate fibers can reduce it [9]. Another source of birefringence arises when the fiber is bent to wrap it around the conductor. Curving the fiber breaks the symmetry of its structure because the outer part of the fiber supports stress tension and, on the contrary, the inner part supports compressive stress. The linear birefringence effect is more significant when the fiber is wrapped in many turns around the conductor for higher sensitivity, but it becomes less significant if the coil has a sufficiently large diameter [57]. If the state of polarization (SOP) of the propagating light remains perfectly linear or circular, the sensor response is not affected. But, if the linear birefringence effect causes the SOP to change from linear to elliptical, the behavior of the light polarization angle in the sensing fiber will be affected not only by the Faraday effect but also by the SOP. Thus, sensor performance is degraded, as its response is affected by the polarization evolution due to the linear birefringence [58]. Li, Wang, and Liu [59] demonstrated the effect of linear birefringence on the response of a fiber optic current sensor that used a spun LoBi fiber as the sensing element. For the purpose of evaluating the measurement uncertainty of optical current sensors, the birefringence extrinsic to a sensor's optical material, such as that caused by the fiber bending process or by external influence quantities such as temperature and vibration, is more important, since the effects of intrinsic birefringence will be present at the time of sensor calibration and will not vary over time. Cheng et al. [60] analyzed the effect of linear birefringence by varying the radius of the sensor's fiber optic coil. The variation found was ±0.45% for radii between 5 cm and 30 cm and decreased to <±0.2% for radii between >30 cm and 50 cm. Madaschi [58] experimentally evaluated the effect of coil radius and number of turns on the response of a standard silica fiber optic sensor. When the coil diameter varied from 6 cm to 14 cm, the effective Verdet constant of the sensor varied from approximately 0.18 rad/(T·m) to approximately 0.7 rad/(T·m) for a wavelength of 1310 nm. In this situation, the length of the sensor was the same, which meant that for each diameter evaluated, the number of turns was different. When evaluating the number of turns, when it increased from 5 to 16, keeping the coil diameter at 13 cm, the effective Verdet constant reduced from 0.64 rad/(T·m) to 0.53 rad/(T·m). To reduce measurement uncertainty, it is recommended to calibrate the fiber optical current sensor with the number of turns and coil diameter equal to or close to the configuration in which it will be used.

Temperature and vibration effects also induce birefringence in fiber optic sensors. When temperature fluctuates, the different coefficients of thermal expansion between the fiber core and cladding lead to axial deformation, thus inducing linear birefringence [61]. Zhang et al. [62] presented the results of experiments on the influence of temperature and vibration on sensor performance using three different reflection schemes and three distinct types of fiber. The 90° Faraday rotating mirror showed the best results, regardless of the sensor fiber type. Sima et al. [63] proposed a new sensing scheme to improve the robustness of fiber optic current sensors using a polarization detection scheme against the effects of birefringence caused by vibration and temperature by placing a fiber polarization rotator (FPR) at the end of the sensing fiber. In the temperature influence tests, the error variance ratio was less than ±2% in the temperature range from −20 °C to 80 °C, while the maximum type A uncertainty was 0.12%. In the vibration tests, the maximum ratio error variation was less than ±0.4%, with the type A uncertainty of 0.029%. Li et al. [10] analyzed the effect of the temperature variation, and consequently of the linear birefringence induced by the curvature of the fiber variation, on the error of an optical current sensor for various values of the radius of the optical fiber coil and for various numbers of turns. In a temperature range from −40 °C to 80 °C, the error variation was approximately 0.25% for a radius of 20 cm and 0.7% for a radius of 10 cm. In the same temperature range, when the number of fiber turns was 25, the error varied by approximately 0.3%, while for 35 turns, the variation was greater than 1% Tantaswadi et al. [64] analyzed the effects of induced linear birefringence due to vibration on the performance of a reciprocal fiber optic current sensor with polarimetric detection. The deviation due to this effect was within ±0.1%. Huang et al. [65] measured the influence of vibrations in the low-frequency range from 10 Hz to 20 Hz on the response of a fiber optic sensor. The response varied by up to 250%.

In bulk OCSs, temperature variation can cause mechanical stress due to the difference between the thermal expansion coefficients of the optical material and the encapsulation material, leading to the appearance of induced or extrinsic birefringence.

3.5. Other Contributions

Sensors may present variations in their response depending on the electrical current being measured, which is called linearity, and these variations must be estimated as a source of uncertainty for the measurement uncertainty. Shao et al. [66] developed and tested a fiber optic current sensor from 50 kA to 1000 kA, and the measured linearity varied by ±0.03%. Shui et al. [32] developed and tested a bulk OCS at low currents, and its linearity was about ±0.28% from 0.2 A to 2.4 A.

For bulk sensors, the position of the sensing element in relation to the current conductor also influences its response. Mihailovic et al. [67] conducted experiments with a bulk sensor, varying its position in relation to the current conductor, and found a variation of approximately 2.5% in the sensor's response. To reduce measurement uncertainty, it is recommended that the sensor be used at a distance from the current conductor equal to the distance at which it was calibrated. This concern becomes less important in bulk sensors that have an encapsulation to facilitate their installation next to the conductor, as this will ensure that the position is always the same.

Sensors that do not completely surround the current conductor are influenced by the magnetic field generated by currents from nearby conductors. Shenwang et al. [68] analyzed the influence of neighboring magnetic fields on optical current sensors, varying the distance and length of the sensor. Perciante and Ferrari [69] proposed a modified architecture of a three-phase system, where the sensor's optical path was parallel to adjacent current conductors and perpendicular to the current conductor to be measured, minimizing the effect of stray magnetic fields. Unfortunately, this solution does not seem to be very practical for existing systems. Xiaojun et al. [70] proposed an architecture for a bulk sensor in order to minimize the effect of stray magnetic fields in a three-phase power system, such that the sensor reaches an accuracy class of 0.2%. Samimi et al. [51] demonstrated that the magnetic fields of neighboring conductors in a three-phase network can cause variations in the response of a bulk sensor of up to 10%. This variation reduces if the distance between the phases is large enough and if the sensors are close to the conductor of the electrical current being measured. If the phase distance is 2 m or more, the sensor can achieve an accuracy class of 0.2%. In outdoor high-voltage substations, the distance between phases is usually greater than 2 m. For example, in the Tucuruí Substation in Brazil, shown in Figure 5, in the 500 kV section, the distance between phases is approximately 9 m. Considering the phase arrangement, the central phase will be more affected by external magnetic fields than the other phases.

The resolution of an instrument or measuring system is the smallest variation that can be observed in its indication and is associated with its capacity in terms of accuracy and precision. An uncertainty is associated with the finite resolution of the instrument or system. Depending on certain conditions of a given measurement process, the contribution of uncertainty associated with resolution is significant in relation to the other contributions. Similar situations can occur in the analog-to-digital conversion process, where electronic resolution impacts the repeated sampling of a signal, leading to digital samples that can differ by just one or two bits. Usually, for a digital indication instrument, its resolution is divided by two to calculate the resolution uncertainty [71,72]. This is the case for sensors that present their measurements in the format of digital indications, such as on a computer or display.

Figure 5. The 9 m distance between phases at the 500 kV Tucuruí Substation.

Measurement repeatability is the precision of a measurement taken under repeatable conditions. The repeatable conditions include the same operating conditions, same location, same operator, same measuring systems, and replicated measurements for the measurement of the same measurand (the quantity intended to be measured) [73,74]. In the case of OCSs, performing measurements under these repeatable conditions can be useful to estimate the uncertainty contributions due to the influence of temperature fluctuations and due to the short-term stability of sensor components.

4. Uncertainty Analysis

Reference [75] sets out general rules for evaluating and expressing the measurement uncertainty that are intended to be applicable to a wide range of measurements. Within the scope of this work, it is worth highlighting some rules and good practices that will be used in evaluating the uncertainty contributions of optical current sensors. When dealing with systematic effects, it is assumed that the result of a measurement has been corrected for all effects that are recognized as significant, but it is also assumed that corrections are considered imperfect due to the lack of knowledge about them, so there is always uncertainty associated with the correction of a systematic effect. Sometimes it may be impractical to correct for a systematic effect, so it is necessary to model this effect as a source of uncertainty. However, the best option to improve measurement uncertainty is to perform the correction and estimate an uncertainty contribution due to this correction [76].

Still, according to reference [55] in relation to the evaluation method, whether by statistical analysis of a set of measurements or by other methods, uncertainty contributions can be classified into two types: type A, which is estimated by statistical methods, or type B, which is evaluated by other methods. A type A uncertainty is estimated from the statistical evaluation of a set of observations. Some forms of analysis may be calculating the standard deviation or variance of a series of independent measurements or using a linear regression method to fit a curve. A type B uncertainty can be estimated by means other than statistical evaluation and can be based on scientific evaluations of all available data, such as data from calibration certificates, uncertainties attributed to reference data from publications, effects of environmental conditions on measurement, data from previous measurements, and data published in datasheets or manuals [77]. In some cases, uncertainty contributions that could be assessed as type B are included in a type A assessment when a reasonable number of observations are able to be performed in the measurement process.

When classifying the uncertainty contributions of optical current sensors, uncertainties arising from short-term effects that are difficult to correct are best quantified through a series of observations and are, therefore, defined as type A. This is the case, for example, for uncertainty contributions due to the effects of temperature fluctuation on a light source and other components, as well as the short-term stability of these same components. The combined effect of these uncertainties will be observed if the measurement procedure provides a reasonable number of readings that cover all conditions related to these uncertainty contributions.

On the other hand, effects that allow corrections to be applied to the mathematical model, such as the effect of performing measurements at a temperature different from that at which the sensor components were calibrated, give rise to uncertainty contributions that can be classified as type B. In the case of temperature difference, after applying the correction, the uncertainty due to this correction must be estimated. Other uncertainty contributions that can be classified as type B are those whose assessment using a statistical method is difficult, impractical, or economically unfeasible. This is the case for uncertainties due to the long-term stability of the sensor components and the resolution of the A/D board.

Figure 6 and Table 2 show a summary of the identified relevant uncertainty contributions to the measurement uncertainty of OCSs. Table 2 also presents a suggestion for their classification as type A or type B evaluation. These uncertainties are discussed below.

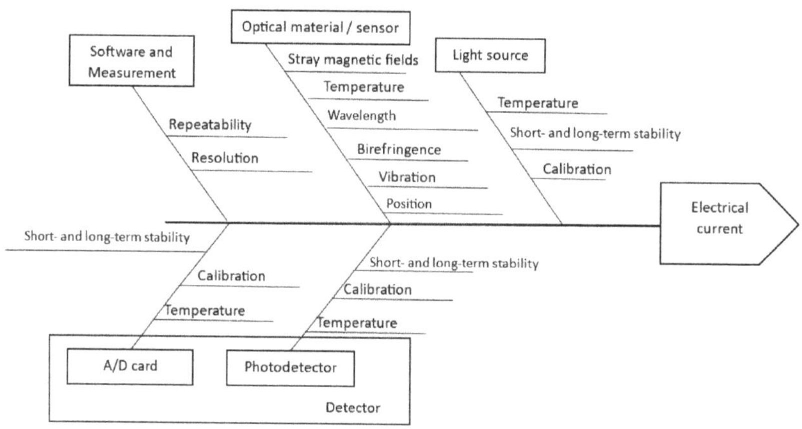

Figure 6. Main OCS uncertainty sources.

The uncertainty associated with the calibration of the OCS can be estimated from its last calibration report. Calibration may have been performed on the entire sensor or on each of its parts, according to Figure 3. The OCS calibration can be performed at various current levels according to the measurement range of interest to enable the assessment of the sensor's linearity and its uncertainty.

The variations in the Verdet constant of the optical material, the light source, the photodetector, and the A/D card caused by use of the OCS at a temperature different from that at which it was calibrated can be estimated by measuring the OCS temperature and the characteristics of each of the OCS components, which can be found in reference data and in the technical documentation provided by their manufacturers. These variations may give rise to correction factors that must be applied to the sensor response; the uncertainties of these corrections can be estimated through the uncertainties declared in the technical documentation and the uncertainty of the sensor temperature measurement.

Table 2. Summary and classification of OCS uncertainty sources.

Uncertainty Source	Type
Calibration of the OCS [1]	B
OCS linearity	B
Verdet constant variation due to temperature fluctuations	A
Correction of Verdet constant drift due to use of the sensor at a temperature far from that at which it was calibrated	B
Correction due to birefringence	B
Correction due to vibration	A
Light-source wavelength short-term stability	A
Correction due to light-source wavelength long-term stability	B
Light-source wavelength variation due to temperature fluctuations	A
Light-source wavelength variation due to driving current fluctuations	A
Correction of the light source wavelength drift due to use of the sensor at a temperature far from that at which it was calibrated	B
Correction of the light-source wavelength drift due to use of the sensor at a driving current far from that at which it was calibrated	B
Position of the sensor in relation to the current conductor	B
Stray magnetic field influence	A
Correction due to the drift since the last photodetector calibration	B
Photodetector short-term stability	A
Photodetector performance variation due to temperature fluctuations	A
Correction of photodetector performance drift due to use of the sensor at a temperature far from that at which it was calibrated	B
Correction of the drift since the last A/D card calibration	B
A/D card short-term stability	A
A/D card performance variation due to temperature fluctuations	B
Correction of the A/D card performance drift due to use of the sensor at a temperature far from that at which it was calibrated	B
Finite resolution of the measured current indication	B
Repeatability of current measurements	A

[1] Calibration can be of the entire sensor or of each of its components.

Variation in the wavelength of a light source due to its use in a driving current different from that used in its calibration may give rise to a correction factor, which can be obtained through the technical documentation of the light source or through experiments to determine the variation in the wavelength of the source as a function of the driving current. This correction will be associated with an uncertainty that can be estimated from the uncertainty of the measurement of the driving current and the characteristics of the light source.

Uncertainties related to the long-term stability or drift since the last calibration of the light source, photodetector, and A/D card can be estimated from the technical documentation provided by their manufacturers. When this information is not available in the technical documentation, this stability or drift can be assessed experimentally, provided that resources are available. If these uncertainties are significant, they can be reduced by calibrating the sensor or its components immediately before and after use. However, this alternative may be inadequate for practical or economic reasons.

To reduce uncertainty due to variations caused by the birefringence effect, it is important to control the factors causing this birefringence. For example, for a fiber OCS, the same radius and number of turns should be used whenever possible. Even with these precautions, a residual uncertainty will be present (such as due to variations in the length of the coil radius) and must be estimated. The influence of temperature on the variation in birefringence is more difficult to assess and must be estimated for each sensor, since it will depend strongly on its peculiar characteristics.

In estimating the uncertainty due to variation of the position of an open-core OCS relative to the conductor or busbar where the current to be measured propagates as a result of vibration, the value of the radial vibration of the conductor or busbar in displacement units (meter or its submultiples, peak-to-peak) can be used, if this is available. Vibration is expected to be much less significant in busbars than in conductors, as the latter are more susceptible to natural phenomena that cause vibration, such as wind.

Uncertainties arising from temperature fluctuations during measurement of the Verdet constant of the optical material, wavelength of the light source, and performance of the photodetector and A/D card, and also uncertainties arising from the short-term stability of the light source (inherent and due to the driving current), the photodetector, and the A/D card, can be statistically estimated together by performing an adequate number of measurements or observations during the measurement. The joint uncertainty assessed through this series of observations will be associated with the repeatability of the OCS.

To estimate the uncertainty due to the position of the OCS in relation to the conductor of the electrical current to be measured, the sensor can be calibrated by setting one or more distances according to the conditions of use of the OCS in the field. The uncertainty will be quantified through the variation of the sensor response obtained for the various distances used in the calibration and also through the uncertainty in the measurement of the distance between the sensor and the conductor.

The uncertainty due to stray magnetic fields caused by adjacent phases in three-phase networks can be evaluated analytically through knowledge of the characteristics of the electrical network where the sensor will be used. Using the distance between the phases and considering that the three-phase network is reasonably balanced, which is the most common situation encountered, uncertainty can be estimated using Equation (2). This estimate can also be made from results found in the literature referring to similar situations, since it will depend mainly on only two characteristics—distance between phases and whether the system is reasonably balanced.

Another point to consider is the fact that some sources of uncertainty can become significant according to certain characteristics of optical current sensors, such as the type of sensing element (fiber or bulk), the enclosure of the current conductor (total or partial), or the detection scheme (polarimetric or interferometric). For example, uncertainty due to vibration will be more significant for sensors that partially enclose the current conductor than for those that enclose it completely, and uncertainty due to birefringence will be more significant in fiber sensors than in bulk sensors. The magnitude of the measurement uncertainty of an optical current sensor, and by how much each uncertainty contribution is more or less significant, will depend on knowledge of the effects that affect its performance, whether they are corrected for, and the form of that correction. As mentioned previously, an uncorrected systematic effect will give rise to a contribution of uncertainty, thus degrading the quality of the sensor. For example, if the effect of extrinsic birefringence of a fiber optic current sensor is not known, its performance will be compromised and there will be an increase in measurement uncertainty.

In the bulk sensor tested by Costa et al. [78], shown in Figure 7, the low thermal coefficient vinyl encapsulation completely encloses the current conductor, but the sensor itself is just a block of SF-57 glass with a polarimetric detection scheme. Its design ensures that the variation in the position of the sensor in relation to the conductor is no more than the diameter of its opening, d = 4.7 cm. Thus, the uncertainty due to vibration can be determined by measuring its response at distances from zero to d. For this sensor, all uncertainties listed in Table 2 apply, with the exception of the uncertainty due to birefringence. On the other hand, for a fiber optic sensor that wraps the electrical current conductor in several turns, the sources of uncertainty due to correction due to vibration, the position of the sensor relative to the conductor, and stray magnetic field influence will be less significant than the others. Some sources of uncertainty will be common and will always be significant in any optical current sensor scheme, such as corrections due to temperature variations in the Verdet constant of the optical material and in the wavelength of the light source and aging of the components.

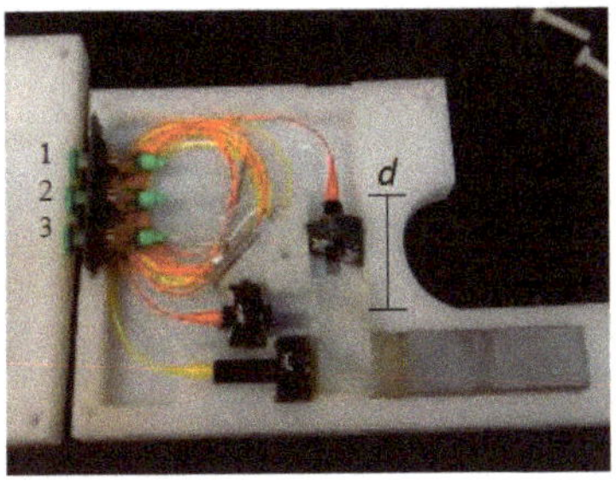

Figure 7. Experimental bulk OCS constructed with SF-57 glass—numbers 1–3 indicate optical connections [78].

5. Conclusions

This article presented a review of the main uncertainty contributions that affect the accuracy of optical current sensors based on the Faraday effect, including the classification of their estimation as type A or type B. It was observed that contributions of uncertainty arising from influencing quantities such as temperature variation, material birefringence, and aging of components can be quite significant for sensor measurement uncertainty, even when schemes are used to compensate or correct the effects of some of these factors, as it is considered that these compensations or corrections are imperfect and that there is always a contribution from associated uncertainty. Some of these corrections, if not conducted, will take a sensor's measurement uncertainty to values that are unsuitable for its use. Some uncertainty contributions are significant in any sensor configuration, while other contributions are more significant in a certain configuration and not significant in other configurations. These configurations include the type of sensing element, whether the sensor surrounds the current conductor completely, and the detection scheme.

The evolution of this paper involved the quantification of the identified uncertainties, so the guide proposed in this paper can be complemented with examples where the measurement uncertainty of one or more types of optical current sensors can be obtained.

Author Contributions: Conceptualization, M.M.C.; methodology, M.M.C.; supervision, J.C.W.A.C.; validation, J.C.W.A.C. and M.A.G.M.; formal analysis, J.C.W.A.C. and M.A.G.M.; project administration, J.C.W.A.C.; writing—original draft preparation, M.M.C.; writing—review and editing, J.C.W.A.C. and M.A.G.M. All authors have read and agreed to the published version of the manuscript.

Funding: This research received no external funding.

Data Availability Statement: Data are contained within the article.

Conflicts of Interest: The authors declare no conflict of interest.

References

1. Pong, P.W.T.; Annaswamy, A.M.; Kroposki, B.; Zhang, Y.; Rajagopal, R.; Zussman, G.; Poor, H.V. Cyber-enabled grids: Shaping future energy systems. *Adv. Appl. Energy* **2021**, *1*, 100003.
2. Larruskain, M.; Valverde, V.; Torres, E.; Buigues, G.; Santos-Mugica, M. Protection Systems for Multi-Terminal HVDC Grids. *Renew. Energy Power Qual. J.* **2018**, *1*, 310–315.
3. Lenner, M.; Frank, A.; Yang, L.; Roininen, T.M.; Bohnert, K. Long-Term Reliability of Fiber-Optic Current Sensors. *IEEE Sens. J.* **2020**, *20*, 823–832.
4. Kucuksari, S.; Karady, G. Experimental Comparison of Conventional and Optical Current Transformers. *IEEE Trans. Power Deliv.* **2010**, *25*, 2455–2463. [CrossRef]
5. Schon, K. *High Voltage Measurement Techniques, Fundamentals, Measuring Instruments, and Measuring Methods*, 1st ed.; Springer International Publishing: Basel, Switzerland, 2019.
6. Wu, J.; Zhang, X. Recent Progress of All Fiber Optic Current Transformers. In Proceedings of the 7th International Forum on Electrical Engineering and Automation (IFEEA), Hefei, China, 25–27 September 2020.
7. Laurano, C.; Toscani, S.; Zanoni, M. A Simple Method for Compensating Harmonic Distortion in Current Transformers: Experimental Validation. *Sensors* **2021**, *21*, 2907. [CrossRef] [PubMed]

8. N'cho, J.S.; Fofana, I. Review of Fiber Optic Diagnostic Techniques for Power Transformers. *Energies* **2020**, *13*, 1789. [CrossRef]
9. Samimi, M.H.; Akmal, A.A.S.; Mohseni, H. Optical Current Transducers and Error Sources in Them: A Review. *IEEE Sens. J.* **2015**, *15*, 4721–4728. [CrossRef]
10. Li, Y.; Liu, X.; Zhang, W.; Liu, J. Error characteristic analysis and experimental research on a fiber optic current transformer. *Appl. Opt.* **2018**, *57*, 8359–8365. [CrossRef]
11. FOCS—Fiber Optic Current Sensors Enabling Smart Grids and Digital Substations. Available online: https://library.e.abb.com/public/a0c2264f26b94c748f7ef2d54bd0bea2/ABB_FOCS%20Kit%20Flyer_20190723.pdf?x-sign=LYr+kJPJwQ5xVj5qzJbKZJh4FzKJcbujknH4FAtts1Vn3ZBMi2r3CYgbxrCrjws8 (accessed on 1 December 2023).
12. Rahmatian, F.; Ortega, A. Applications of Optical Current and Voltage Sensors in High-Voltage Systems. In Proceedings of the IEEE/PES Transmission & Distribution Conference and Exposition, Caracas, Venezuela, 15–18 August 2006. [CrossRef]
13. Ahmed, S. Performance Evaluation of Optical Fiber Current Transformer based Over-current Protection. *FES J. Eng. Sci.* **2019**, *8*, 40–47. [CrossRef]
14. Tzelepis, D.; Psaras, V.; Tsotsopoulou, E.; Mirsaeidi, S.; Dyśko, A.; Hong, Q.; Dong, X.; Blair, S.; Nikolaidis, V.; Papaspiliotopoulos, V.; et al. Voltage and Current Measuring Technologies for High Voltage Direct Current Supergrids: A Technology Review Identifying the Options for Protection, Fault Location and Automation Applications. *IEEE Access* **2020**, *8*, 203398–203428. [CrossRef]
15. Jogalekar, A.; Katti, P. An overview on renovation and modernization of existing substations. In Proceedings of the International Conference on Nascent Technologies in Engineering (ICNTE), Vashi, India, 27–28 January 2017. [CrossRef]
16. Rose, A.H.; Hudson, M.C.; Hadley, E.; Perkins, P. Optical Inline-Sagnac Current Sensor, Part II: Performance and Applications. *IEEE Trans. Power Deliv.* **2022**, *37*, 5394–5401. [CrossRef]
17. Azad, S.P.; Leterme, W.; Van Hertem, D. A DC grid primary protection algorithm based on current measurements. In Proceedings of the 17th European Conference on Power Electronics and Applications (EPE'15 ECCE-Europe), Geneva, Switzerland, 8–10 September 2015. [CrossRef]
18. Lei, C.; Zhizhong, G.; Yue, C.; Guizhong, W. Measurement of inrush current in transformer based on optical current transducer. *J. Phys. Conf. Ser.* **2019**, *1187*, 022004. [CrossRef]
19. Ivanov, N.; Vorobev, P.; Bialek, J.; Kanafeev, R.; Yanin, M. Fiber Optic Current Transformers for Transformer Differential Protection during Inrush Current: A Field Study. In Proceedings of the IEEE Madrid PowerTech, Madrid, Spain, 28 June–2 July 2021. [CrossRef]
20. Li, C.; Shao, H.; Zhao, W.; Wang, J.; Zhang, H. Fiber-Optic Current Sensor Utilized for DC High Current Calibration. In Proceedings of the 2016 Conference on Precision Electromagnetic Measurements, Ottawa, ON, Canada, 10–15 July 2016. [CrossRef]
21. Zhao, J.; Shi, L.; Sun, X. Design and Performance Study of a Temperature Compensated ±1100-kV UHVdc All Fiber Current Transformer. *IEEE Trans. Instrum. Meas.* **2021**, *70*, 7001206. [CrossRef]
22. Zhou, S.; Tang, H.; Pan, B.; Zhang, W. Abnormal handling mechanism and improvement measures of optical DC current transformer in smart grid environment. *PeerJ Comput. Sci.* **2022**, *8*, 1132. [CrossRef]
23. Bohnert, K.; Gabus, P.; Nehring, J.; Brandle, H. Temperature and vibration insensitive fiber-optic current sensor. *J. Lightwave Technol.* **2002**, *20*, 267–276. [CrossRef]
24. Temkina, V.; Medvedev, A.; Mayzel, A.; Sivolenko, E.; Poletaeva, E.; Dudnik, I. Experimental Study of Temperature Impact on Fiber Optic Current Sensor Elements. In *International Conference on Next Generation Wired/Wireless Networking*; Springer International Publishing: Cham, Switzerland, 2022; Volume 13158. [CrossRef]
25. Oppegård, A.; Katoulaei, M.K.; Oleinikova, I. Complete FEM-based model of a bulk-glass optical current transformer. In Proceedings of the International Conference on Smart Energy Systems and Technologies (SEST), Eindhoven, The Netherlands, 5–7 September 2022. [CrossRef]
26. Short, S.X.; Tantaswadi, P.; Carvalho, R.T.; Russell, B.D.; Blake, J. An experimental study of acoustic vibration effects in optical fiber current sensors. *IEEE Trans. Power Deliv.* **1996**, *11*, 1702–1706. [CrossRef]
27. Tantaswadi, P.; Tangtrongbenchasil, C. Numerical analysis of localized vibration effects and its compensation technique on reciprocal fiber-optic polarimetric current sensor. In Proceedings of the IEEE International Conference on Industrial Technology, Bangkok, Thailand, 11–14 December 2002. [CrossRef]
28. Leung, F.Y.C.; Chiu, W.C.K.; Demokan, M.S. Fiber-optic current sensor developed for power system measurement. In Proceedings of the International Conference on Advances in Power System Control, Operation and Management, APSCOM-91, Hong Kong, China, 5–8 November 1991.
29. Morello, R.; Mukhopadhyay, S.C.; Liu, Z.; Slomovitz, D.; Samantaray, S.R. Advances on Sensing Technologies for Smart Cities and Power Grids: A Review. *IEEE Sens. J.* **2017**, *17*, 7596–7610. [CrossRef]
30. Samimi, M.H.; Akmal, A.A.S.; Mohseni, H.; Jadidian, J. Open-Core Optical Current Transducer: Modeling and Experiment. *IEEE Trans Power Deliv.* **2016**, *31*, 2028–2035. [CrossRef]
31. Shui, H.; Rong, Y.; Dong, T.; Sun, Y.; Yan, W. Current measurement method based on integral reconstruction of magnetic rotation angle. *Int. J. Energy* **2023**, *2*, 52–56. [CrossRef]
32. Silva, R.M.; Martins, H.; Nascimento, I.; Baptista, J.M.; Ribeiro, A.L.; Santos, J.L.; Jorge, P.; Frazão, O. Optical Current Sensors for High Power Systems: A Review. *Appl. Sci.* **2012**, *2*, 602–628. [CrossRef]

33. Shao, J.; Liu, W.; Liu, C.; Xu, D. A passive optical fiber current sensor based on YIG. In Proceedings of the 1st Asia-Pacific Optical Fiber Sensors Conference, Chengdu, China, 7–9 November 2008; pp. 1–4. [CrossRef]
34. Wang, H.; Guan, Y. Study on Long-Term Operation Stability of Fiber Optical Current Transformer Based on Faraday Effect. In Proceedings of the International Conference on Intelligent Transportation, Big Data and Smart City, Halong Bay, Vietnam, 19–20 December 2015. [CrossRef]
35. Xu, Z.; Yue, C.; Yan, X.; Fan, J.; Pang, F. Multi-Scale Anomaly Detection in Fiber Optic Current Transformers. In Proceedings of the IEEE 6th International Electrical and Energy Conference (CIEEC), Hefei, China, 12–14 May 2023. [CrossRef]
36. Müller, G.M.; Gu, X.; Yang, L.; Frank, A.; Bohnert, K. Inherent temperature compensation of fiber-optic current sensors employing spun highly birefringent fiber. *Opt. Express* **2016**, *24*, 11164–11173. [CrossRef]
37. Liu, C.; Wang, D.; Li, C.; Su, K.; Li, D.; Yu, D.; Wang, L.; Si, L.; Jin, J. Fiber optic current sensor temperature compensation through RBF neural network. In Proceedings of the Applied Optics and Photonics China, Beijing, China, 30 November–2 December 2020. [CrossRef]
38. Petricevic, S.J.; Mihailovic, P.M. Compensation of Verdet Constant Temperature Dependence by Crystal Core Temperature Measurement. *Sensors* **2016**, *16*, 1627. [CrossRef] [PubMed]
39. Müller, G.M.; Frank, A.; Yang, L.; Gu, X.; Bohnert, K. Temperature Compensation of Interferometric and Polarimetric Fiber-Optic Current Sensors with Spun Highly Birefringent Fiber. *J. Lightwave Technol.* **2019**, *37*, 4507–4513. [CrossRef]
40. Madden, I.; Michie, W.; Cruden, A.; Niewczas, P.; McDonald, J. Temperature compensation for optical current sensors. *Opt. Eng.* **1999**, *38*, 10. [CrossRef]
41. Liang, Z.; Han, Q.; Zhang, T.; Tang, Y.; Jiang, J.; Cheng, Z. Nonlinearity Compensation of Magneto-Optic Fiber Current Sensors Based on WOA-BP Neural Network. *IEEE Sens. J.* **2022**, *22*, 19378–19383. [CrossRef]
42. Li, Y.; Zhang, W.; Liu, X.; Liu, J. Characteristic Analysis and Experiment of Adaptive Fiber Optic Current Sensor Technology. *Appl. Sci.* **2019**, *9*, 333. [CrossRef]
43. Jia, Q.; Han, Q.; Liang, Z.; Cheng, Z.; Hu, H.; Wang, S.; Ren, K.; Jiang, J.; Liu, T. Temperature Compensation of Optical Fiber Current Sensors with a Static Bias. *IEEE Sens. J.* **2022**, *22*, 352–356. [CrossRef]
44. Hoffmann, H.J. A survey of isotropic and anisotropic modifications of the refractive index. *J. Phys. IV Fr.* **1992**, *2*, 21–30. [CrossRef]
45. Chen, X.; Jiang, L.; Yang, M.; Yang, J.; Shu, X. Research on the mean-wavelength drift mechanism of SLD light source in FOG. *J. Phys. Conf. Ser.* **2019**, *1300*, 012047. [CrossRef]
46. Wang, Z.; Li, Q.; Qi, Y.; Huang, Z.; Shi, J. Wavelength dependence of the sensitivity of a bulk-glass optical current transformer. *Opt. Laser Technol.* **2006**, *38*, 87–93. [CrossRef]
47. Müller, G.M.; Quan, W.; Lenner, M.; Yang, L.; Frank, A.; Bohnert, K. Fiber-optic current sensor with self-compensation of source wavelength changes. *Opt. Lett.* **2016**, *41*, 2867–2870. [CrossRef]
48. Williams, P.A.; Rose, A.H.; Day, G.W.; Milner, T.E.; Deeter, M.N. Temperature dependence of the Verdet constant in several diamagnetic glasses. *Appl. Opt.* **1991**, *30*, 1176–1178. [CrossRef] [PubMed]
49. Schott Optical Glass—Datasheets. Available online: https://refractiveindex.info/download/data/2017/schott_2017-01-20.pdf (accessed on 18 November 2023).
50. Temkina, V.; Medvedev, A.; Mayzel, A.; Mokeev, A. Compensation of Fiber Quarter-Wave Plate Temperature Deviation in Fiber Optic Current Sensor. In Proceedings of the IEEE International Conference on Electrical Engineering and Photonics, St. Petersburg, Russia, 17–18 October 2019. [CrossRef]
51. Samimi, M.H.; Bahrami, S.; Akmal, A.A.S.; Mohseni, H. Effect of Nonideal Linear Polarizers, Stray Magnetic Field, and Vibration on the Accuracy of Open-Core Optical Current Transducers. *IEEE Sens. J.* **2014**, *14*, 3508–3515. [CrossRef]
52. He, X.; Wang, G.; Gao, W.; Wang, Y.; Gao, H. The effect analysis of impact on a fiber optic current sensor. *Optik* **2021**, *238*, 166724. [CrossRef]
53. Sawale, B.A. Effect of induced linear birefringence on Faraday current sensor using Ultra-low birefringence Optical Fiber. In Proceeding of the International Conference on Electrical Power and Energy Systems (ICEPES), Bhopal, India, 14–16 December 2016. [CrossRef]
54. Ulmer, E. A High-Accuracy Optical Current Transducer for Electric Power Systems. *IEEE Trans. Power Deliv.* **1990**, *5*, 892–898. [CrossRef]
55. Tabor, W.; Chen, F. Electromagnetic Propagation through Materials Possessing Both Faraday Rotation and Birefringence: Experiments with Ytterbium Orthoferrite. *J. Appl. Phys.* **1969**, *40*, 2760–2765. [CrossRef]
56. Forman, P.; Jahoda, F. Linear birefringence effects on fiber-optic current sensors. *Appl. Opt.* **1988**, *27*, 3088–3096. [CrossRef]
57. Madaschi, A.; Brunero, M.; Ferrario, M.; Martelli, P.; Boffi, P.; Martinelli, M. Experimental Evaluation of the Birefringence Effect on Faraday-Based Fiber-Optic Current Sensors. *IEEE Sens. Lett.* **2022**, *6*, 1–4. [CrossRef]
58. Hotate, K.; Konishi, Y. Formulas describing error induced by fiber linear-birefringence in fiber-optic current sensors. In Proceedings of the 15th Optical Fiber Sensors Conference Technical Digest, Portland, OR, USA, 10 May 2002. [CrossRef]
59. Li, Y.; Wang, B.; Liu, J. Modeling and analysis on the sensing characteristic of fiber optical current transformer. In Proceedings of the IECON 2017—43rd Annual Conference of the IEEE Industrial Electronics Society, Beijing, China, 29 October–1 November 2017. [CrossRef]
60. Cheng, S. Optimization design of sensing coil of all-fiber optical current transformer. In Proceedings of the 7th International Power Electronics and Motion Control Conference, Harbin, China, 2–5 June 2012. [CrossRef]

61. Song, C.; Zhizhong, G.; Guoqing, Z.; Wenbin, Y.; Yan, S.; Shixin, J. The Analysis on Temperature Characteristic of Fiber Optic Current Sensor Ratio Error. In Proceedings of the Fifth International Conference on Instrumentation and Measurement, Computer, Communication and Control (IMCCC), Qinhuangdao, China, 18–20 September 2015. [CrossRef]
62. Zhang, H.; Qiu, Y.; Huang, Z.; Jiang, J.; Li, G.; Chen, H.; Li, H. Temperature and Vibration Robustness of Reflecting All-Fiber Current Sensor Using Common Single-Mode Fiber. *J. Lightwave Technol.* **2014**, *32*, 4311–4317. [CrossRef]
63. Sima, W.; Zeng, L.; Yang, M.; Yuan, T.; Sun, P. Improving the Temperature and Vibration Robustness of Fiber Optic Current Transformer Using Fiber Polarization Rotator. *IEEE Trans. Instrum. Meas.* **2022**, *71*, 1–12. [CrossRef]
64. Tantaswadi, P.; Tangtrongbenchasil, C.; Maheshwari, S. Study of current measurement error due to vibration in reciprocal fiber-optic polarimetric current sensor. In Proceedings of the 4th IEEE International Conference on Power Electronics and Drive Systems, Denpasar, Indonesia, 25 October 2001. [CrossRef]
65. Huang, J.; Liu, B.; Chen, Y.; Tian, J.; Xia, L. Effect of Low Frequency Vibration of Sensing Coil on Fiber Optic Current Sensors. In Proceedings of the 21st International Conference on Optical Communications and Networks (ICOCN), Qufu, China, 31 July–3 August 2023. [CrossRef]
66. Shao, H.; Li, C.; Lin, F.; Wang, L.; Liang, B.; Wang, J.; Zhao, W. Development and Applications of Wideband High Current Transducers. In Proceedings of the 2020 Conference on Precision Electromagnetic Measurements, Online, 20–24 August 2020. [CrossRef]
67. Mihailovic, P.; Petricevic, S.; Stojkovic, Z.; Radunovic, J. Development of a portable fiber-optic current sensor for power systems monitoring. *IEEE Trans. Instrum. Meas.* **2004**, *53*, 24–30. [CrossRef]
68. Shenwang, L.; Zhizhong, G.; Guoqing, Z.; Wenbin, Y.; Yan, S.; Zhiqing, L.; Dong, Y.; Guopeng, W. Research on Magnetic Crosstalk Characteristics of Straight Light Path Type Optical Current Transformer. In Proceedings of the Sixth International Conference on Instrumentation & Measurement, Computer, Communication and Control (IMCCC), Harbin, China, 21–23 July 2016. [CrossRef]
69. Perciante, C.; Ferrari, J. Magnetic Crosstalk Minimization in Optical Current Sensors. *IEEE Trans. Instrum. Meas.* **2008**, *57*, 2304–2308. [CrossRef]
70. Xiaojun, Z.; Zhuohang, L.; Yansong, L.; Jun, L. Clustered magneto-optical current sensor to eliminate the interference of a phase-to-phase magnetic field. *Opt. Contin.* **2022**, *1*, 197–214. [CrossRef]
71. Hogan, R. How to Determine Resolution Uncertainty. *Cal Lab Mag.* **2017**, *24*, 20–26.
72. Phillips, S.D.; Toman, B.; Estler, W.T. Uncertainty Due to Finite Resolution Measurements. *J. Res. Natl. Inst. Stand. Technol.* **2008**, *113*, 143–156. [PubMed]
73. De Bièvre, P. Measurement uncertainty is not synonym of measurement repeatability or measurement reproducibility.... *Accredit. Qual. Assur.* **2008**, *13*, 61–62. [CrossRef]
74. Alvarez-Prieto, M.; Jiménez-Chacón, J.; Montero-Curbelo, Á. Do we need to consider metrological meanings of different measurement uncertainty estimations? *Accredit. Qual. Assur.* **2009**, *14*, 623–634. [CrossRef]
75. Evaluation of Measurement Data—Guide to the Expression of Uncertainty in Measurement—JCGM 100:2008—GUM 1995 with Minor Corrections. Available online: https://www.bipm.org/documents/20126/2071204/JCGM_100_2008_E.pdf/cb0ef43f-baa5-11cf-3f85-4dcd86f77bd6 (accessed on 19 December 2023).
76. Cohen, E.R. Error and Uncertainty in Physical Measurements. In *Modeling Complex Data for Creating Information. Data and Knowledge in a Changing World*; Dubois, J.E., Gershon, N., Eds.; Springer: Berlin/Heidelberg, Germany, 1996. [CrossRef]
77. Farrance, I.; Frenkel, R. Uncertainty of Measurement: A Review of the Rules for Calculating Uncertainty Components through Functional Relationships. *Clin. Biochem. Rev.* **2012**, *33*, 49–75.
78. Costa, M.M.; Reis, S.M.L.; Brígida, A.C.S.; Gonçalves, C.S.L.; Costa, J.C.W.A.; Jorge, P.A.S.; Martinez, M.A.G. High Voltage Compact Optical Current Sensor Experimental Tests. In Proceedings of the Nineteenth Brazilian Symposium on Microwaves and Optoelectronics and Fourteenth Brazilian Congress on Electromagnetism, Online, 8–12 November 2020.

energies

MDPI

Article

Electrical Transients in Industrial Facilities

Attila Kovács [1,*], Judit Somogyiné Molnár [1] and Károly Jármai [2]

[1] Department of Electrical and Electronic Engineering, Institute of Physics and Electrical Engineering, University of Miskolc, H-3515 Miskolc, Hungary; judit.somogyine@uni-miskolc.hu
[2] Department of Mechanical Engineering, Institute of Energy and Chemical Engineering, University of Miskolc, H-3515 Miskolc, Hungary; karoly.jarmai@uni-miskolc.hu
* Correspondence: attila.kovacs@zf.com

Abstract: It is essential to protect control, regulation and data storage units used in automated robotic manufacturing from the consequences of harmful electrical grid harmonics. In this paper, the effectiveness of active/passive harmonic filters built into LED lamps, test benches, and industrial robots was investigated (in the latter two cases, light and heavy load states were applied). Based on network analysis, it was found that the built-in passive harmonic filters of the luminaries were ineffective because the THD of the current was approximately two or three times the permissible value according to the IEEE 519-2022 standard. It was proved that the built-in active harmonic filter of the test bench at a heavy load worked properly, but at a light load it was not effective, while the magnitude of harmonic distortion of the industrial robot exceeded the allowed level (three times the limit at a light load, seven times the limit at a heavy load). Further measurements were also performed at three locations: an engineering room, a gearbox house machine tool, and an office room (where there were no built-in filters). The results proved that in each case, there was a significant exceedance of the limit value. Therefore, protection against harmonics needs to be installed, and then the measurements must be repeated.

Keywords: electrical transients; harmonic distortion; electrical power disturbances; voltage sags; total harmonic distortion (THD); built-in active/passive harmonic filter

1. Introduction

Today, there are almost no mechanical movement or electrical energy conversion devices that do not rely on electronic power control. Since the electricity network is constantly changing, the main directions of research are forecasting and prognosis. Due to the drastic increase in the number of non-linear loads, it is possible to estimate the magnitude of harmonic distortions and the level of losses in residential, industrial, and office environments; one method for this is the Norton equivalent modelling technique [1]. Semiconductor production requires very high-quality electricity, and thus some energy providers create special strategies to improve the quality, which contribute to ensuring that its quality does not deteriorate either now or in the future [2]. In the processing industry, e.g., paper production, specific research has been carried out on the possible consequences of poor energy quality, what losses can be caused by inadequate electricity quality, and the effects of high harmonic injection equipment on the operation of other equipment [3]. The effect of harmonics was also investigated in a large steel factory and a forecasting program was developed to estimate the expected harmonic loads, which can also be used in the installation of new factories [4]. Since the load of IT equipment cannot be precisely determined in advance, research is being conducted to predict the harmonic injection of a large number of single-phase computer equipment in the case of offices and office buildings [5]. Currently, the impact of renewable energy sources on energy quality is also a major problem, since the basic component of these devices is a DC/AC inverters containing non-linear elements; therefore, due to their mass distribution, they must be treated as a high or even priority hazard [6]. The EMS (Energy Management System ISO 50001) standard [7] also requires the use of power-controlled drives above certain power limits, but different types of induction heating devices can also be classified as electrical energy conversion equipment. These non-linear devices have electrical transducer feedback. The possible theoretical errors caused by distorted waveforms are well known; the sine waves of current and voltage under normal conditions are pure sine waves, but they are distorted in systems containing harmonics [8].

Researchers often deal with the comprehensive modelling of the components of an energy supply system, as well as the simulation of wind and solar power plants and the in-depth description of the effects of harmonics. One of the possible methods for reducing harmonics is the design of passive filters [9]. In the electricity system, photovoltaic (PV) energy belonging to renewable energy sources is connected to the electricity distribution network via an inverter which can generate harmonics near the switching frequency. Therefore, losses due to harmonics must also be considered in this case. The harmonic impedance can be modelled, and the harmonic losses can also be calculated in advance [10]. Problems caused by the distortion of waveforms generally include the distortion of power supply voltage waveforms, resonance risks in systems at high frequencies, the failure of sensitive electronic devices, PLC and CNC devices, malfunctions in protection and control systems, and overheating in electric motors and transformers [11]. Studying the effect of current harmonics on distribution transformers is crucial for the network planning and maintenance activities of energy providers. Therefore, the effects of harmonics on eddy current and other stray losses, the temperature of the hottest point, and the lifetime of distribution transformers are investigated [12]. The frequency and magnitude of the problem depends on factors such as the type of installation, the operating voltage level of the installation, the type/intensity of the load and the short-circuit power of the network [13].

Practical experience shows that in electricity supply, it is difficult to recognize whether the size of the harmonics in the grid is already harmful. Unfortunately, even nowadays, suppliers carry out direct network analysis only for large consumers and there is an ongoing debate about what amount is produced by the industrial installation and what is delivered to the consumer from the supplier's network. Companies seldom carry out audits of their internal electrical network and these analyses are mostly only aimed at detecting and documenting "blackouts". Direct testing of low-power equipment [14] such as office and production control equipment, controls, light sources, dimmers, etc., is often left out of the audit. Today's largest sources of harmonic disturbance are electronic power controls and LED light sources [15,16]. Several types of LED lamps produce different amounts of harmonic current depending on the ballast configuration. This harmonic distortion can be reduced by combining different types of LED lamps, although this is not common practice in industrial halls in order to maintain the homogeneity of lighting. The voltage sag tests show that all tested lamps are sensitive to the magnitude and duration of the voltage sag [17]. The high harmonic content of the phase currents and the current of the neutral conductor can be detected in the case of LED lamps. LED lamps with controls can cause excessive heat production in the wires during operation. In circuits loaded with such current harmonics, it is necessary to apply the methodology of cable selection, primarily because of the possible high neutral conductor current [18]. Of course, the development of models of LED lamps and their analysis makes it possible to calculate and predict the harmonic emission. The parameterization of these model is determined based on experimental measurements [19].

Plug-in hybrid and pure electric vehicle chargers are sources of harmonic disturbance as well. For predictions, the effects of EV charging load on cable current and temperature in the presence of harmonic currents of different magnitudes and orders are modelled for two typical modes of EV charging, peak-time and off-peak home charging [20]. In the case of plug-in hybrid electric vehicles (PHEV), not only is their harmonic injection investigated, but also the effect of their charging electric load on the transformer loss-of-life, since the accurate prediction of charging load is almost impossible due to its stochastic nature. Therefore, a stochastic model can be used to examine the effect of real-time charging rates on the charging load and the resulting effect on transformer derating. By using a stochastic model, the effect of real-time charging rates on the charging load and the resultant impact on transformer derating can be investigated [21]. When designing the load control of intelligent distribution systems, the effect of generated current harmonics by electric vehicle battery chargers must be considered. Research shows that charging EV batteries in plug-in electric vehicles increases power losses more than harmonic currents. It was proved that charging EV batteries in plug-in electric vehicles increases power losses rather than the harmonic currents [22].

2. Harmonic Disturbance Sources

Any distorted, non-sinusoidal periodic function can be analyzed using Fourier's theorem and represented as the sum of terms consisting of the following components [9,23,24]:

- The DC component, if applicable;
- A sinusoidal term at the fundamental frequency;
- Sinusoidal members (harmonics) with a frequency represented by a multiple of the fundamental frequency.

The nth-order harmonic (usually called the nth harmonic) in the signal is the sinusoidal component with a frequency n times the fundamental frequency. A distorted periodic current waveform expanded into a Fourier series is expressed as follows [9]:

$$i_t = I_0 + \sum_{n=1}^{\infty} I_n \sin(n\omega t + \Phi_n) \tag{1}$$

where I_0 is the DC component, I_n is the maximum value (amplitude) of the nth harmonic, ω is the angular frequency (2π times the fundamental frequency), while Φ_n is the initial phase of the harmonic component at time t = 0. That is, a distorted periodic signal has a fundamental frequency component, namely 50 Hz, plus a second harmonic component of 100 Hz, plus a third harmonic component of 150 Hz, and so on. Harmonic distortion of the voltage waveform affects both electricity distribution networks and consumer equipment. Harmonic distortion levels have increased rapidly in electricity networks in recent years, mainly due to the increasing use of non-linear semiconductor devices, which are responsible for a large portion of the harmonic distortion [25,26].

Figure 1 shows a linear consumer, while Figure 2 a non-linear consumer based on signal shapes.

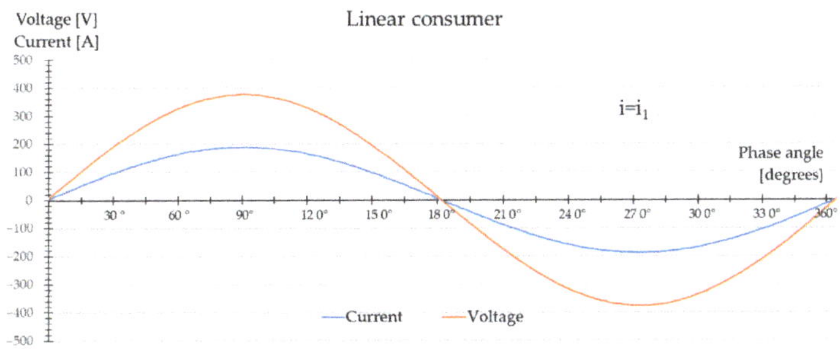

Figure 1. Linear consumer waveforms.

One can see that the current of linear consumers is proportional to the voltage applied to them with the same waveform. In the case of non-linear consumers, the resulting current is calculated by the RMS (root mean square) [27] formula, i.e., the square root of the sum of the squared harmonic currents. If their current in the frequency domain is analyzed (the spectrum of the signal is obtained by Fourier transform), one can see that, in addition to the fundamental harmonic of the mains (50 Hz), there are also integer multiples of it

(100, 150, 200, 250 Hz, etc.). The harmonic distortion of the mains voltage is due to the harmonic currents of non-linear loads. These currents flow through the impedances of the electrical network, creating harmonic voltages. This effect is called the harmonic feedback of consumers [28].

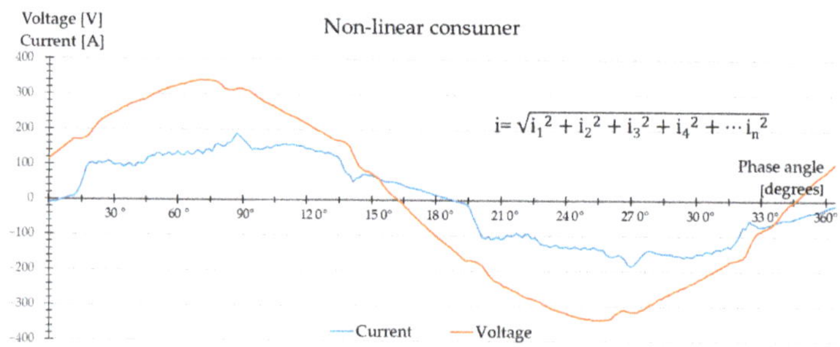

Figure 2. Non-linear consumer waveforms.

It should be noted that not all types of loads inject harmonics into the supply system. This depends on the type of load characteristics. While in the previous figures the line current and its harmonic components are represented by time-domain waveforms, it is simpler and more common to represent harmonics using the current spectrum. In Figure 3, the fundamental harmonic of the current (50 Hz) and the third (150 Hz), fifth (250 Hz), seventh (350 Hz) and ninth (450 Hz) harmonics of the current are plotted in the time domain over a period T.

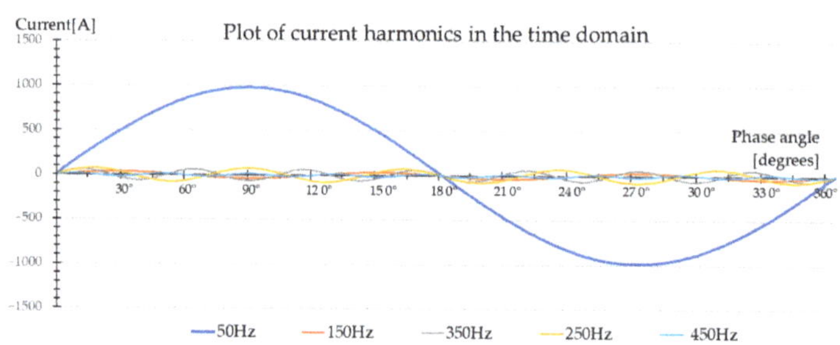

Figure 3. Plot of the current harmonic components of a distorted current waveform.

Summing these sinusoidal waveforms gives the true waveform of the current, as illustrated in Figure 4.

Waveforms can be produced from a basic sine component and the sum of superimposed sinusoidal harmonics. The generation of the waveform of Current 1 is illustrated in Figure 4. Current 1 consists of a 50 Hz fundamental component and harmonics of different orders and magnitudes with respect to the fundamental harmonic. In this case, harmonics are only denoted to the ninth odd order, but they can be denoted up to 256 orders (of course, it is worth investigating this as far as useful information can be obtained).

The frequency domain spectra of these harmonic currents are illustrated in Figure 5 as amplitude (magnitude) values in reference to the fundamental harmonic (50 Hz). This spectrum shows the amplitudes of the harmonic orders for each harmonic frequency. As shown in Figure 4, the fifth and seventh harmonics have a greater amplitude. Since the plots are based on our own real measurements, the ratios of the time- and frequency-domain currents are the same.

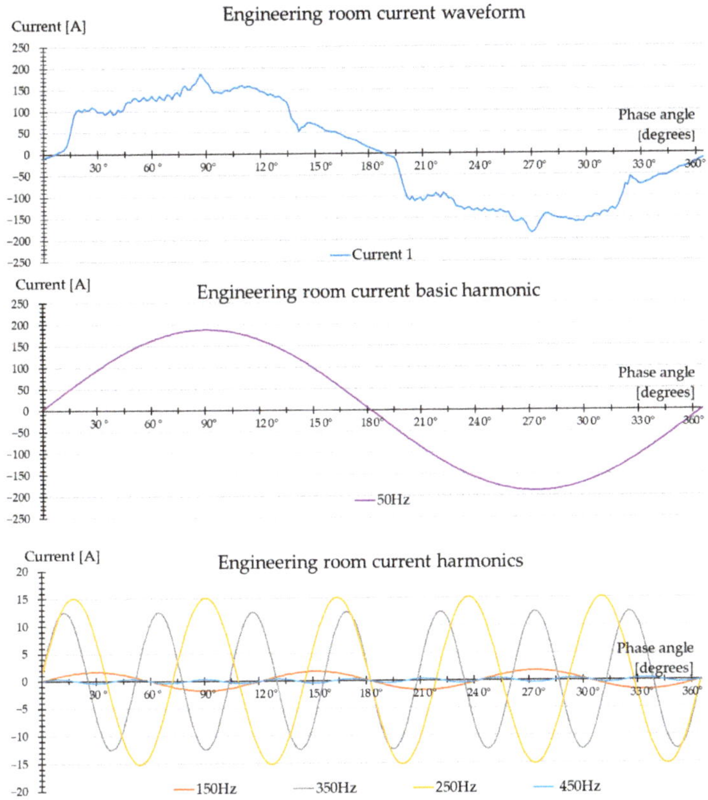

Figure 4. Distorted Current 1 waveform with fundamental and harmonic components.

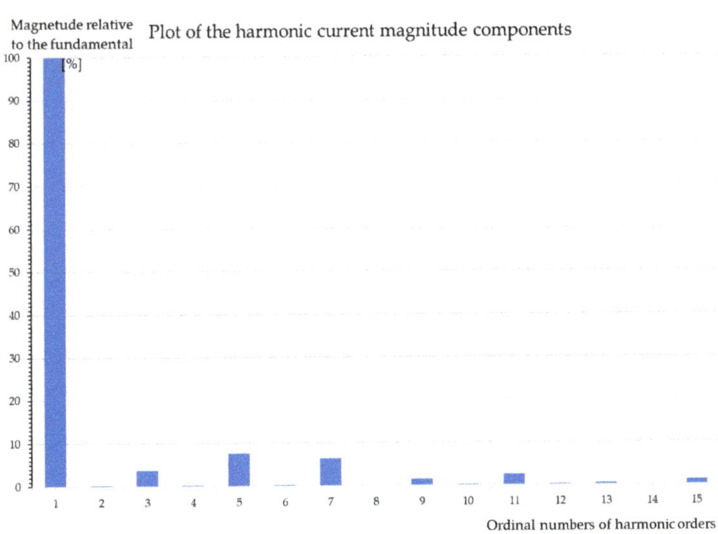

Figure 5. Plot of the harmonic current magnitude components of a distorted current waveform.

It is well known that harmonics are sinusoidal voltages or currents whose frequency is a multiple of the fundamental frequency at which the power supply is designed to operate (e.g., 50 Hz). An electrical supply system is inherently very passive, and the source outputs and the voltages measured across the system would be close to sinusoidal at the declared frequency if there was no load connected to the system. It is well known that for a pure sinusoidal voltage wave with a frequency of 50 Hz and an amplitude of 230 V, the spectrum would be zero at all frequencies except 50 Hz, where the value would be 230 V. For a distorted voltage wave, the spectrum contains harmonic frequencies characteristic of the nature of the distortion. The basic frequency is the component of the spectrum at which the network is designed to operate. It is usually the first and largest component of the spectrum, as can be seen in Figure 5. The term "total harmonic distortion" (THD) describes the effective sum of the voltages of all harmonic frequencies present relative to the fundamental.

The two most commonly used distortion indicators are the individual and the total harmonic distortion indices, which are usually described by Equations (2) and (3). The former shows the ratio of the component observed at a given ordinate h to the fundamental harmonic, while the latter relates the resulting harmonic content of the signal to the basic harmonic. The two indicators can be determined for both voltage and current as follows [29]:

$$D_h = \frac{U_h}{U_1} \tag{2}$$

$$THD_U = \frac{\sqrt{\sum_{h=2}^{\infty} U_h}}{U_1} \tag{3}$$

where U_h is the harmonic voltage of order h, U_1 is the fundamental harmonic voltage, D_h is the individual harmonic distortion, while THD_U is the total harmonic distortion. During the measurements, the true RMS [27] value for any signal was measured. This true effective value of the current returns the effective value that has the same effect and is capable of the same performance as the direct current of the same voltage or current. Thus, in the case of a non-linear consumer, according to Figure 2, the true effective value of the current is equal to the square root of the square sum of the current of the fundamental harmonic and harmonics described by Equation (4).

$$i = \sqrt{{i_1}^2 + {i_2}^2 + {i_3}^2 + {i_4}^2 + \ldots {i_n}^2} \tag{4}$$

3. Switching Power Supplies

Over the last 20 years, switching power supplies have increasingly replaced the transformer rectifier power supplies in electronic equipment. These devices generate a large amount of fifth harmonic emissions relative to their rated power consumption, which are randomly phase-directed and thus can add up [30,31]. Consequently, in power systems, at all voltage levels up to 400 kV, the background level of harmonic level distortion is increasing. In three-phase distribution networks, odd harmonics are most common. It is essential to measure harmonics of the orders 3, 5, 7, 9, 11, 13, etc., up to harmonics of the order 25, as practical experience has shown. As the harmonic voltage distortion increases, the risk of widespread problems due to premature ageing of the equipment and overloading of the neutral conductors increases. Generators in the power system usually produce a pure sine waveform at their end points. This sinusoidal waveform is considered to be the pure form of the AC voltage, and deviation from it is called distortion. More and more consumers are drawing a non-sinusoidal current from the electricity system. As a result, as described above, due to sine wave distortion of the grid voltages and currents, several zero crossings can occur within a period [9]. The consequence will be a malfunction of the switching power supplies. As an example, consider the circuit shown in Figure 6.

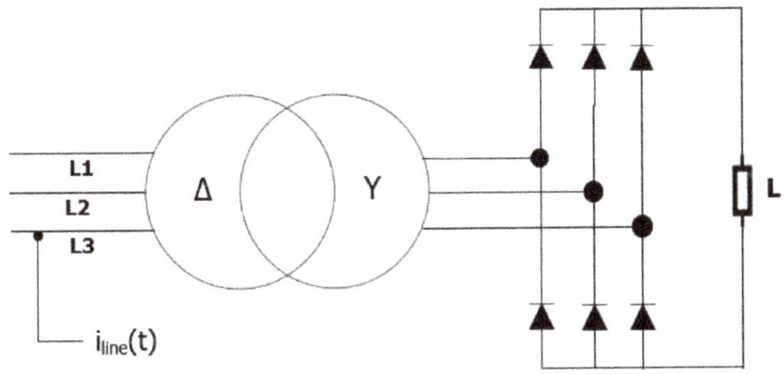

Figure 6. Example of a load that takes a non-sinusoidal load from the network.

The thyristor bridge is connected to a pure sine wave network via a transformer. The frequency of this sine wave is called the fundamental frequency f_1. When the load is connected to the power supply, a line current flows, denoted as $i_{line}(t)$. Figure 7 shows an approximation of the current waveform.

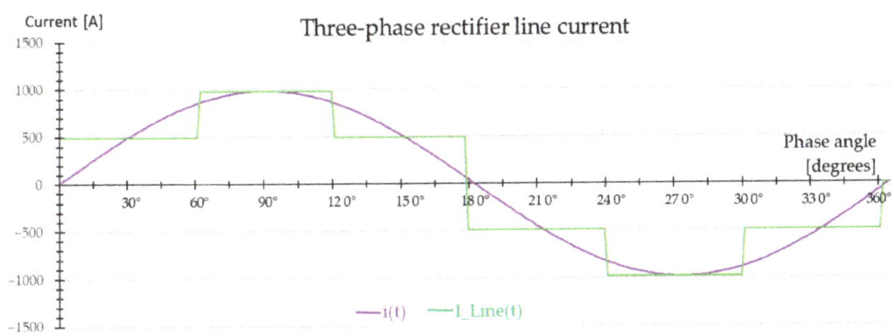

Figure 7. Line current flowing in the branch of a three-phase, two-way, six-stroke (3F, 2W, 6S) diode rectifier power supply.

From this figure, it can be seen that the line current $i_{line}(t)$ is very different from the sinusoidal current waveform, the fundamental component i(t), and therefore something must be added to this sinusoidal component to obtain the actual current flowing. To determine what this "something" is, look at Figure 8, which shows the waveform of the current obtained by adding component i_h5(t) with a frequency equal to five times the fundamental frequency to the i(t) component.

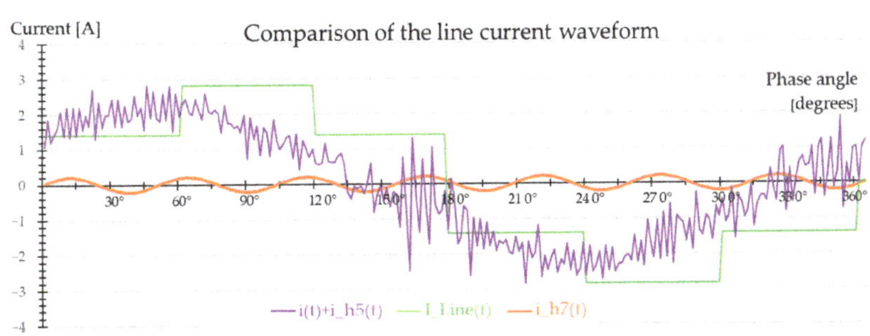

Figure 8. Comparison of the line (t) line current waveform with the i(t) + i_h5 (t) waveform.

Currently, power electronic control-based equipment is the main source of harmonic pollution in low-voltage networks [32]. Usually, in such equipment, the switching semiconductor conducts only in a fraction of the base period, as this is how this type of equipment can gain its main characteristics of energy saving, dynamic performance, and control flexibility. It has been mentioned earlier that generators in the power supply system usually output a pure sinusoidal waveform at their end points; this sinusoidal waveform is the pure form of the AC voltage, and deviation from this is called distortion. Currently, there are more and more consumers with different power electronics solutions—due to the use of semiconductors in switching mode—causing the equipment to draw a non-continuous current from the power system, rather than a sinusoidal current, with a significant amount of distortion. The explanation for the formation of harmonics is as follows. While in linear consumers there is a direct proportionality between voltage and current, in non-linear, non-continuous consumers, it is observed that the current no longer follows the sinusoidal voltage waveform, but so-called positive and negative current peaks occur. Since the semiconductors used in switching mode conduct for only a fraction of the base period of the mains period, the consumers draw pulsating input currents from the mains, and these feedbacks caused by the pulsating currents cause the harmonics in the mains.

The aim of this study is to identify precisely the potential problems caused by harmonics, especially in electronic and IT devices used in industry. Later, an exact mathematical method must be found to identify the harmonics generated and to calculate the network feedbacks for preventive purposes, so that the expected magnitude of harmonics can be predicted at the design stage of electronic circuits. In addition, this information can be used to find specific, local, cost-effective filtering methods for existing harmonics in the electrical network, targeting to the sensitivity of the equipment concerned.

4. Possible Consequences of the Presence of Harmonics

From the distorted signal shapes shown in Figure 8, it can be seen that if multiple zero crossings are created, the circuit controlling the thyristors, synchronized with the zero crossing, will make faulty connections and thus, due to the increase in the fill factor, the output voltage of the switching power supply may increase—if no overvoltage protection is installed—and destroy the circuits it supplies. This process can be seen in Figure 9.

The presence of harmonics can also have other consequences, such as the skin effect [33]. DC current is homogeneously distributed across the cross-section of the conductor, so the current density is the same over the entire surface of the conductor. In the case of alternating current, at high frequencies, this is no longer true because the alternating current always tends to flow near the outer surface of the conductor, caused by the electro-magnetic field surrounding the conductor and varying with frequency, which creates a current in the conductor similar to an eddy current. According to Lenz's law, this current is not eddy-like, but is in the opposite direction to the current generating the magnetic field. As the two currents repel each other, the collateral magnetic field forces the conductor current to the surface, which becomes more and more pronounced at increasing frequencies.

In the vast majority of cases, this is not considered at the mains frequency of 50 Hz, as the effect is negligible at this frequency. However, from 250 Hz onwards, i.e., from the fifth harmonic upwards, at higher frequencies, the skin effect is significant, as it creates increasing collateral losses. This should be considered especially when sizing single-core cables. Harmonics also influence plated busbars, reducing the skin effect, but because of their sizing in relation to the resonance, harmonics must be taken into account.

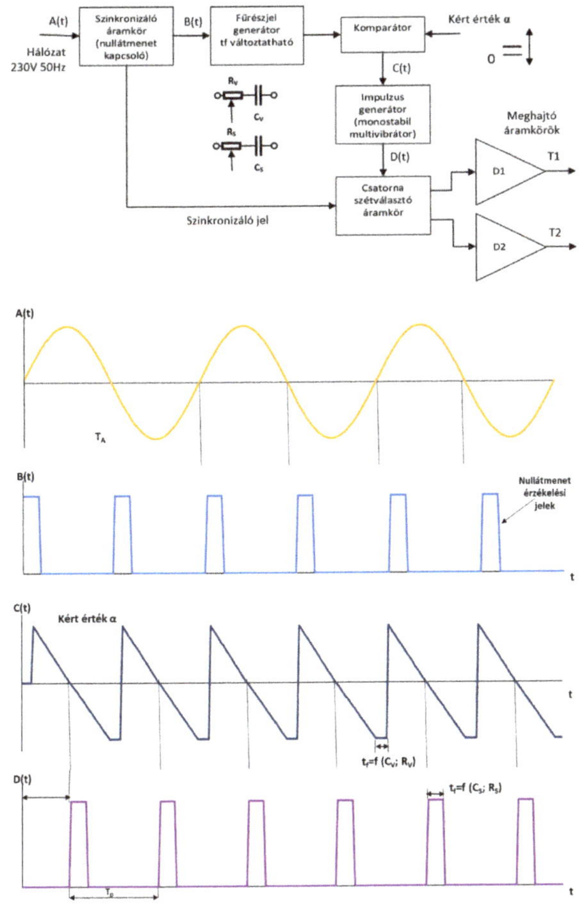

Figure 9. Block diagram and control of a network zero-crossing synchronized thyristor IC.

5. Filtering Harmonics

If the distribution network is "polluted" with harmonics, the quality parameters of the energy service are also significantly degraded [30]. The true RMS value of the current in the neutral conductor increases. If a fully symmetrical load is assumed in the three-phase power system, i.e., loads of equal magnitude are connected between each phase and the neutral, then the fundamental, positive-order, and negative-order harmonics are eliminated in the summation, while the zero-order ones are summed algebraically. For this reason, the current in the neutral conductor will not be zero even with a symmetrical load since the sum of the zero-sequence currents in the three phase conductors flows in the neutral conductor in the case of a four-wire Y connection. The effective value of the current of the neutral conductor can be calculated as the square root of the sum of the squares of the effective value of the zero-order currents. In the case of symmetrically connected non-linear loads, this results in a significant additional load in the neutral conductor. This results in a substantial excess load in the neutral conductor for symmetrically coupled non-linear loads. In locations where most loads are non-linear, e.g., office buildings, two to three times the phase conductor current may flow in the neutral [34]. Harmonics may cause overloading in generators, motors, and transformers [35]. In a harmonics-contaminated network, electrical equipment may be subject to unwanted vibration and premature ageing, as well as the hum and unwanted sound effects of transformers are also increased, not to mention the overloading of phase correction capacitors, which can also lead to their premature ageing and premature replacement, generating significant costs for users.

In a three-phase symmetrical system, if there are no current harmonics, of course no current flows in the neutral conductor. However, in the presence of harmonics, significant current flows in the neutral conductor, which can be several times that of the phase current. The protections may be unduly tripped, reducing the security of the power supply and the life of the protection circuit breakers. The reactive current also causes disturbances in the operation of compensation systems, generating resonant voltages in the resonant circuit formed by the inductive nature of the network and the capacitive system of the reactive current compensation.

In reality, the resulting voltage distortion is proportional (to a limited extent) to the impedance of the power supply, which proves that the harmonic source is a current generator. The impedance of the supply network is usually very small, so the voltage distortion caused by the harmonic current will be small, often so small that it is lost in the background noise. This can be misleading because it gives the impression that there is no harmonic problem, when in fact there are large harmonic currents. This problem can be compared in an attempt to find circulating currents in the ground by measuring the voltage. So, if one suspects or tries to prove that there are no harmonics, the current has to be measured.

One of the simplest solutions for filtering out the third harmonic is the transformer [32,36], which is provided by default in all medium-voltage/low-voltage transformers (MV/LV transformers), so that in the case of wye/delta grounded transformers, the 3rd harmonic current is not transferred to the delta side. This is not only applicable for the MV/LV voltage conversion, but can also be used as a decoupling transformer, thus ensuring that the third harmonic is filtered out. However, in the case of high-power transformers, due to the capacitance between the primary and secondary windings, which can be very high, high power can be used to transfer harmonic voltages from the primary side to the secondary side. Also problematic are the harmonics from the power grid coming through the neutral and the earth conductor, of which the transformer does not filter out any harmonics. Therefore, when choosing the cross-section of the neutral conductor, it is essential to consider the harmonic potential content of the grid.

With active harmonic filters, the consumer current signal is made as perfect as possible by injecting an opposite-signature harmonic current in "real time" into the harmonic-contaminated current drawn by the consumer, thus providing a clean (50 Hz) sinusoidal signal waveform to the mains, which does not distort the mains voltage. Today's active harmonic filters can track this with a response time of 300–500 μs, which is comparable to the 50 Hz mains frequency, i.e., the 20 ms mains T period. Compensation is carried out up to the 50th harmonic, but in practice currents above the 25th harmonic are not often encountered. IGBT-insulated gate bipolar power transistors with a switching frequency of 10 to 15 kHz are used to generate the signal waveforms.

Another power quality improvement technique is APLC (active power line conditioning) [37,38], which is an active filter type that compensates for distortion of the power system waveform. The objective was to develop and present a procedure to calculate the amount of APLC-injected current required to minimize voltage harmonic distortion in the entire power system. The procedure is intended to be applied with APLC frequency domain correction in networks where periodic harmonic distortion is experienced. The injection currents are determined using non-linear optimization theory. The main solution lies in the development of a simple procedure for finding the Fourier series of the optimal APLC injection current waveform. The procedure can be applied in any of the following situations: 1. single-phase APLC in a single-phase network; 2. single-phase APLC in a three-phase network; or 3. three-phase APLC in a balanced three-phase network.

An active harmonic filter with a series design is shown in Figure 10.

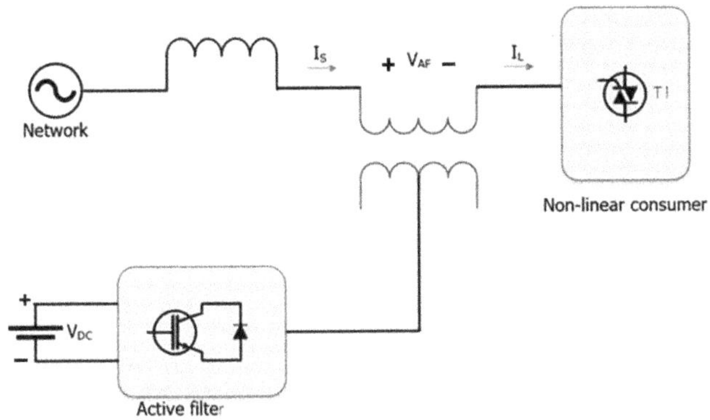

Figure 10. Active filter with in-line design.

This type of filter is connected to the mains through a dedicated transformer. It has the advantage of compensating for the effects of voltage drops and directly influencing the harmonic content of the voltage. Its major disadvantage is its high production cost due to its complex design and, as can be seen from the wiring diagram, the fact that it is very difficult to integrate into the existing electrical network. Therefore, instead of this type of filter, the shunt-type filter has become more widespread in practice and is almost the only one used today.

An example of an active harmonic filter with a shunt design is shown in Figure 11.

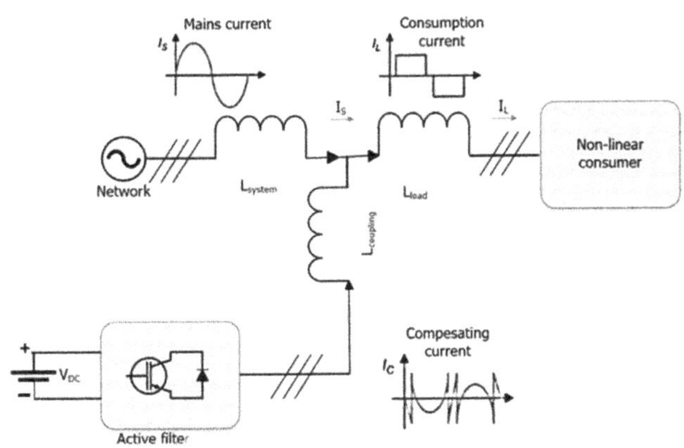

Figure 11. Shunt design active filter.

This type of filter injects a current of the same magnitude as the measured harmonic current generated by the non-linear consumer, but in phase opposition, into the network, resulting in a current flowing to the incoming network containing only fundamental harmonics [30]. This is described by Equation (5).

$$I_{network} = I_{filter} + I_{load} \tag{5}$$

The filter uses PWM (pulse-width modulation) to generate the expected signal shape. This filters is capable of filtering harmonics up to 50 orders. For this purpose, an inverter with an NPC (neutral point clamped) [39] topology is used, using IGBT upstream drivers [40].

6. The IEEE 519-2022 Standard Recommends

In order to evaluate the measured data, it is essential to know the relevant standards. When examining the standards, a distinction must be made between the maximum permissible voltage and current distortion values recommended for the power supply's electricity distribution network, devices and equipment connected to the so-called residential network, and the internal electrical network of industrial facilities. The MSZ EN 61000-3-12 or IEC 61000-3-12 standards must be used to test a product under laboratory conditions, while the IEEE 519-2022 standard (not yet a Hungarian standard) recommends maximum permissible harmonic distortion values for equipment installation. In cases of current harmonics, the creators of the IEEE 519 standard decided to limit the magnitude of current harmonics depending on the power of the voltage source. This is logical since a network with a high electrical power can suppress the effect of current harmonics to a much greater extent without affecting its voltage than a network with a lower power. The purpose of creating current harmonic limits was to limit the harmonic injection of individual consumers into the network so that they do not cause unacceptable voltage sine wave distortion [41]. The IEEE 519 standard specifies limits for order-by-order harmonic distortion and total current demand distortion (TDD), not total harmonic distortion (THD). THD specifies the maximum possible values only for the voltage and causes its magnitude to depend on the voltage level of the common connection point (Point of Common Coupling—PCC). A significant difference is that in the IEEE 519 standard, the limit values of voltage harmonics are constant at all frequencies, while in IEC 519, the permissible magnitude of voltage harmonics decreases with frequency. Among other things, the IEEE 519-2022 standard imposes two requirements on harmonics—the absolute maximum voltage distortion (THD), i.e., the previously mentioned THD_U, as well as the harmonic distortion according to the variable maximum total current demand (TDD). In the case of the former, one can see the relevant limits in Table 1, while for the latter, they can be seen in Table 2. TDD depends on the ratio of the short-circuit current (I_{SC}) of the common electrical network connection point (PCC) and the load current (I_L) of the connected device. In this way, the effect of current harmonics produced by individual equipment on the network can be distinguished. The reason for this is that the harmonics produced by consumers with lower current consumption are less able to influence the quality of electricity on the network compared to large consumers. The TDD values allowed in the standard refer to the given common connection point (PCC), which ensures the connection between the high-current supply point and the connected equipment and can be used at any voltage level. In industrial conditions, the PCC is considered an internal point of the own electrical network or electrical distribution system. IEEE 519-2022 differs from the original intent of IEEE 519 in that the latter considers only the point of connection between the power supplier and the consumer. In contrast, IEEE 519-2022 already allows the internal electrical network to be divided into several connection points (PCC). Thus, in a large-scale industrial facility, the limit values of the standard can be applied separately to several common connection points with different power currents protected by circuit breakers.

Table 1. Recommended possible voltage distortion limits by the IEEE 519-2022 standard.

Bus Voltage V at PCC (kV)	Individual Harmonic (%)	Total Harmonic Distortion (%)
$V \leq 1.0$	5.0	8.0
$1 < V \leq 69$	3.0	5.0
$69 < V \leq 161$	1.5	2.5
$161 < V$	1.0	1.5 [1]

[1] High-voltage systems can have up to 2.0% THD, where the cause is an HVDC terminal whose effects will have attenuated at points in the network where future users may be connected.

Table 2. Recommended possible current distortion limits for systems rated 120 V through 69 kV by IEEE 519-2022 standard.

I_{SC}/I_L	Individual Harmonic Order (Odd Harmonics) [1,2] Maximum Harmonic Current Distortion in Percent of I_L					Total Current Demand Distortion (%)
	$2 \leq h < 11$	$11 \leq h < 17$	$17 \leq h < 23$	$23 \leq h < 35$	$35 \leq h < 50$	
<20 [3]	4.0	2.0	1.5	0.6	0.3	5.0
20–50	7.0	3.5	2.5	1.0	0.5	8.0
50–100	10.0	4.5	4.0	1.5	0.7	12.0
100–1000	12.0	5.5	5.0	2.0	1.0	15.0
>1000	15.0	7.0	6.0	2.5	1.4	20.0

[1] Even harmonics are limited to 25% of the odd harmonic limits above. [2] Current distortions that result in a DC offset, e.g., half-wave converters are not allowed. [3] All power generation equipment is limited to these values of current distortion, regardless of actual Isc/I_L.

7. Network Analysis

It is essential to investigate the sine-wave distortion effect of non-linear loads on the network under real measurement conditions [42,43]. Our analyses were carried out in 2023 and 2024 at an industrial facility. The aim of the measurements was to test the harmonic injection of individual equipment and equipment groups at the PCC of the electrical network. This allowed us to obtain information on the extent to which building services and manufacturing equipment, office buildings, and halls can be considered as the potential sources of harmonic disturbances and their impact on power quality. These measurement results can be used in comparative analyses in our future research. For example, whether the principle of superposition can be applied to current harmonics in grouped installations, whether the magnitude of harmonics increases linearly with the number of identical installations, or whether the magnitude of harmonics of the same order decreases for certain harmonics can be assessed. For similar machines, the questions of do they add together in orders and does the amplitude of the current harmonics increase proportionally with the frequency can be addressed. During the analyses, the effectiveness of the passive and active harmonic filters installed by the manufacturers both in the loaded and unloaded states of the tested equipment were validated. For the measurements, the network analyzer type PQA824 from the manufacturer HT Italia SRL Faenza (RA)-ITALY [44] was used. The main features of the instrument were recording MIN/MAX/AVG, voltage anomaly detection, harmonic analysis, patch analysis, flicker analysis, recording of high-speed voltage transients from 5.5 µs, and recording of turn-on currents. The instrument was parameterized according to the user's manual, while the recorded data were read out by using HT Italia's Topview software version 2.2.2.3 [45]. The measurement results were evaluated using fast Fourier transform (FFT). Measurements were performed for three different consumers: an LED lamp, a test bench, and an industrial robot.

7.1. LED Lamp Built-In Passive Harmonic Filter Validation

Currently, LED lighting is used in almost all industrial facilities, halls, and offices, but also in the case of workplace lighting. It can be predicted that these electricity consumers can also have a serious impact on the quality of electricity [46]. The aim of this study is to determine the effectiveness of the passive harmonic filters built into the luminaires. This was investigated through network analysis, in the case of a single LED luminaire, and by measuring the lighting of an entire hall at the feed point. First, one HIAL 2000/840/W/F type LED lamp was examined with 230 V voltage and 140 W electrical power. These fixtures have a built-in passive harmonic filter, according to the distributor.

The voltage and current waveforms for one period are shown in Figure 12. The waveform of the current has been distorted to such an extent that its sinusoidal waveform cannot even be recognized. One can also see multiple zero crossings. The total harmonic distortion (THD) of the current was 15.20%. The harmonic distortion according to the total current demand (total current demand distortion—TDD) was also determined. It was

16 A/1 A = 16. According to Table 2 the permissible distortion could only ne 5%, so the difference is more than three times the permissible level, namely 10.20%.

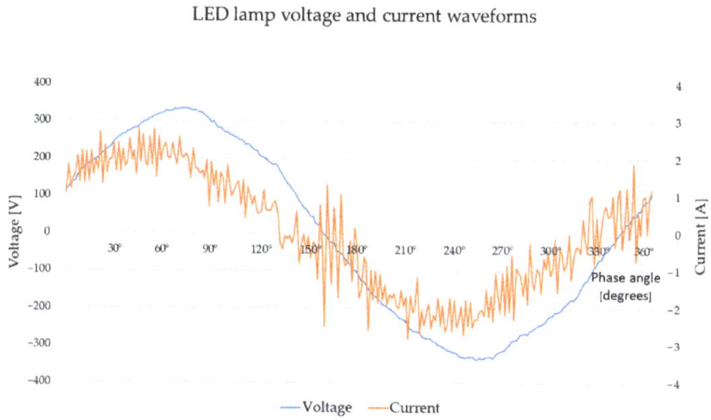

Figure 12. LED lamp's distorted current and voltage waveform.

Figure 13 shows the current spectrum. According to Table 2, the maximum possible value per harmonic order up to the 11th harmonic recommended in the standard is 4%, but these current amplitude values are 7.82% and 7.45% for the 3rd and 7th harmonics. Since the waveform of the current in this case also indicates that there may be harmonic emissions at higher frequencies, the spectrum of the current in the supraharmonic range (2–150 kHz) was also examined, which can be seen in Figure 14. From the 79th order (3.95 kHz) to the 127th (6.35 kHz), it approaches and in many cases even significantly exceeds the 0.6% allowed magnitude of the amplitude relative to the fundamental harmonic according to Table 2. This is important because in practice, the harmonic pollution of the equipment above the 15th harmonic order is no longer examined and, as one can see, it would also be necessary to carry out this in the case of higher orders.

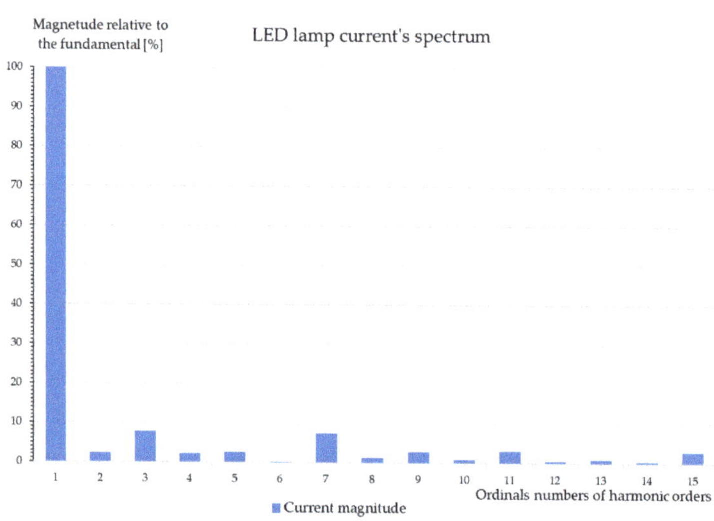

Figure 13. LED lamp's harmonic current magnitude components of the distorted current waveform.

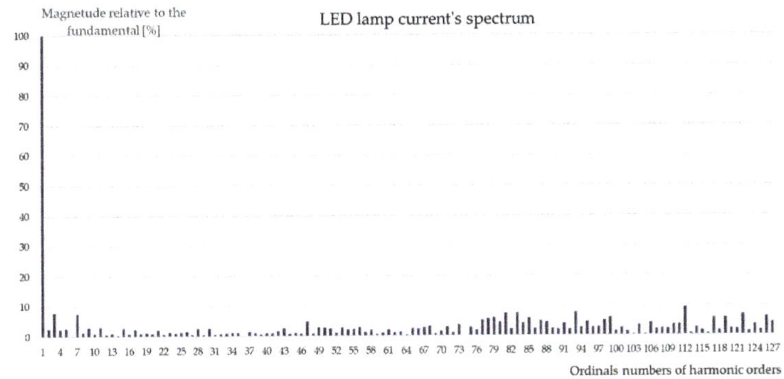

Figure 14. LED lamp's supraharmonics current magnitude components of the distorted current waveform.

This measurement was performed with one and two LED lamps, and there were no significant differences in the signal shapes and the degree of distortion. The table provided by the TopView software version 2.2.2.3 displaying the voltage, current and THD measurement results of the analyzer is shown in Figure 15. From the measurement results based on the graphs, it can probably be concluded that the built-in passive harmonic filter is either not included or is ineffective.

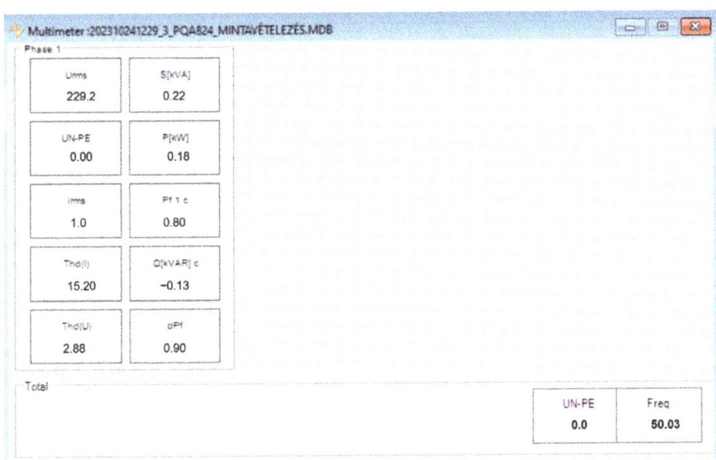

Figure 15. LED lamp's measured voltage, current, THD, and power factor provided by TopView.

7.2. LED Lamp Built-In Passive Harmonic Filter Validation in the Case of Hall Lighting

It was found that the larger these distortions are in the case of lighting an entire production area, the larger the harmonic injection of the hall lighting towards the electrical distribution network is. The hall was empty, meaning that there was no production equipment or other office equipment that could influence the measurement. Figure 16 shows the measurement results of this hall. Four hundred 140 Watt electric power lamps were installed, with a total of 56 kW of maximum effective electric power consumption. Based on Figure 16, it can be said that the voltage waveform is almost pure sinusoidal, and the distortion of the current is not significant. Since the magnitude of the currents of each phase and the THD of the currents were almost the same, it made sense to analyze only the current of phase 2.

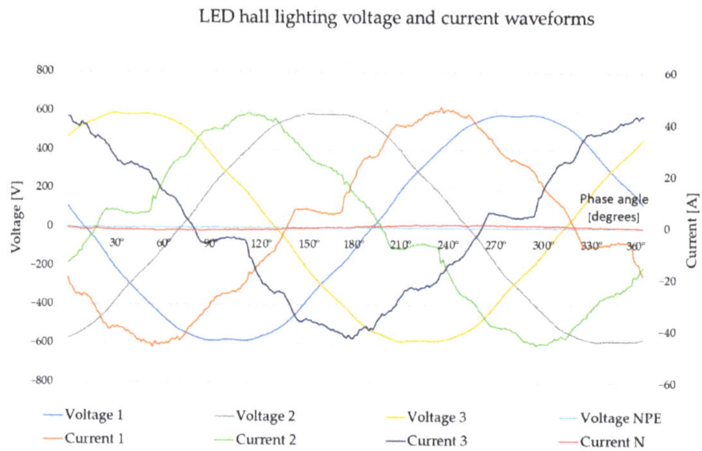

Figure 16. LED hall lights' distorted current and voltage waveform.

The value of the current's total harmonic distortion was 14.71% for the i_2 current. The harmonic distortion according to the total current demand (TDD) is 2300 A/28.3 A = 81.27. According to Table 2, the permissible distortion can be only 8%, so the deviation from the limit value is 6.71%. Figure 17 shows the current spectrum. One can see that instead of the 7% allowed value up to the 11th harmonic in Table 2, the values of the 3rd and 7th harmonics are 10.98% and 6.73%, respectively. The table provided by the TopView [24] software displaying the voltage, current, THD, and reactive energy measurement results of the analyzer is shown in Figure 18.

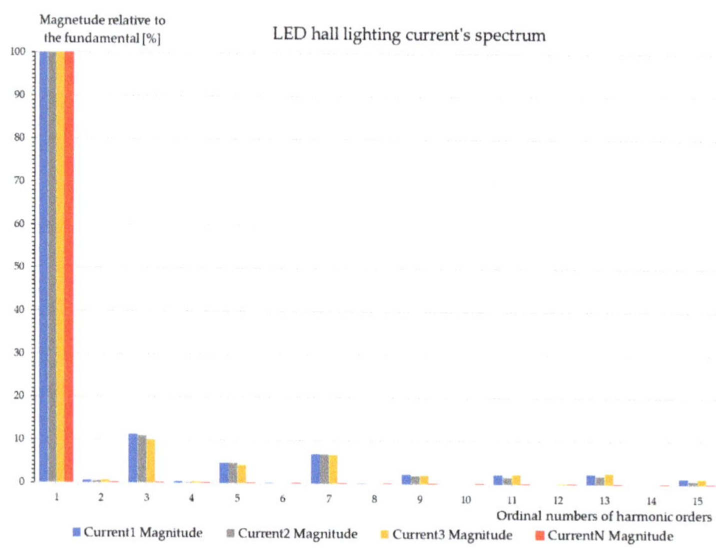

Figure 17. LED hall lights' harmonic current magnitude components of the distorted current waveform.

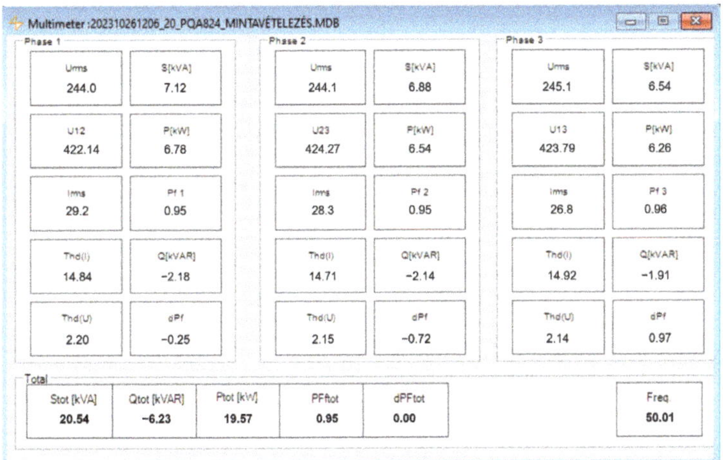

Figure 18. LED hall lights' measured voltage, current, THD, and power factor provided by TopView.

Comparing the results of the individual LED lamp and the group LED hall lighting, the difference is clearly visible. According to Figure 16, the current sine wave distortions of the same magnitude as in the case of a single LED lamp do not appear. But, by looking at the spectra in Figure 17, one can see that the magnitude of the amplitudes per order is very similar to the individual luminaire measurement in Figure 13. This proves that the harmonic distortions of the individual LED luminaires are preserved even in the case of grouping a large number of units distributed by phase, and they do not compensate for each other's distortions. However, the harmonics of the individual luminaires do not add up according to the principle of superposition. This will be an important research result for sizing the central harmonic filter of the lighting distribution network.

7.3. Test Bench Built-In Active Harmonic Filter Validation

In the following, the measurement results performed in a modern factory are presented. The efficiency of an active harmonic filter installed in a test bench equipped with a high-performance frequency converter at full load and during use in the more typical low-load or idling state was examined.

7.3.1. Test Bench at a Heavy Load

In this case as well, the magnitude of the currents of each phase and the total harmonic distortion of the currents was almost the same (see Figure 19); therefore, only Current 3 was analyzed here. The maximum value of the current of the neutral conductor was 5.15 A, which is negligible compared to the maximum phase currents of 108.2 A. The network analysis was carried out with the machine in the extended state under normal production conditions. The applied current varied between 103.3 A and 108.2 A per phase, so it can be considered symmetrical. On the other hand, there was a high level of harmonic current pollution, and therefore the current waveform was also highly distorted. Thus, the THD of the current was 4.35% for Current 3. As was mentioned before, the proposed maximum allowable value of THD as defined in IEEE 519-2022 is given by the TDD. This ratio is now 630 A/108.2 A = 5.82, and according to Table 2, the permissible distortion can only ne 5%. In the case of Current 3, this value is 4.35%, which corresponds to the requirements of the standard, and the built-in active harmonic filter works properly.

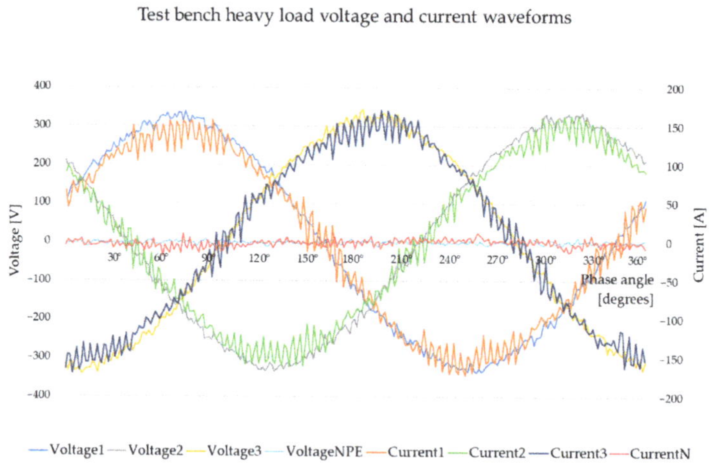

Figure 19. Distorted current and voltage waveforms measured on the test bench at a heavy load.

Note that the current of the neutral conductor is 5.15 A, but it should be zero in normal operation. The current in the neutral conductor is caused by the not perfectly symmetrical (slightly asymmetrical) load, which was negligible in this case. But in the case of larger neutral conductor currents when its magnitude approaches the magnitude of the phase conductor current, this can cause overloading and overheating in the line in the transformer, which can lead to failures [34]. Electricity distribution networks are in most cases TN (three-phase neutral systems) [47], and in this system, the secondary windings of the transformer are connected to a star and the star points of the windings are grounded and the neutral conductor is connected to this point. All harmonics are found in the phase voltages, and hence harmonics appear in the phase and line currents. The neutral current is the sum of the phase currents. The effective value of this current will be the square root of the square sums of the zero sequence currents. This current causes an additional load in the neutral conductor for symmetrically connected non-linear loads, the magnitude of which may be as high as two to three times the phase current. If this current is not considered in network design and the current in the neutral conductor due to harmonics—which causes potential shifts due to voltage drops in individual conductor sections—is not taken into account, it can cause serious disturbances in the operation of electronic equipment. Later, the possible direct consequences of harmonic disturbances on the power sources of equipment operating on the electrical network will be discussed. Figure 20 shows the spectrum of the phase of currents.

As before, on the horizontal axis, the line numbers represent the harmonic orders, the integer nth multiples of the fundamental frequency (50 Hz), while on the vertical axis, the amplitude is plotted against the fundamental harmonic. Based on the measurements, it can be said that the magnitude of harmonic distortion does not exceed the level allowed by the standard in the case of harmonic orders and the active filter works very well at a heavy load.

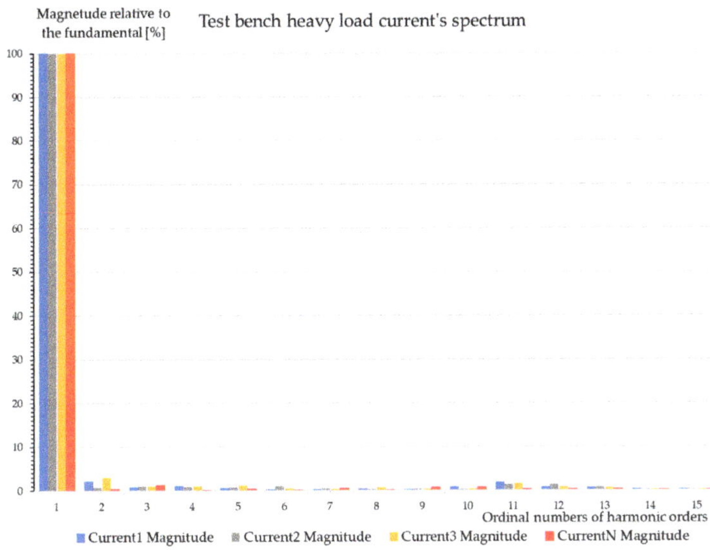

Figure 20. Harmonic current magnitude components of the distorted current waveforms measured on the test bench at a heavy load.

The table provided by the TopView software displaying the voltage, current, THD, and reactive power measurement results of the analyzer is shown in Figure 21.

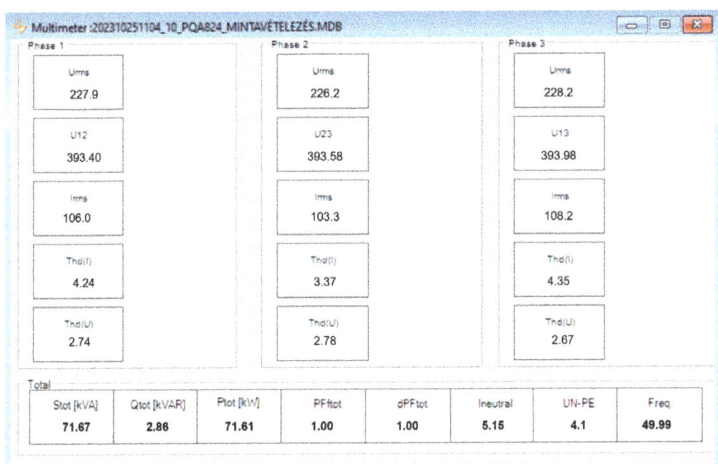

Figure 21. Measured voltage, current, THD, and power factor of test bench at heavy load provided by TopView.

7.3.2. Test Bench at Light Load

Figure 22 shows the voltage and current waveforms at a light load state. The voltage sine is almost free of distortion, but the current sine cannot be recognized; accordingly, the current THD value was 104.54%. As was mentioned before, the standard specifies the recommended maximum value of the total harmonic distortion of the current with the value of the total current demand harmonic distortion (TDD), which depends on the ratio of the short-circuit current of the common electrical network connection point and the load current of the given device. This ratio is now 630 A/10.4 A = 60.57, which is between 50 and 100, so the permissible distortion would be only 12%. The deviation from the limit value is 92.54%.

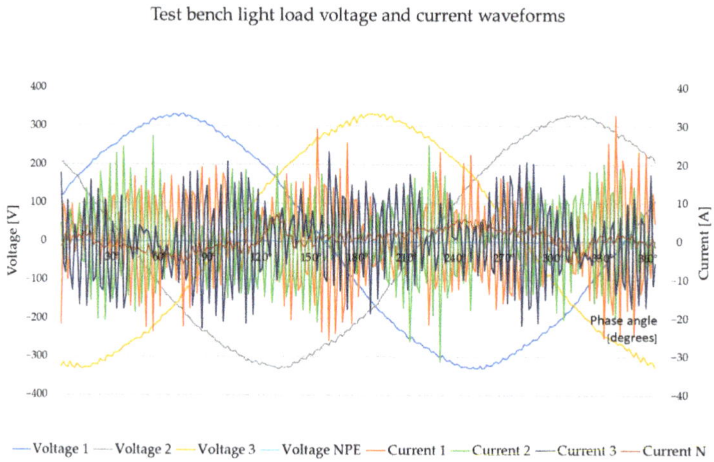

Figure 22. Distorted current and voltage waveforms measured on the test bench at a light load.

In the current spectrum shown in Figure 23, one can see that there is no harmonic order whose value relative to the fundamental harmonic does not exceed the maximum possible value of 4% up to the 11th harmonic written in the standard. Current amplitude values vary between 10% and 88%. Based on the above, it can be stated that the built-in active filter is not effective in a small or unloaded state and allows serious signal shape distortions in the network. It is advisable to increase the efficiency of the filtering as soon as possible, because it can cause serious problems in the various data recording and control units.

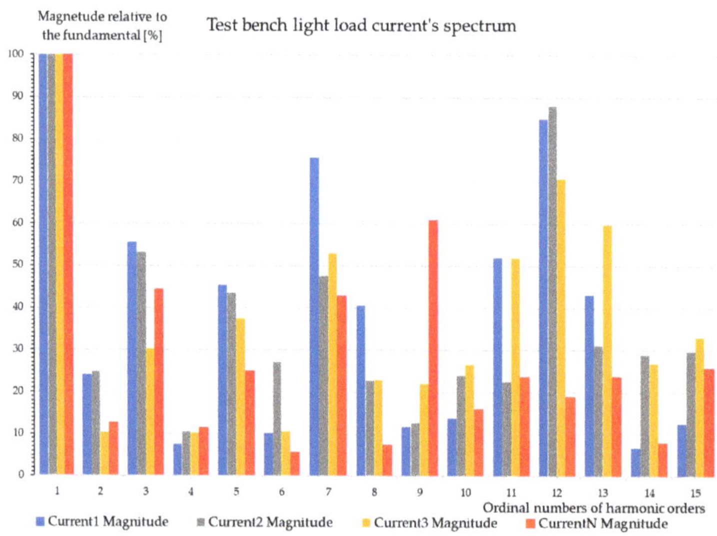

Figure 23. Harmonic current magnitude components of the distorted current waveforms measured on the test bench at a light load.

The table provided by the TopView software displaying the voltage, current, THD, and reactive power measurement results of the analyzer is shown in Figure 24.

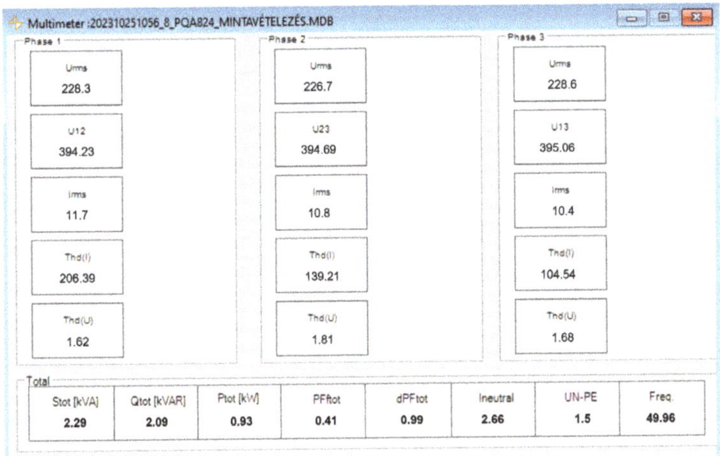

Figure 24. Measured voltage, current, THD, and power factor of test bench at light load provided by TopView.

7.4. Industrial Robot Built-In Active Harmonic Filter Validation

Currently, there is almost no industrial facility where robots, robot cells, and cobots are not used for various technologies. In their variety of applications, they are considered as very significant consumers of electricity, especially if we take into account the automated equipment directly connected to them, such as robot cells or machining centers. Since it can be said that such devices are highly automated, their control and regulation electronics can be highly sensitive to electrical network disturbances and possible voltage dips. In this section, the magnitude of the harmonic injection of such an industrial robot and the efficiency of the built-in active harmonic filter at light load and heavy load states of the robot are examined.

7.4.1. Industrial Robot at Light Load

Figure 25 shows the measured voltage-current waveforms. Since the magnitude of the currents of each phase and the total harmonic distortion of the currents were almost the same, only Current 2 was analyzed. The voltage sine is almost free of distortion, but the current sine cannot be recognized. The measured current varies between 1.6 A and 2 A per phase, so it can be considered symmetrical. In Figure 25, one can see that the current sine waveform is very distorted, and the THD of the current was 59.42% for Current 2. The proposed maximum allowable value of THD according to IEEE 519-2022 is given by the TDD, which depends on the ratio of the PCC termination current I_{SC} to the load current I_L of the device, as was previously presented. This ratio is now 630 A/2 A = 315; therefore, the permissible distortion according to Table 2 can be only 15%, so the difference is more than three times, namely 44.42%.

Figure 26 shows the frequency spectrum of the measurement, i.e., the magnitudes of the harmonics with respect to the fundamental harmonic. However, it can also be concluded from the measurement that according to Table 2, instead of the recommended 12% up to the 11th harmonic, the maximum harmonic current distortions in the case of Current 2 were 33.96%, 21.16%, and 7.63% for the 5th, 7th, and 9th harmonics, respectively. Instead of the recommended 5.5% from the 11th to the 17th harmonic, for the 15th harmonic, it was 7.68%. In the case of higher harmonic orders, their magnitudes were already negligible; therefore, they are not included in the analysis.

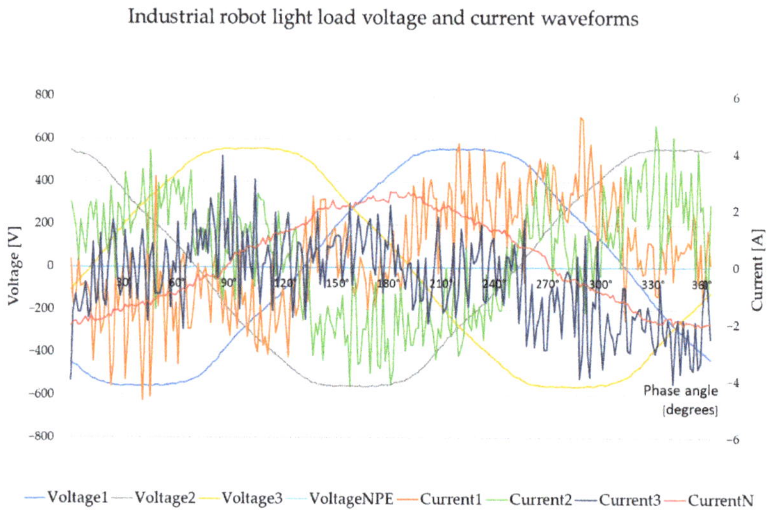

Figure 25. Distorted current and voltage waveforms measured on an industrial robot at a light load.

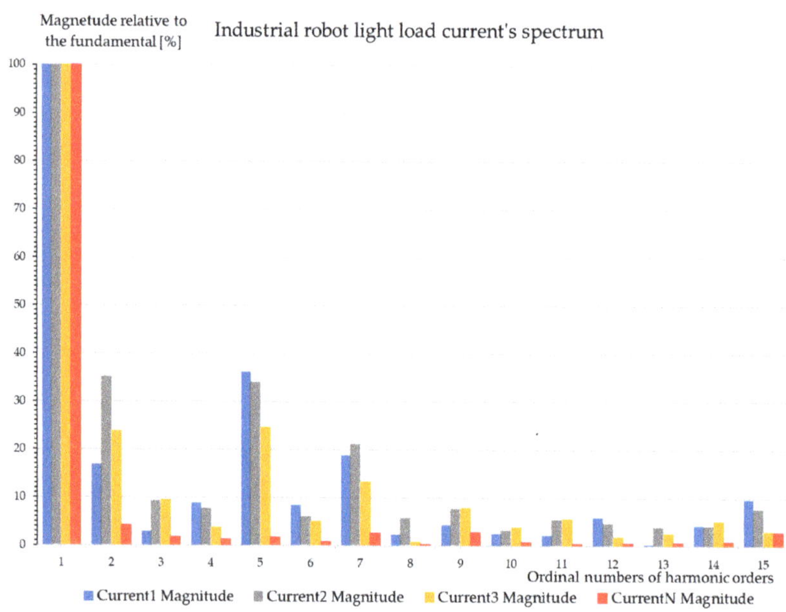

Figure 26. Harmonic current magnitude components of the distorted current waveforms measured on an industrial robot at a light load.

The measured data provided by the network analyzer (voltage, current, THD and reactive power) can be seen in Figure 27.

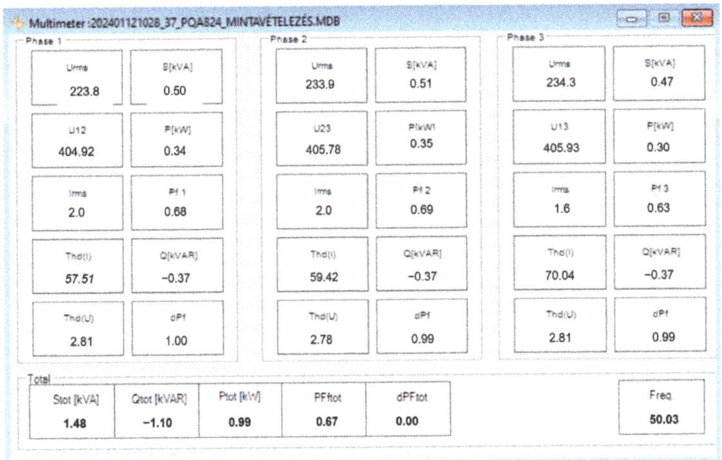

Figure 27. Measured voltage, current, THD, and power factor of industrial robot at light load provided by TopView.

7.4.2. Industrial Robot at Heavy Load

During the analysis, measurements were also performed under heavy load. After all, the more robot axes operate at the same time, the greater the simultaneous electric power consumption of the drive cards that operate them. This therefore represents the greatest opportunity to measure the greatest distortion of electric signals and the injection of harmonics into the network. The measured voltage and current waveforms are illustrated in Figure 28. It can be stated that the voltage sine wave is slightly distorted, while the current sine wave is already significantly distorted, and the shape of the sine waves of the current clearly demonstrates the non-linear current draw from the network. After all, the direction of the current changes four times instead of once within one period. If one looks at the significant number of zero crossings within one period, it is clear that this must be taken into account not only for the robot, but also for the robot cell equipment when designing their switch mode power supplies.

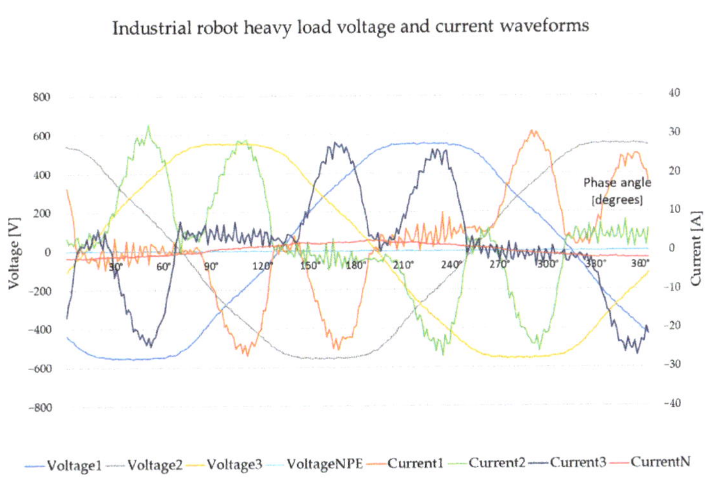

Figure 28. Distorted current and voltage waveforms measured on an industrial robot at a heavy load.

Figure 29 shows the current spectrum. The harmonic distortion of the total current measured at the common connection point (PCC) of the robot cell is THD(I), and in the case of Current 3 it was 83.94%. In this case, the permissible maximum value of the total harmonic distortion of the current is also given by the TDD. The short-circuit current of the common electrical network connection point (PCC) is 800 A, and the load current of the robot cell is 8.3 A. The required ratio is 800 A/13 A = 61.53. The maximum allowable total current demand harmonic distortion (TDD) recommended by the IEEE 519-2022 standard is 12%. Therefore, THD = 83.94% is seven times the permissible value. Based on the ratio number 61.53, the permissible harmonic current distortion is 10% for the 2nd–11th harmonics, and 4.5% for the 11th–17th harmonics. These limits are exceeded multiple times for every odd harmonic order. The value for the 5th harmonic (250 Hz) is 66.06%, that for the 7th (350 Hz) is 40.74%, and that for the 11th (550 Hz) 7.61%.

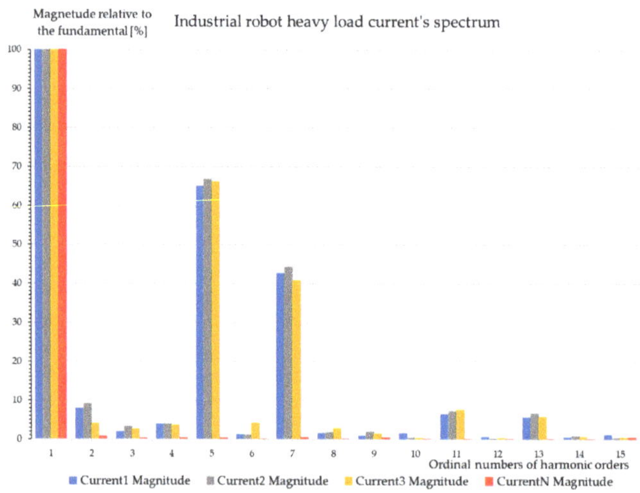

Figure 29. Harmonic current magnitude components of the distorted current waveforms measured on an industrial robot at a heavy load.

The table provided by the TopView software displaying the measurement results of the analyzer is shown in Figure 30.

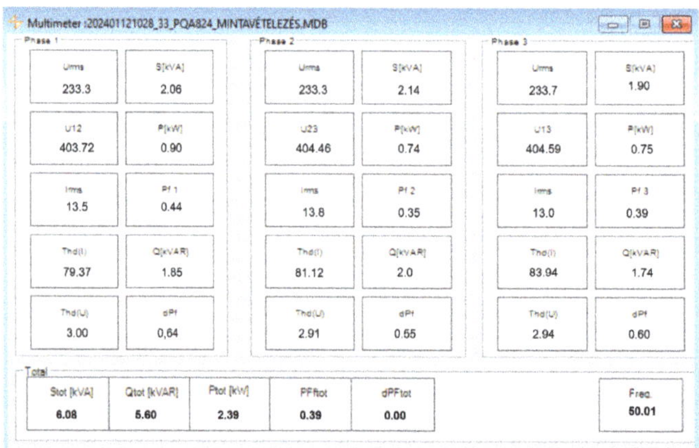

Figure 30. Measured voltage, current, THD, and power factor of industrial robot at heavy load provided by TopView.

8. Network Analysis of Different Locations of the Industrial Facility

In the previous section, the efficiency of the built-in filters was investigated. Measurements were also performed at three different locations within an industrial facility: the engineering room, the gearbox house machine tool, and an office room. In these cases, there are no built-in filters. These measurement results will contribute to the subsequent development of what kind of protection against harmonics needs to be installed for each location. After installing the proposed active filters, network analysis will be performed again, and thus in the future a similar validation process will be carried out like in Section 7.

8.1. Engineering Room

Figure 31 shows a voltage–current waveform diagram in a mechanical room. Since the magnitude of the currents of each phase and the total harmonic distortion of the currents were almost the same, only Current 2 was analyzed in this paper. It can be seen that the current of the neutral conductor was almost zero, while all voltage and current waveforms are distorted. These network analysis results correspond to the literature description of harmonic distortions. The measurement also shows how the current waveform is distorted by harmonics in the building services room, where the electronic speed regulation of electric motors and the use of switching power supply units to power the control electronics are significant.

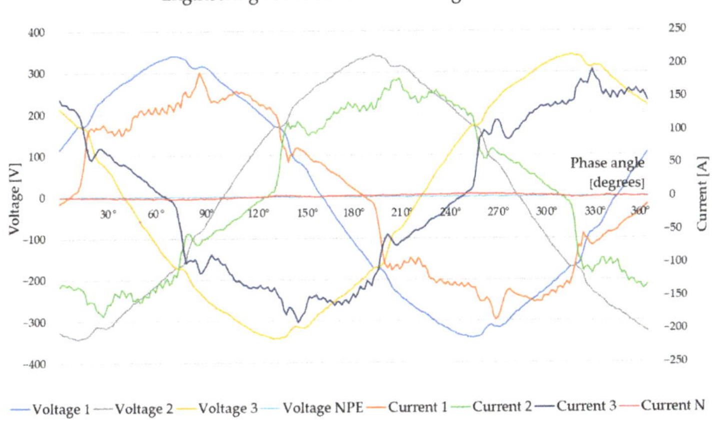

Figure 31. Distorted current and voltage waveform measured in the engineering room.

Figure 32 shows the frequency spectrum of the measurement.

The table provided by the TopView software displaying the measurement results of the analyzer is shown in Figure 33.

The scale shows the magnitudes of the harmonics with respect to the fundamental harmonic. One can observe that the harmonics of the neutral conductor were negligible. Therefore, the current of the neutral conductor was not analyzed either. The recorded current varies between 113.5 A and 116 A per phase, so it can be said to be symmetrical. But in Figure 6, one can also see that the current sine waveform is very distorted, and the THD of the current was 14.85% for Current 2. The proposed maximum allowable value of THD according to IEEE 519-2022 is given by the TDD, which depends on the ratio of the PCC termination current I_{SC} to the load current I_L of the device, as was previously presented. This ratio is now 2000 A/112,8 A = 17.73; therefore, the permissible distortion according to Table 2 would be only 5%, so the THD of 14.85% is almost three times the recommended maximum value, and the limit value was exceeded by 9.85% in the case of Current 2. However, it can also be concluded from the measurement that according to Table 2, instead of the recommended 4% up to the 11th harmonic, the maximum harmonic current distortion for Current 2 was 8.53%, 6.19%, and 7.2% for the 5th, 7th, and 11th harmonics, respectively (these values are almost twice the recommended value). Instead of the recommended 2% from the 11th harmonic to the 17th harmonic, the distortion was 4.00% for the 13th harmonics. In the case of higher harmonic orders, their magnitudes were already negligible; therefore, they are not included in the analysis. Of course, further measurements should also be carried out in this case to determine the exact sources of the main harmonic disturbances.

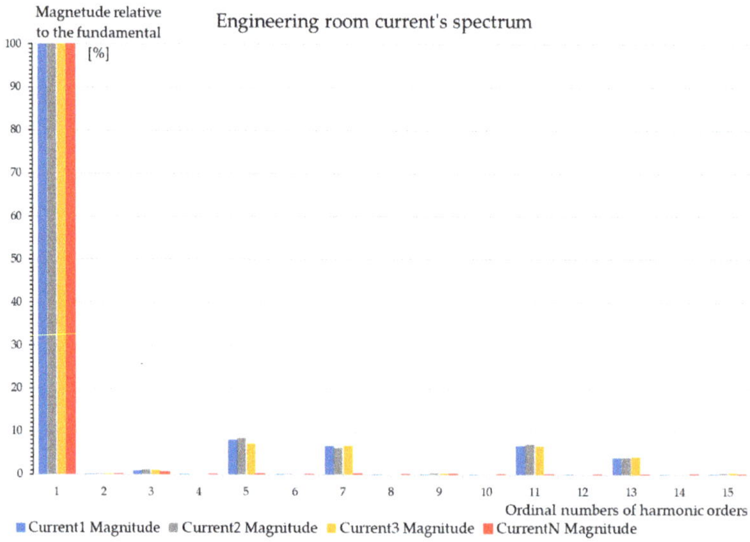

Figure 32. Harmonic current magnitude components of the distorted current waveform measured in the engineering room.

Figure 33. Measured voltage, current, THD, and power factor of engineering room provided by TopView.

8.2. Gearbox House Machine Tool

Measurements were also performed in a gearbox house machine tool. The voltage and current waveforms of a special machine tool can be seen in Figure 34.

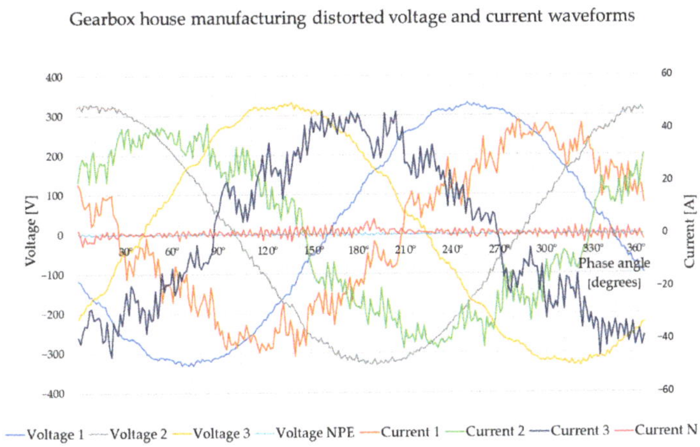

Figure 34. Gearbox house manufacturing distorted voltage and current waveforms.

In this case as well, the magnitude of the currents of each phase and the total harmonic distortion of the currents were almost the same; therefore, only Current 3 was analyzed here. The maximum value of the current of the neutral conductor was 1.56 A, which is negligible compared to the maximum phase currents of 43 A.

The network analysis was carried out with the machine in the extended state under normal production conditions. The applied current varied between 26.7 A and 28.2 A per phase, so it can be said to be symmetrical. On the other hand, there was a high level of harmonic current pollution and therefore the current waveform was also highly distorted. Thus, the THD of the current was 17.3% for the Current 3. As was mentioned before, the proposed maximum allowable value of THD as defined in IEEE 519-2022 is given by the TDD. This ratio is now 630 A/28.2 A = 22.34; according to Table 2, the permissible distortion should only be 8%. The limit value was thus exceeded by 9.85% in the case of Current 3. The other problem is that the current of the neutral conductor is 1.56 A, whereas it should be zero under normal operation. The current in the neutral conductor is caused by the asymmetrical load, and the current flowing in the neutral conductor can cause overloading and overheating in the line in the transformer, which can lead to failures. Electricity distribution networks are in most cases TN (three-phase neutral systems), where the secondary windings of the transformer are connected to a star and the star points of the windings are grounded and the neutral conductor is connected to this point. All harmonics are found in the phase voltages, and hence harmonics appear in the phase and line currents. The neutral current is the sum of the phase currents. The effective value of this current will be the square root of the square sums of the zero-sequence currents. This current causes an additional load in the neutral conductor for symmetrically connected non-linear loads, the magnitude of which may be as high as two to three times the phase current. If this current is not considered in network design and the current in the neutral conductor due to harmonics—which causes potential shifts due to voltage drops on individual conductor sections—is not taken into account, it can cause serious disturbances in the operation of electronic equipment. Later, the possible direct consequences of harmonic disturbances on the power sources of equipment operating on the electrical network will be discussed.

Figure 35 shows the spectrum of the phase of currents in Figure 34.

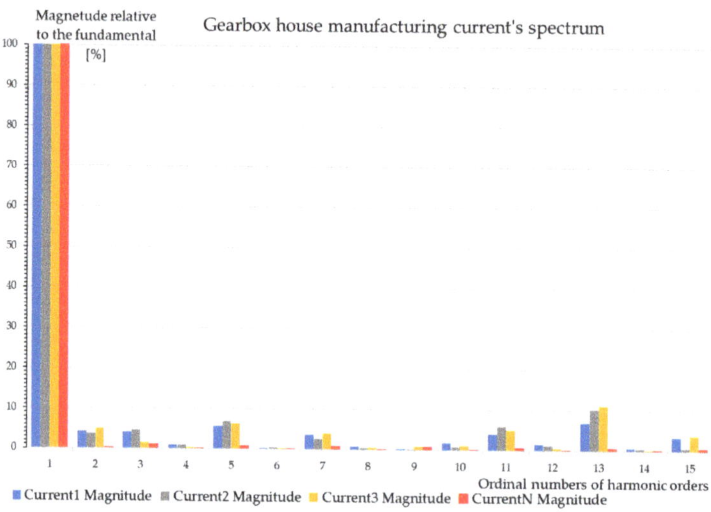

Figure 35. Gearbox house manufacturing harmonic current magnitude components of the distorted current waveform.

The table provided by the TopView software displaying the measurement results of the analyzer is shown in Figure 36.

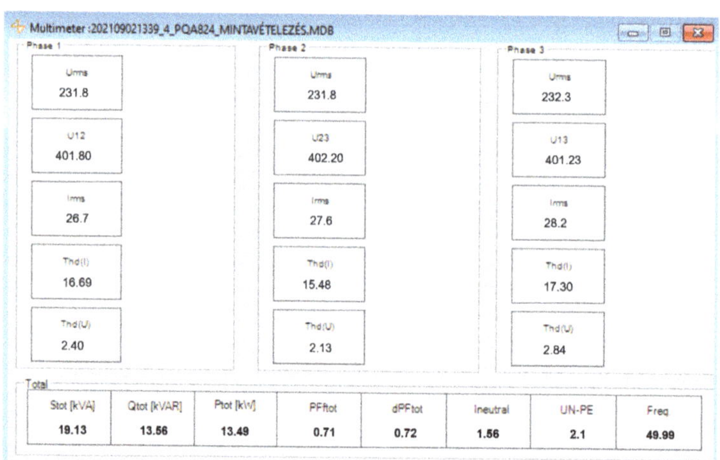

Figure 36. Measured voltage, current, THD, and power factor of gearbox house provided by TopView.

As before, on the horizontal axis, the line numbers represent the harmonic orders, the integer nth multiples of the fundamental frequency (50 Hz), while on the vertical axis, the amplitude is plotted against the fundamental harmonic. It can be observed that the harmonics of the neutral conductor were also negligible in this case. Therefore, the current of the neutral conductor was not analyzed either.

One can see that the magnitude of the 5th-, 7th- and 11th-order harmonics exceeds the maximum possible 4% value up to the 11th harmonic recommended in the standard, because these current amplitude values were 6.31%, 3.92%, and 5.90% respectively. From the 11th harmonic to the 17th harmonic, instead of the recommended 2%, the magnitude of the 13th harmonic was 11.15%, while that of the 15th harmonic was 3.71%. In the case of higher harmonic orders, their magnitudes were already negligible; therefore, they are not included in the analysis. So, based on the measurements, it can be said here again that the magnitude of harmonic distortion exceeds the level allowed by the standard and it is necessary to filter these values and eliminate the cause of the current flowing in the neutral conductor.

8.3. Office Room

In the following, the results of a single-phase network analysis in an office room are presented. The waveforms of one period of voltage and current are shown in Figure 37.

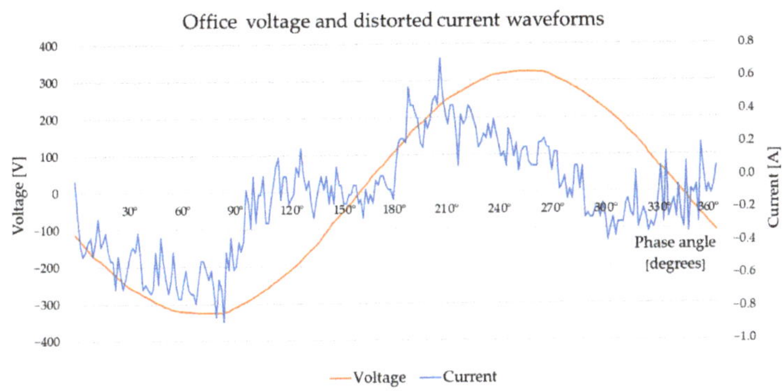

Figure 37. Office distorted voltage and current waveforms.

The current waveform is distorted to such an extent that its sinusoidal waveform is unrecognizable. One can also see the multiple zero transitions. The TDD value was also determined in this case. The value of THD of the current was 59.47%. The total current demand distortion (TDD) is 16 A/0.3 A = 53.33; according to Table 2, the permissible distortion should only be 12%. The limit value was exceeded by 47.47%.

Figure 38 shows the frequency spectrum of the current in Figure 37.

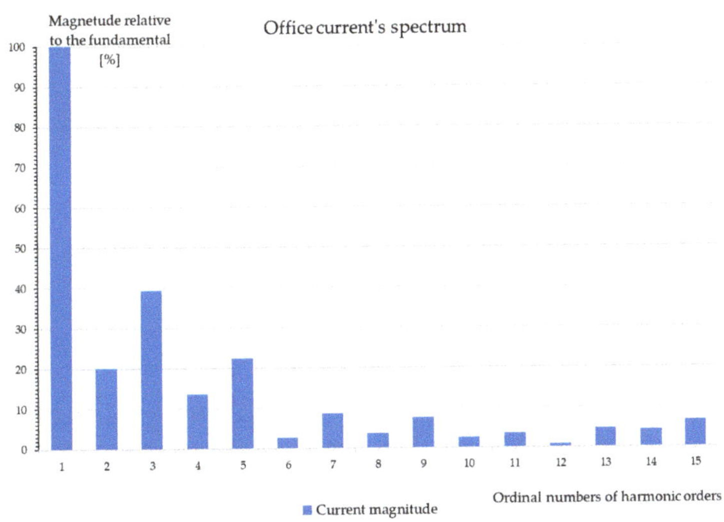

Figure 38. Office distorted current waveform components of harmonic current.

The table provided by the TopView software displaying the measurement results of the analyzer is shown in Figure 39.

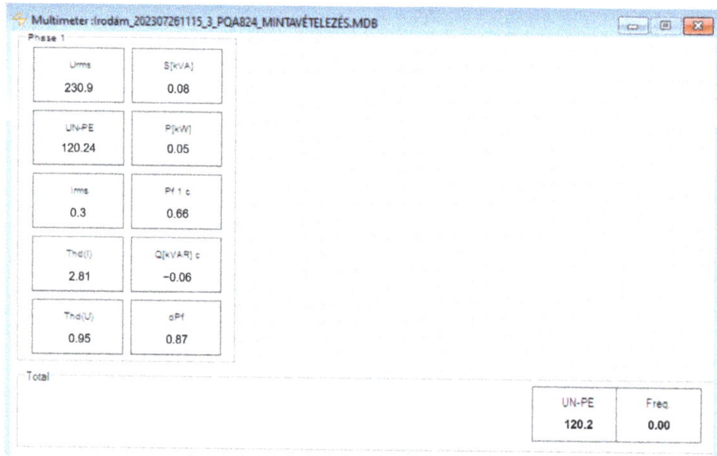

Figure 39. Measured voltage, current, THD, and power factor of office room provided by TopView.

It can be seen that the third harmonic (150 Hz) is very high, which is mainly typical of LED lighting, so the luminaires in the office are not fitted with filters, contrary to the contractor's claim. According to Table 2, the maximum possible value of 4% per harmonic order up to the 11th harmonic is recommended in the standard, but these current amplitude values for the 3rd and 5th harmonics were 39.38% and 22.39%. The measured value is ten times the recommended maximum value. Targeted research is also currently being carried out to identify the devices in office consumers that could cause the current amplitudes shown in this figure. Based on the measurement, it was proved that the built-in harmonic filters claimed by the manufacturers of office equipment and computers are in many cases not true or their quality is not adequate. Our future plan is to install a test filter in the central electricity supply network of the offices and then carry out the measurements again.

9. Conclusions

Virtually all modern electrical and electronic equipment has a switching power supply, controlling the electricity drawn and thus creating a non-linear load. The voltage and frequency of a perfect power supply are within the permissible limits, and its voltage curve is perfectly sinusoidal. Electrical harmonics are a by-product of modern electronics. They are particularly prevalent where there are a large number of computing and electronic power control devices. The technical quality of electricity, especially the value of harmonics is receiving more and more attention currently. A modern way to reduce harmonics is to use active harmonic filtering devices. These devices are designed to fulfill several functions at the same time, performing a complex electrical power shaping task instead of harmonic filtering.

The aim of the first presented measurements was to investigate equipment with in-built active and passive filters. Measurements were performed on three different devices: an LED lamp, a test bench, and an industrial robot. For the appropriate validation in the latter two cases, light and heavy load states were both investigated. The measured total harmonic current distortion (THD) values were compared to the IEEE 519-2022 standard. In the case of a single LED lamp, the total harmonic distortion of the current was 15.20%, while the permissible distortion should be only 5%, so the difference is more than three times (similar results were obtained for two luminaires). In the case of the LED hall lighting, the limit value should be only 8%, so the THD of 14.71% is almost twice the recommended maximum value. In the case of the test bench at a light load, the THD value was 139.21%, which is more than twelve times the permissible value of 12%. However, when the test bench was under a heavy load, the harmonic distortion was 4.24%, 3.37%, and 4.35%

THD per phase, which can meet even the strictest requirements (5%) according to the IEEE 512-2022 standard. The THD value of an industrial robot under a heavy load was 90%, which is more than ten times the permissible 8%. Overall, it can be concluded that the built-in passive filters of the LED lamps were not effective, while in the case of the test bench and the robot, the efficiency of the active harmonic filters was not adequate. When sizing the active filters in these cases, the designers did not consider the condition of the machines, which can be significant even in series production due to transitions and machine failures. The solution is to install an active harmonic filter that performs equally well in the cases of low and high current consumption. From the measurement results it can be said that the built-in active harmonic filter of the tested industrial robot was completely ineffective at both low and full load. The effectiveness of the filters in the case of machine tools was also examined, and the same result was obtained. The authors (at the concern level) are in contact with filter manufacturers, sharing the measurement results and experience, and joint research is planned in order to increase active harmonic filtering efficiencies.

The aim of the second presented set of measurements was to investigate the current harmonic injection and to examine the compliance with the standard of equipment connected to the internal electrical network of an industrial structure. Measurements were performed in these cases at three different locations: an engineering room, a gearbox house machine tool, and an office room. The current harmonic distortion per unit current and the total harmonic current distortion were measured during network analysis. The values were compared to the IEEE 519-2022 standard. For the machine room, the allowed distortion could be only 5%, so the THD, at 14.85%, was almost three times the recommended maximum value; for the gearbox manufacturing machine, the THD was 17.3%, which is higher than the allowed 12%. For the office room, the THD was 59.47%, almost five times the permitted value of 12%.

As it can be seen, the amount of permissible harmonic distortion [40] depends on the short-circuit current PCC of the electrical connection point, but this varies between installation locations. Therefore, in industrial facilities where networks have smaller short-circuit currents, the same harmonic injection will have a much greater impact on the quality of electricity. In the future, the possibilities of using current harmonics for predictive maintenance, machine learning, and anomaly detection will be investigated.

Author Contributions: Conceptualization, A.K. and J.S.M.; methodology, A.K.; software, A.K.; validation, A.K., J.S.M. and K.J.; formal analysis, A.K.; investigation, A.K.; resources, A.K.; data curation, A.K.; writing—original draft preparation, A.K and J.S.M.; writing—review and editing, A.K. and J.S.M.; visualization, A.K.; supervision, J.S.M.; project administration, J.S.M.; funding acquisition, A.K. All authors have read and agreed to the published version of the manuscript.

Funding: This research received no external funding.

Data Availability Statement: The raw data supporting the conclusions of this article will be made available by the authors on request.

Conflicts of Interest: The authors declare no conflict of interest.

Glossary

THD	Total harmonic distortion
EMS	Energy Management System ISO 50001
PHEV	Plug in hybrid vehicle
PLC	Programmable logic controller
CNC	Computer numerical control
LED	Light-emitting diodes
DC	Direct current
RMS	Root mean square
TDD	Total current demand distortion
PCC	Point of common coupling

FFT	Fast Fourier transform
TN	Three-phase neutral system
AC	Alternating current
MV	Medium voltage
LV	Low voltage
APLC	Active power line conditioning
PWM	Pulse-width modulation
NPC	Neutral point clamped
IGBT	Insulated-gate bipolar transistor

References

1. Ghorbani, M.J.; Mokhtari, H. Impact of harmonics on power quality and losses in power distribution systems. *Int. J. Electr. Comp. Eng.* **2015**, *5*, 166–174. [CrossRef]
2. Hunter, I. Power quality issues-a distribution company perspective. *Power Eng. J.* **2001**, *15*, 75–80. [CrossRef]
3. Radmehr, M.; Farhangi, S.; Nasiri, A. Effects of power quality distortions on electrical drives and transformer life in paper industries: Simulations and real time measurements. In Proceedings of the Pulp and Paper Industry Technical Conference, Appleton, WI, USA, 18–23 June 2006.
4. Yan, Y.H.; Chen, C.S.; Moo, C.S.; Hsu, C.T. Harmonic analysis for industrial customers. *IEEE Trans. Ind. Appl.* **1994**, *30*, 462–468. [CrossRef]
5. Mansoor, A.; Grady, W.M.; Staats, P.T. Predicting the net harmonic currents produced by large numbers of distributed single-phase computer loads. *IEEE Trans. Power Deliv.* **1995**, *10*, 2001–2006. [CrossRef]
6. Masoum, M.A.S.; Fuchs, E.F. *Power Quality in Power Systems and Electrical Machines*, 2nd ed.; Academic Press: Cambridge, MA, USA, 2015.
7. ISO (International Organization for Standardization). ISO Central Secretariat. ISO 50001:2018. Vernier, Geneva, Switzerland. 2018. Available online: https://www.iso.org/files/live/sites/isoorg/files/store/en/PUB100400.pdf (accessed on 20 August 2024).
8. Pinyol, R. Harmonics: Causes, Effects and Minimization. Salicru White Papers. 2015. Available online: https://www.salicru.com/files/pagina/72/278/jn004a01_whitepaper-armonics_(2).pdf (accessed on 2 August 2024).
9. Das, J.C. *Power System Harmonics and Passive Filter Designs. Power System Harmonics and Passive Filter Designs*, 1st ed.; Wiley-IEEE Press: Hoboken, NJ, USA, 2015.
10. Wenqian, Y.; Xiang, Y.; Longwei, X.; Chao, Z.; Xinsheng, M. Harmonic loss analysis of low-voltage distribution network integrated with distributed photovoltaic. *Sustainability* **2023**, *15*, 4334. [CrossRef]
11. Sumaryadi, D.; Gumilang, H.; Suslilo, A. Effect of power system harmonic on degradation process of transformer insulation syste. In Proceedings of the IEEE 9th International Conference on the Properties and Applications of Dielectric Materials, Harbin, China, 19–23 July 2009.
12. Shafiee-Rad, M.; Kazerooni, M.; Ghorbany, J.; Mokhtari, H. Analysis of the grid harmonics and their impacts on distribution transformers. In Proceedings of the IEEE Power and Energy Conference, Champaign, IL, USA, 24–25 February 2012.
13. Aktif. Problems Caused by Harmonics in Electrical Plants. Available online: https://aktif.net/en/problems-caused-by-harmonics-in-electrical-plants/# (accessed on 2 August 2024).
14. Szűcs, V.; Fodor, A.; Görbe, P. Examination of nonlinear distortion of IT power supplies in low voltage grid from the current quality point of view. *Acta Phys. Pol.* **2018**, *134*, 307–310. [CrossRef]
15. Caicedo, E.J.; Romero, A.A.; Zini, C.H. Assessment of the harmonic distortion in residential distribution networks. *Ing. Investig.* **2017**, *37*, 72–84. [CrossRef]
16. Bonner, A.; Grebe, T.; Gunther, E.; Hopkins, L.; Marz, M.B.; Mahseredjian, J.; Xu, W. Modeling and simulation of the propagation of harmonics in electric power networks. Part I: Concepts, models, and simulation techniques. *IEEE Trans. Power Deliv.* **1996**, *11*, 452–465.
17. Uddin, S.; Shareef, H.; Mohamed, A. Power quality performance of energy-efficient low-wattage LED lamps. *Measurement* **2013**, *46*, 3783–3795. [CrossRef]
18. Poplawski, T.; Kurkowski, M. Nonlinear Loads in lighting installations—Problems and threats. *Energies* **2023**, *16*, 6024. [CrossRef]
19. Molina, J.; Mesas, J.J.; Mesbahi, N.; Sainz, L. LED lamp modelling for harmonic studies in distribution systems. *IET Gener. Transm. Distrib.* **2017**, *11*, 1063–1071. [CrossRef]
20. Wang, H.; Zhou, W.; Qian, K.; Meng, S. Modelling of ampacity and temperature of MV cables in presence of harmonic currents due to EVs charging in electrical distribution networks. *Int. J. Electr. Power Energy Syst.* **2019**, *112*, 127–136. [CrossRef]
21. Jouybari-Moghaddam, H.; Alimardani, A.; Hosseinian, S.H. Influence of electric vehicle charging rates on transformer derating in harmonic-rich battery charger applications. *Arch. Electr. Eng.* **2012**, *61*, 483–497. [CrossRef]
22. Heidarian, T.; Joorabian, M.; Reza, A. The effect of plug-in electric vehicles on harmonic analysis of smart grid. *Int. J. Emerg. Electr. Power Syst.* **2015**, *16*, 559–567. [CrossRef]
23. Sallam, A.A.; Malik, P.O. Harmonics in Electric Distribution Systems. In *Electric Distribution Systems*, 2nd ed.; El-Hawary, M.E., Ed.; Wiley-IEEE Press: Piscataway, NJ, USA, 2019; Chapter 14; pp. 379–402.

24. Fourier Analysis. Carl von Ossietzky University Oldenburg—Faculty V—Institute of Physics. Module Introductory Laboratory Course Physics—Part I. Available online: https://uol.de/f/5/inst/physik/ag/physikpraktika/download/GPR/pdf/E_Fourieranalyse.pdf?v=1666712279 (accessed on 2 August 2024).
25. Emanuel, A.E.; Janczak, J.; Pileggi, D.J.; Gulachenski, E.M.; Root, C.E.; Breen, M.; Gentile, T.J. Voltage distortion in distribution feeders with nonlinear loads. *IEEE Trans. Power Deliv.* **1994**, *9*, 79–87. [CrossRef]
26. Teng, J.H.; Liao, S.H.; Leou, R.C. Three-phase harmonic analysis method for unbalanced distribution systems. *Energies* **2014**, *7*, 365–384. [CrossRef]
27. What is True-RMS? Available online: https://www.fluke.com/en-us/learn/blog/electrical/what-is-true-rms (accessed on 2 August 2024).
28. Magadum, P.K. Effects of nonlinear loading on power quality. In Proceedings of the 13th International Conference on Advances in Computing, Control, and Telecommunication Technologies, Hyderabad, India, 27–28 June 2022; Volume 8, pp. 1119–1123.
29. Tarnik, I. Villamos Hálózati Zavarok/Electrical Network Disturbances. 2008. Available online: https://www.muszeroldal.hu/measurenotes/tarmikcikk.pdf (accessed on 2 August 2024).
30. Peng, F.Z. Harmonic sources and filtering approaches. *IEEE Ind. Appl. Mag.* **2001**, *7*, 18–25. [CrossRef]
31. Mustafa, T.I.A.H.; Cabral, S.H.L.; Almaguer, H.A.D.; Meyer, L.H.; Puchale, L.H.B.; Vier, G.B.; Cereja, J.E. Analysis of the Behavior of the 5th Harmonic over Power Transmission Lines. In Proceedings of the IEEE PES Asia-Pacific Power and Energy Engineering Conference (APPEEC), Kota Kinabalu, Malaysia, 7–10 October 2018; pp. 400–405. Available online: https://ieeexplore.ieee.org/document/8566407 (accessed on 11 August 2024).
32. Jewel, W.; Miller, W.L.; Casey, T. Filtering dispersed harmonic sources on distribution. *IEEE Trans. Power Deliv.* **2000**, *15*, 1045–1051. [CrossRef]
33. Chapman, D. *Application Note—Harmonics: Causes and Effects*, 2nd ed.; European Copper Institute: Brussels, Belgium, 2011; pp. 1–21.
34. Calvas, L. *Electrical Disturbances in LV*, 1st ed.; Cahier Technique No. 141; Schneider Electric: Claix, France, 2000; Available online: https://www.tekdok.dk/files/cahiers_techniques/Electrical_disturbances_in_LV.pdf (accessed on 2 August 2024).
35. Wagner, V.E.; Balda, J.C.; Griffith, D.C.; McEachern, A.; Barnes, T.M.; Hartmann, D.P.; Phileggi, D.J.; Emannuel, A.E.; Horton, W.F.; Reid, W.E.; et al. Effects of harmonics on equipment. *IEEE Trans. Power Deliv.* **1993**, *8*, 672–680. [CrossRef]
36. Emanuel, A.E.; Orr, J.A.; Cyganski, D.; Gulachenski, E.M. A survey of harmonic voltages and currents at distribution substations. *IEEE Trans. Power Deliv.* **1991**, *6*, 1883–1890. [CrossRef]
37. Grady, W.M.; Samotyj, M.J.; Noyola, A.H. Minimizing network harmonic voltage distortion with an active power line conditioner. *IEEE Trans. Power Deliv.* **1991**, *6*, 1690–1697. [CrossRef]
38. Ketabi, A.; Farshadnia, M.; Malekpour, M.; Feuillet, R. A new control strategy for active power line conditioner (APLC) using adaptive notch filter. *Int. J. Electr. Power Energy Syst.* **2013**, *47*, 31–40. [CrossRef]
39. Imperix: Neutral Point Clamped Inverter (NPC). Available online: https://imperix.com/doc/implementation/neutral-point-clamped-inverter (accessed on 2 August 2024).
40. Wu, J.-C.; Jou, H.-L.; Hsaio, H.-H.; Xiao, S.-T. A new hybrid power conditioner for suppressing harmonics and neutral-line current in three-phase four-wire distribution power systems. *IEEE Trans. Power Deliv.* **2014**, *29*, 1525–1532. [CrossRef]
41. *IEEE 519-2022*; IEEE Standard for Harmonic Control in Electric Power Systems. IEEE: Piscataway, NJ, USA, 2022. Available online: https://standards.ieee.org/ieee/519/10677/ (accessed on 2 August 2024).
42. Caramia, P.; Carpinelli, G.; Rossi, F.; Verde, P. Probabilistic iterative harmonic analysis of power systems. *IEE Proc.-Gener. Transm. Distrib.* **1994**, *141*, 329–338. [CrossRef]
43. Medina, A.; Segundo, J.; Ribeiro, P.; Xu, W.; Lian, K.L.; Chang, G.W.; Watson, N.R. Harmonic analysis in frequency and time domain. *IEEE Trans. Power Deliv.* **2013**, *28*, 1813–1821. [CrossRef]
44. HT Italia. Available online: https://www.ht-instruments.com/en/ (accessed on 2 August 2024).
45. HT Italia: Management Software Topview. Available online: https://www.ht-instruments.com/en/accessories/topview/ (accessed on 2 August 2024).
46. Karim, F.A.; Ramdhani, M.; Kurniawan, E. Low pass filter installation for reducing harmonic current emissions from LED lamps based on EMC standard. In Proceedings of the International Conference on Control, Electronics, Renewable Energy and Communications, Bandung, Indonesia, 13–15 September 2016; pp. 132–135.
47. Lakervi, E.; Holmes, E.J. *Electricity Distribution Network Design*, 2nd ed.; Peter Peregrinus LTD, Institution of Engineering and Technology: London, UK, 2003.

energies

MDPI

Article

Analysis of the Impact of Volt/VAR Control on Harmonics Content and Alternative Harmonic Mitigation Methods

Krzysztof Lowczowski *, Jaroslaw Gielniak, Zbigniew Nadolny and Magdalena Udzik

Institute of Electric Power Engineering, Poznan University of Technology, Piotrowo 3A, 60-965 Poznan, Poland
* Correspondence: krzysztof.lowczowski@put.poznan.pl

Abstract: This article presents information about harmonic distortion and resonance in distribution networks. The theory behind harmonics and resonance is presented. Examples from the literature and the results of power quality measurements, as well as the authors' experiences connected with significant changes in harmonic distortions, are presented. The harmonic resonance phenomenon is explained, and the risk of resonance in a distribution system network is highlighted. Attention is paid to the connection of a new power plant to the network; however, other risks, e.g., those connected to network reconfiguration, are also highlighted. Further simulation case studies are presented to show interactions between volt/VAR control and harmonic distortion. Simulations consider a few scenarios: impact of voltage change on impedance characteristics and resulting harmonics amplitudes, the impact of a capacitor on impedance characteristics, and the impact of network expansion on harmonic distortion. The final part presents alternative, low-cost harmonics mitigation methods. The concept of the utilization of phase-shifting transformers for two twin-type 1 MW plants located next to each other is verified by on-site measurement. The concept of adapting the harmonics spectrum of new devices to impedance characteristics is presented. Finally, an alternative concept for active mitigation of harmonics under resonance conditions is provided. The concept is based on the reactive power correction in order to change the harmonics phase shift. A comparison of harmonic mitigation methods and general recommendations are provided. Further research is outlined.

Keywords: harmonics/distortion; THD; resonance; PQ quality; phase shift

1. Introduction

The development of renewable energy sources allows for reducing the utilization of fossil fuels and limiting CO_2 emissions. Unfortunately, the development of renewable energy sources is connected to many technical issues, which arise in all fields of electrical power engineering, e.g., protection relay coordination, voltage control complexity, power quality issues unwanted islanding, and many more [1]. At the same time, however, there is significant growth in technology that allows for compensating technical issues. Among many improvements, one can mention better observability of network, better coordination of devices, development of new monitoring technologies, and control algorithms [2]. This paper focuses on problems connected to power quality issues, which can be divided into a few categories [3,4]:

- Level of voltage;
- Slow and fast voltage changes, rapid voltage changes;
- Asymmetry of voltage;
- Voltage distortions.

Each category requires a detailed explanation. For the sake of simplicity, one can say that the level of voltage is the most common issue. The voltage level is easy to observe using even the simplest meters and has a direct impact on devices, e.g., inverters switch off above the voltage threshold. Voltage changes have an important impact on the operation of devices, since voltage changes result in varying currents drawn by devices and therefore varying outputs, e.g., light emission [5]. Voltage asymmetry is connected to higher energy losses and faster aging, which have a particularly large impact on three-phase motors [6].

Voltage distortions are categorized into subgroups depending on the frequency range of the distortion harmonics up to 2 kHz and supraharmonics in the range of 2 kHz to 150 kHz [7]. Harmonics are calculated using Fourier transformation in sine or cosine form (1, 2) [8]. Harmonics are classified into groups and subgroups, as marked in yellow in Figure 1 [9]. Harmonics groups exhibit a wide frequency spectrum, while harmonics subgroups include only ±5 Hz (±1 bar in Figure 1). Interharmonic groups and subgroups include frequencies between harmonics. Interharmonic groups exhibit a wider frequency band ±5 Hz compared with interharmonic subgroups.

$$i(t) = I_{dc} + \sum_{h=1}^{\infty} I_h sin(h\omega_1 t + \Theta_k) \quad (1)$$

$$i(t) = I_{dc} + \sum_{h=1}^{\infty} I_h cos(h\omega_1 t + \Theta_k - 90^\circ) \quad (2)$$

where Θ_k is the harmonic phase angle.

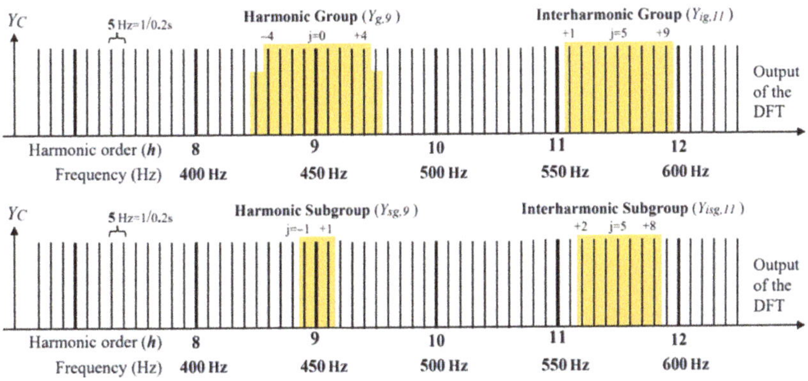

Figure 1. Illustration of the harmonic group of order 9, *Yg,9*, the interharmonic group of order 11, *Yig,11*, the harmonic subgroup of order 9, *Ysg,9*, and the centered interharmonic subgroup of order 11, *Yisg,11*, for the case of f1 = 50 Hz [9].

Typically, the literature presents selected power quality problems separately; however, nowadays, in networks, one may expect more and more interactions between power quality parameters because of the high variation of voltage, which is the result of the integration of renewable energy sources and storages as high power loads, like charging stations. Variations in voltage have a great impact on reactive power flow and capacitive and reactive power (3, 4), which in turn has an impact on the impedance characteristics of the power system and harmonics resonance frequencies. The goal of this paper is to present the interactions between voltage, VAR, and harmonics changes, as well as potential issues and risks connected with harmonics. The simulation results show that voltage changes shift the harmonic resonance band and resulting harmonic flow, the modification of reactive power obtained by capacitor tap change has an impact on the resonance band, whereas harmonic mitigation under resonance conditions is based on reactive power change, which has an impact on voltage level. Further, it is worth underlining that all the modifications have an impact on the loading level.

$$Q_c = U \cdot I = U^2 2\pi f C \quad (3)$$

$$Q_L = U \cdot I = \frac{U^2}{2\pi f L} \quad (4)$$

where $Q_{L/c}$ is the reactor/capacitor power, L is the inductance, and C is the capacitance.

Current and voltage distortion may be connected to a great number of issues—additional energy losses, extensive heating and thermal aging, risk of sudden failure particularly in the context of resonances, maloperation of control or protection algorithms, malfunction of consumer loads, specific heating of cable terminal, and others [7,10]. The high spectrum of negative consequences of voltage distortions requires monitoring, analysis, and proper utilization of mitigation methods. The monitoring is carried out using smart meters and power quality meters. Smart meters, the main task of which is to calculate power consumption and generation, typically allow for the monitoring of basic parameters, like asymmetry or THD (5), which express general harmonic distortions [11]. Power quality meters are much more advanced devices. PQ meters allow for observing the harmonic spectrum up to 40 or 50, and the harmonics angle, with additional indicators like TDD (6) and supraharmonics. TDD is total demand distortion, where harmonics are normalized to the maximum current. TDD allows us to observe when the harmonics have the greatest impact on the power system. As a result of the presented features, smart meters can be considered a simple PQ indicator, whereas PQ analyzers allow for gathering data for detailed analysis. PQ analyzers are both fixed and portable. Regulations require installing fixed PQ meters at the point of renewable energy source coupling, which ensures that all relevant data are available.

$$THD = \frac{\sqrt{\sum_{h=2}^{hmax} (Y_h)^2}}{Yh1} \quad (5)$$

where Uh is the content of the harmonic row and Uh1 is the content of the first (fundamental 50 Hz) harmonic.

$$TDD = \frac{\sqrt{\sum_{h=2}^{hmax} I_h^2}}{Id} \quad (6)$$

where Id is the maximum peak demand load current and I_h is the content of the harmonic row.

Based on the situation and results of analysis, one can utilize mitigation methods, among which one can mention probably the most popular passive and active filters, proper arrangement of devices, and minimization of harmonic emissions by the optimization of the hardware and control algorithms e.g., use of the variable switching frequency of inverters [12].

The modification of the inverter control algorithm is typically beyond the possibilities of network engineers due to the lack of access to configuration software, high complexity, and an enormous amount of different inverter topologies [13]. Power electronic engineers typically have limited knowledge about power systems, which consist of hundreds or thousands of elements; therefore, the improvement of inverters is generally focused on the general THD level, not on specific harmonics. More advanced inverters may be equipped with adaptive mitigation algorithms; however, such algorithms are often still at the research stage [14–16]. As a result of the presented problems, typical harmonic mitigation methods require special considerations at the design stage or extra devices afterward, which increase costs. The mitigation methods proposed in this paper can be introduced in a relatively simple way without additional costs; however, the effective operation can be observed only for resonance band mitigation. The solution may be important in case of hostile actions where the structure of the power system may be outside the design specifications and a dangerous resonance band could occur. The risk of resonances is also increasing under normal operating conditions due to the increasing number of devices and their characteristic harmonics, e.g., energy storages being charged from PV panels. A detailed description of conventional harmonic mitigation methods including advanced STATCOM devices can be found in [17–19].

This paper presents requirements connected to harmonics limits and results of power quality measurement taken in a 110 kV and MV network. Examples of sudden changes in harmonics and resulting THD are presented. Factors affecting harmonic distortion are explained, and general recommendations are provided. The next paragraph presents simulation case studies. Simulations present the impact of distribution system extension on harmonic distortions. Additionally, the impact of volt/VAR control on harmonic amplitude is provided. The next paragraph presents proposed passive and active harmonic mitigation methods—utilization of transformer phase shift to reduce harmonics or analysis of the energy source's harmonics spectrum to avoid the resonance band. Finally, the concept of reactive power regulation in order to reduce harmonic distortion under resonance conditions is provided and a comparison of different harmonic mitigation methods is performed. The summary contains the most important findings and outlines further research.

2. Harmonics in the Distribution System Network

2.1. Basics of Harmonics Load Flow

Harmonics analysis requires knowledge about voltage harmonics (existing distortions) in power networks, and knowledge about the structure of power systems and different devices. Further, it requires knowledge of harmonics summation or cancellation effects [20–22]. The voltage harmonics describe distortions resulting from the operation of all devices connected to the analyzed network. The structure of the power system is needed in order to develop a simulation or mathematical model of the analyzed network, which allows for detailed analysis, e.g., planning of new devices or analysis of failures causes [23].

According to common knowledge, the voltage harmonics in high voltage networks are negligibly small, which is often true. However, in the case of high penetration of renewables and industrial devices, e.g., rectifiers, and inverters, the voltage distortions can be noticeable. An example of the harmonics content is shown in Figure 2. Typically, the 5th, 7th, and 11th harmonics have the largest amplitude; it is explained that those are characteristic harmonics emitted by conventional rectifying units [24]. Figure 2 presents voltage distortions expressed as THD (5). The THD varies between 0 and 1% over time.

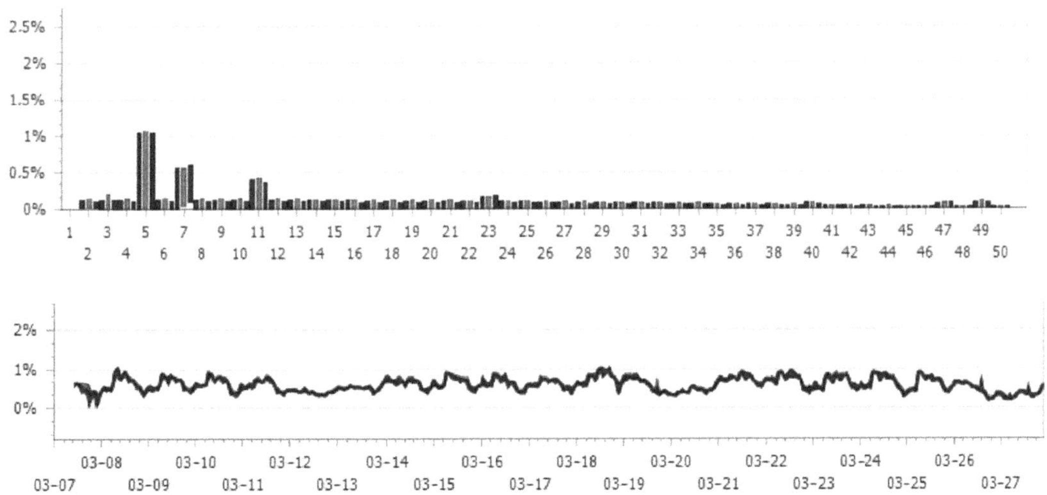

Figure 2. Exemplary voltage harmonic spectrum (expressed as percentage of fundamental component) in 110 kV network with a high share of wind farms and industrial devices. Voltage distortion expressed as THD (%) over time; red—phase 1, green—phase 2, and blue—phase 3.

Analyzing the harmonics presented in Figure 2 (measurement using A class quality meter) allows us to conclude that the amplitudes of the harmonics are below the thresholds specified in norms and regulations. According to [25], 95% of 10 min average values of harmonics must be below the permissible values specified in Table 1. There are different permissible values depending on the type of connection group [26]:

- I group—entities whose devices, installations, and networks are connected directly to networks with a rated voltage higher than 110 kV;
- II group—entities whose devices, installations, and networks are connected directly to the network with a rated voltage of 110 kV;
- III group—entities whose devices, installations, and networks are connected directly to networks with a rated voltage higher than 1 kV but lower than 110 kV;
- IV group—entities whose devices, installations, and networks are connected directly to the network with a rated voltage not higher than 1 kV and a connection power higher than 40 kW or the rated current of the pre-meter protection in the current path higher than 63A;
- V group—entities whose devices, installations, and networks are connected directly to the network with a rated voltage not higher than 1 kV, a connection power not higher than 40 kW, and a rated current of the pre-meter protection not higher than 63A;
- VI group—entities whose devices, installations, and networks are connected to the network through a temporary connection, which will be, under the terms specified in the contract, replaced by a target connection, or entities whose devices, installations, and networks are connected to the network for a specified period, but not longer than a year.

Table 1. Permissible values of higher harmonics.

For I and II Connection Group					
Odd Harmonics			**Even Harmonics**		
Not Being a Multiplicity of 3		**Being a Multiplicity of 3**	**Harmonic Order**	**Relative Voltage Value in Percent of the Fundamental Component (uh)**	
Harmonic Order	**Relative Voltage Value in Percent of the Fundamental Component (uh)**	**Harmonic Order**	**Relative Voltage Value in Percent of the Fundamental Component (uh)**		
5	2%	3	2%	2	1.5%
7	2%	9	1%	4	1%
11	1.5%	15	0.5%	>4	0.5%
13	1.5%	>15	0.5%		
17	1%				
19	1%				
23	0.7%				
25	0.7%				
>25	$0.2 + 0.5 \cdot 25/h$				
for III to V connection group					
5	6%	3	5%	2	2%
7	5%	9	1.5%	4	1%
11	3.5%	15	0.5%	>4	0.5%
13	3%	>15	0.5%		

Table 1. *Cont.*

For I and II Connection Group				
Odd Harmonics			**Even Harmonics**	
Not Being a Multiplicity of 3		**Being a Multiplicity of 3**	**Harmonic Order**	**Relative Voltage Value in Percent of the Fundamental Component (uh)**
Harmonic Order	**Relative Voltage Value in Percent of the Fundamental Component (uh)**	**Harmonic Order**	**Relative Voltage Value in Percent of the Fundamental Component (uh)**	
17	2%			
19	1.5%			
23	1.5%			
25	1.5%			
>25	0.5 + 25/h			

The general trend is that the higher the harmonic row, the lower the acceptable level of harmonic amplitude. For the connection groups I and II, the permissible value of harmonics above the 25th order is roughly 0.5%. The more detailed analysis of Figure 2 allows us to notice that 47 and 49 harmonics also have significant amplitudes in relation to acceptable values. One can observe an amplitude of ~0.15, which is approximately 33% of the acceptable value (0.02 + 0.5·25/47). Harmonics limits may differ around the world depending on the standards and regulations, e.g., [27].

Harmonics analysis is even more complicated, since the inverters are always delivered with the OEM (original equipment manufacturer) filters inside the inverter [28]. It is also possible that additional dedicated filters are installed on the AC or DC side [29,30]. Unfortunately, producers rarely provide information about the internal filters; therefore, often in practice, one needs to accept that the results of harmonics analysis are connected to errors. To avoid potential errors, on-site measurements are typically recommended. Currently, every time new energy sources are installed, power quality (PQ) analysis devices are also installed, which allows for better monitoring of the whole network [31].

Current Situation in the Distribution System

The voltage harmonic distortions can be strongly influenced by the modification of the power system structure. Figure 3 shows the fifth harmonic amplitude measured in three phases, varying over time. Based on the time of occurrence, one can indicate that distortions are higher during working days and lower during weekends; therefore, the changes are most likely caused by nearby industrial activities. The 5th harmonic, 19.03, reached 1% (marked in red), which was caused by the modification of the power system structure—a disconnection of one of two 110 kV lines, which was closer to the 220/110 station. In the presented case, the amplitudes of the harmonics have a large safety margin from the limits defined in Table 1; at the same time, the presented case shows that one needs to be very careful during the analysis of the impact of new devices on the power system, since the harmonic content may change instantaneously when the power system structure and the resulting impedance characteristic change. The impedance characteristics of exemplary power system elements are presented in Figure 4. The resulting impedance is a combination of elements' impedances, e.g., line impedance is additional to source impedance. The PWM inverter ensures lower impedance than the conventional inverter, which ensures the minimization of harmonic distortion [32].

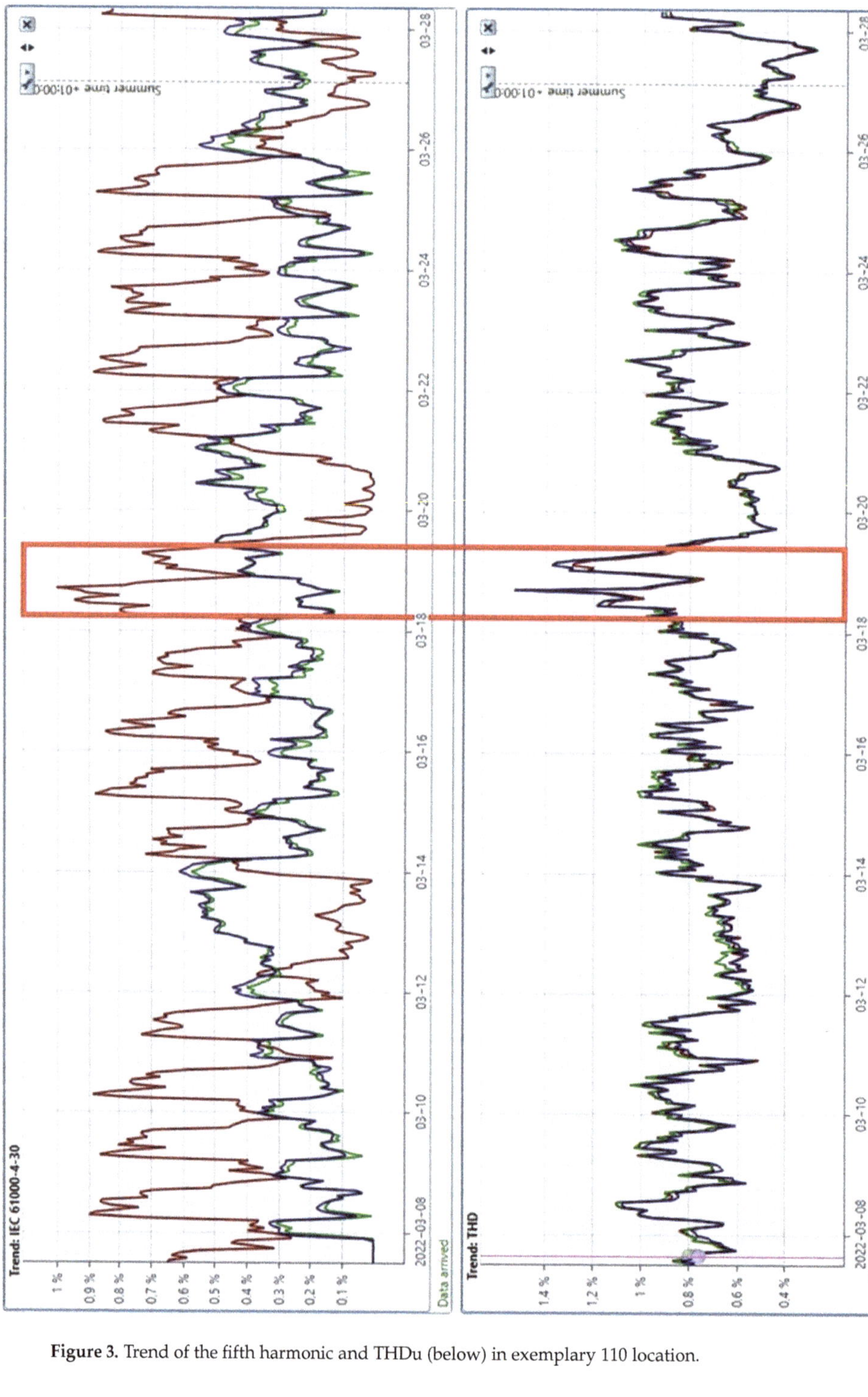

Figure 3. Trend of the fifth harmonic and THDu (below) in exemplary 110 location.

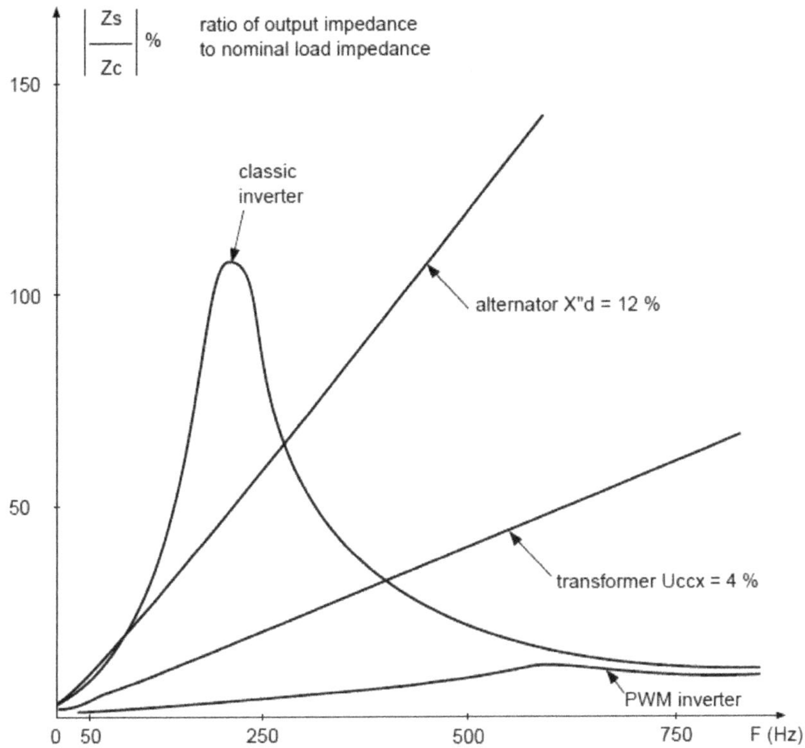

Figure 4. Impedance characteristics of different power system elements [32].

The impedance characteristics show that impedance strongly differs depending on the distortion frequency (harmonic row). As a result, there is a risk that the impedance will be reduced to low/high levels, resulting in dangerously high currents/voltage—series/parallel resonance. Resonance may occur during both steady and transient states and result in devastating surges—transient overvoltage (TOV) [33].

As a result of the presented factors, harmonics analysis is difficult. Moreover, because of the long time frame, preferably a few weeks, numerous configurations of the power system that need to be considered, and operating points, e.g., tap charger positions, harmonics analysis is a time-consuming process. One needs to underline that regulations may state that only normal power system configurations are subject to verification of requirements. In practice, in the case of a 110 kV network, the 5th, 7th, and 11th harmonics are often characterized by the highest amplitudes because both the rectifiers and inverters often emit a large amount of the 5th and 7th harmonics. Furthermore, the harmonics can be within the resonance band of the network [34]. Authorial analysis shows that EHV lines could also be susceptible to the resonance effect and potentially dangerous resonance of low-order harmonics. The analysis is performed in order to verify if, after the modification of the generation module, the NC RFG grid code requirements were fulfilled [35]. Among many different requirements, the reactive power capability requirements have an indirect impact on the harmonics distortions in the power network due to the impact of new devices on the power system impedance characteristics. In some cases, it is difficult to fulfill reactive power capability requirements without modification of the power system structure. The power plant operator is allowed to change the position of transformer tap changers or use conventional reactive power sources, e.g., capacitor banks, in order to fulfill the requirements. As a result, the impedance characteristics of the power system can be changed; therefore, the possibility of resonance for some operational points of the power plant exists [36]. The structure of the power network could also be changed during maintenance activities of power plants, e.g., disconnection of one of the internal feeders of a power plant. Therefore, many different configurations of power plants have to be carefully analyzed to exclude the risk of unwanted resonances and potential resulting failures.

The measurements show that short-term (1s presented) amplitudes of harmonics and THDu in MV networks can be much higher than those in HV networks. Figure 5 presents the 1 s voltage distortion THD. The voltage distortion varied between 0.5 and 2.5% for most of the time (between 15.02 and 18.03). At the end of the measurement period (18.03), the THD increased significantly. The THD level varied between 1 and 3.5% for most of the time; however, values up to 5.5% could be observed relatively often. According to the norms, the limits may be exceeded for a short time, e.g., a few seconds, because the requirements are specified regarding the 10 min average values. Moreover, the PQ norm allows that 5% of the 10 min average values will be above the thresholds specified in 50, 160 [37]. On the other hand, the requirements describing connection possibilities indicate that if any limit, e.g., single harmonic, is exceeded, the decision about the integration of a new device should be negative [38]. New connection requirements indicate that harmonic analysis may be omitted if inverters are certified by PTPIRE [39], which is justified by the great technological progress in inverters over the years. The harmonic spectrum of an exemplary new-generation inverter is presented in Table 2. The omission of analysis creates a risk that potential resonances could be initially missed. Moreover, one needs to note that even the best inverters in high numbers may have a noticeable impact on power quality. Despite new renewables, one may expect that more high-power electronic loads will be introduced to power systems, e.g., electric vehicle charging stations. Therefore, it is expected that the power quality will degrade in the future.

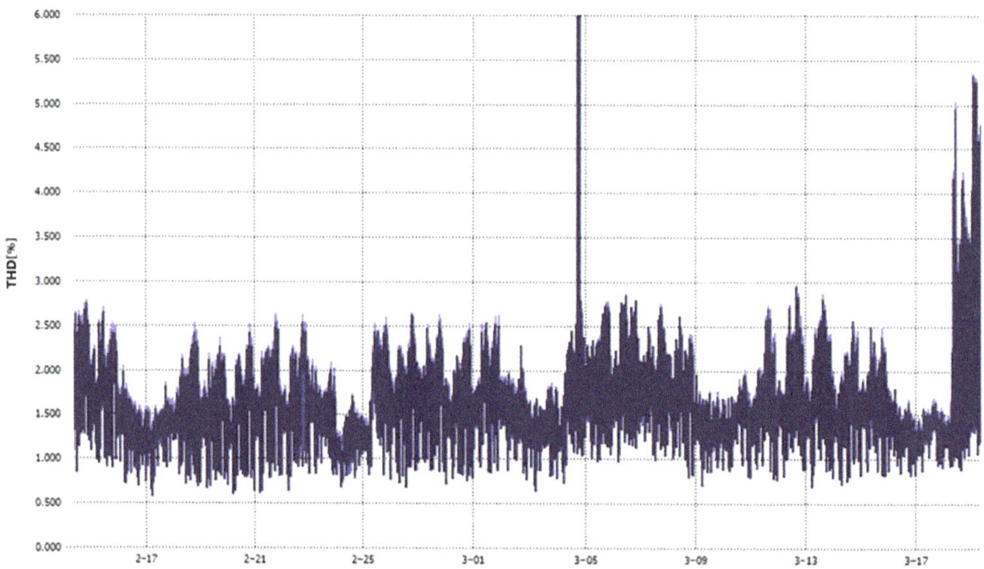

Figure 5. Voltage distortion expressed as THD (%) as a function of time (month–day) in 110/15 kV substation at the 15 kV side.

Power quality issues are a result of the power of electronic converters; however, the susceptibility to distortion is a result of the short-circuit power of the network (and other parameters) at the point of coupling. The lower the voltage level, the smaller the short-circuit power, and therefore the higher the distortions that typically exist in lower-voltage power networks. Typically, the short-circuit power in substations is in the following ranges:

EHV substations 10,000–30,000 MVA, HV (110 kV) 1000–10,000 MVA, and 150–250 MVA in MV networks. Higher short-circuit powers are observed when short-circuit power sources are closer to the substation and are larger, e.g., the higher nominal power of a transformer or synchronous generator.

Table 2. Harmonic spectrum of new generation inverter [40].

Harmonic Row	Current Magnitude [A]			% Of Fundamental		
	L1	L2	L3	L1	L2	L3
1	134,856	134,736	134,840	100,940	100,851	100,928
2	0.252	0.139	0.299	0.187	0.103	0.221
3	0.380	0.209	0.215	0.282	0.155	0.159
4	0.107	0.200	0.095	0.079	0.148	0.071
5	0.278	0.226	0.239	0.206	0.168	0.177
6	0.076	0.066	0.044	0.056	0.049	0.033
7	0.070	0.096	0.083	0.052	0.072	0.062
8	0.047	0.036	0.061	0.035	0.027	0.046
9	0.122	0.039	0.100	0.091	0.029	0.074
10	0.040	0.061	0.071	0.030	0.045	0.052
11	0.224	0.167	0.175	0.166	0.124	0.130
12	0.075	0.055	0.044	0.055	0.041	0.033
13	0.261	0.245	0.286	0.193	0.182	0.212
14	0.042	0.038	0.058	0.031	0.028	0.043
15	0.053	0.102	0.079	0.039	0.076	0.059
16	0.043	0.065	0.073	0.032	0.048	0.054
17	0.235	0.170	0.197	0.175	0.126	0.146
18	0.107	0.066	0.053	0.079	0.049	0.039
19	0.281	0.242	0.229	0.208	0.180	0.170

Recently, the Polish Railway (PLK, PKP) announced a green railway initiative, according to which a high share of electricity is produced using renewables [41]. The integration of renewables with railway networks is even more challenging than distribution networks because the short-circuit power is even lower, and the traction loads emit a large amount of harmonics due to passive 6- or 12-pulse rectifiers. The short-circuit power decreases with the increase in distance from the substation. Therefore, deep in the MV network, the power quality may be significantly lower, e.g., 5 km of 240 mm^2 cable decreases the short circuit power from 200 MVA to 137 MVA, which is often the case for the supply of MV railway substations. As a result, the susceptibility of the network to harmonic distortions increases. The railway substation may also be supplied from the 110 kV network. In this case, the railway substation is equipped with four winding transformers: primary 110 kV, two windings for traction, and one non-traction load winding [42]. From the harmonics perspective, the best point of connection would be a 110 kV network; however, from the techno-economical perspective, the integration of a new source is justified when the power is roughly 20 MW or more. Another option for energy source connection is the non-traction load winding, typically 15 kV. Unfortunately, the non-traction load winding has very little power, in the range of 1 MVA, which results in a great reduction in the short circuit power to approximately 20–30 MVA. The final option for the integration of renewables with the railway network is connection to the DC network, which considers specific regulations and also distortions [43,44]. Harmonic analysis should also include analysis of waveforms, at least for typical transient conditions, e.g., the switching on of the inverter, particularly in locations with a high share of devices activated at similar time periods [45]. An exemplary startup of a high-power PV plant is presented in Figure 6 below (current marked in blue and resulting voltage distortions marked in red). One can see strongly distorted currents and resulting voltage distortions.

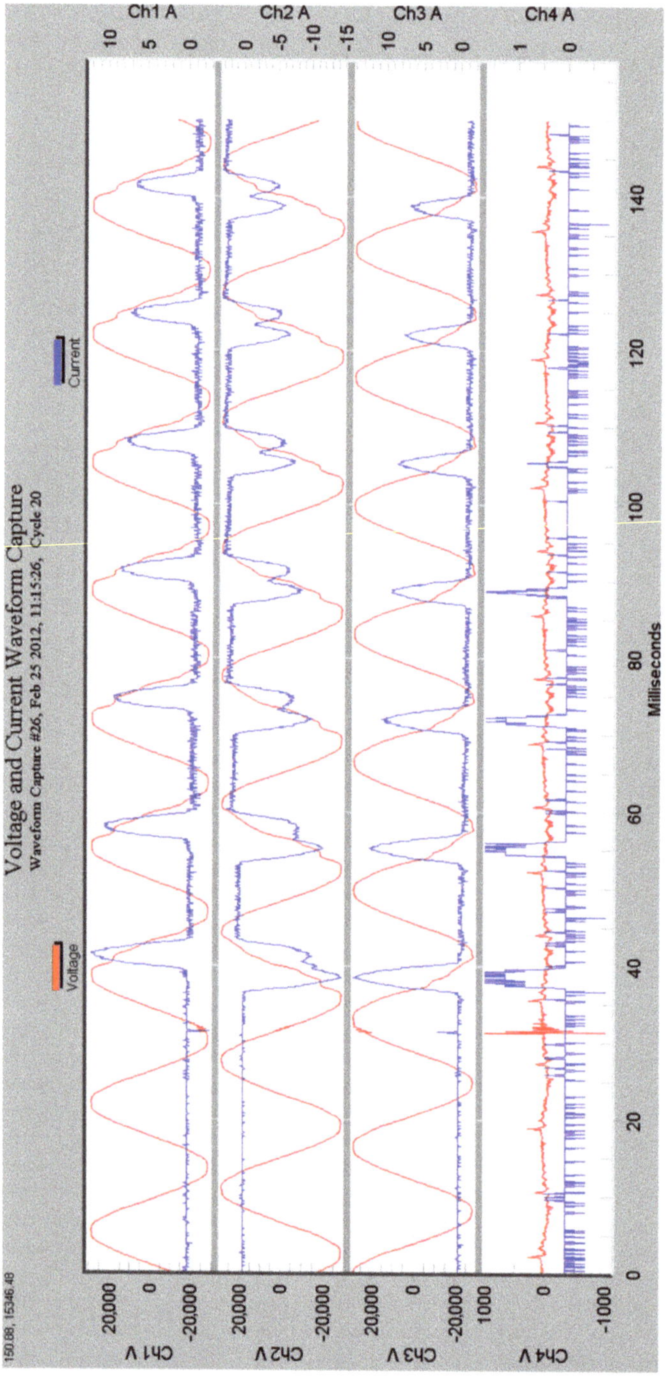

Figure 6. High-power photovoltaic startup current (blue) [46].

3. Harmonic Analysis—Simulations

The purpose of the simulation was to draw attention to changes in the power system during system regulation or development. Attention is paid to the resonance band and its sensitivity to power system development and volt/VAR regulation. Simulations were performed using a modified test model available in the PowerFactory software 2024 SP4—wind farm of 20 kV [47]. The network topology and basic parameters are presented in Figure 7 (parameters in the Appendix A). First, the impact of the voltage level is analyzed. As can be seen in Figure 8, the analyzed model is susceptible to voltage change.

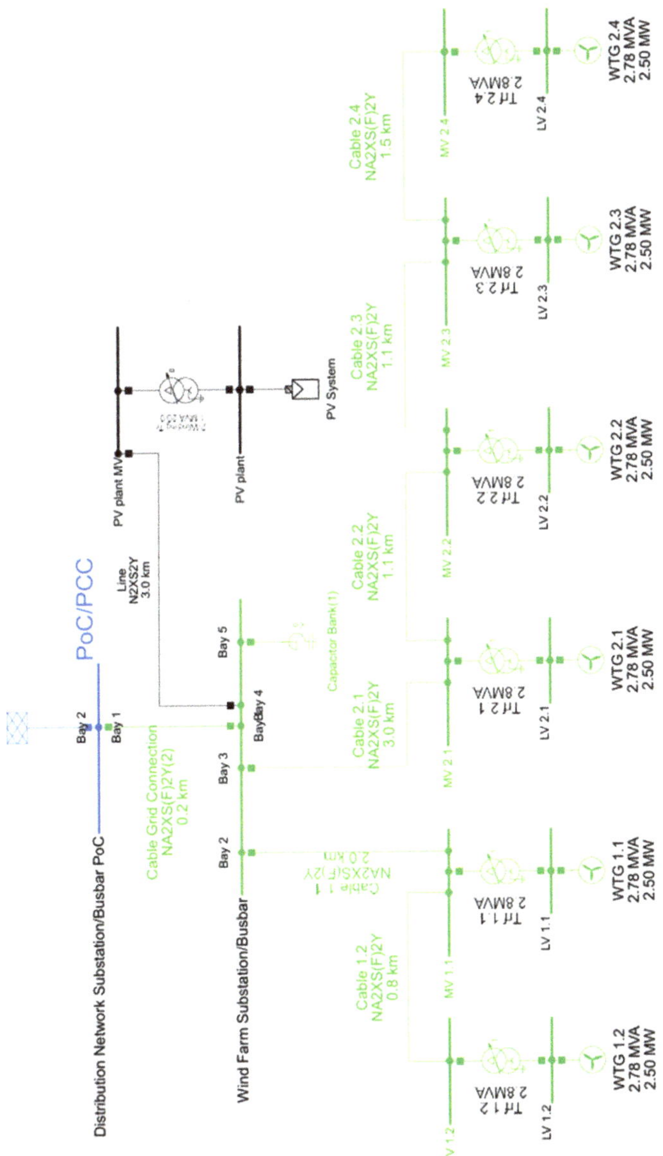

Figure 7. MW diagram of the modified farm elevator network.

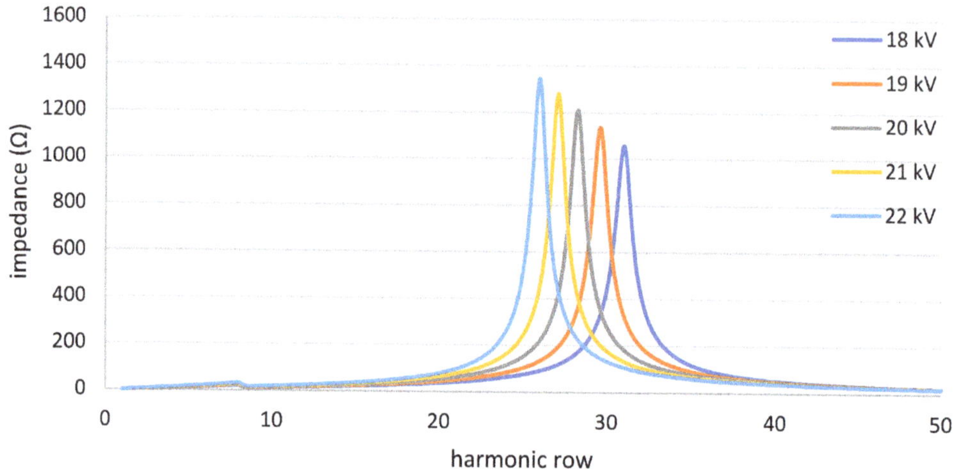

Figure 8. Impact of voltage level on impedance characteristic in point of coupling (PCC).

In the case of a nominal voltage—20 kV—the resonance is observed for the 28th harmonic, whereas a reduction in voltage to 18 kV (10%) shifts the resonance to the 31st harmonic. A voltage increase to 22 kV (10%) shifts the resonance to the 26th harmonic. The shift may trigger the resonance of harmonics, which are negligible under normal operating conditions. The resulting voltage harmonics are presented in Figure 9. The harmonics are the result of wind farm inverters with default settings connected to the network, as defined by the impedance characteristics presented in Figure 8. It can be seen that when a harmonic, e.g., 24, is in the resonance band (22 kV), the distortions are increased.

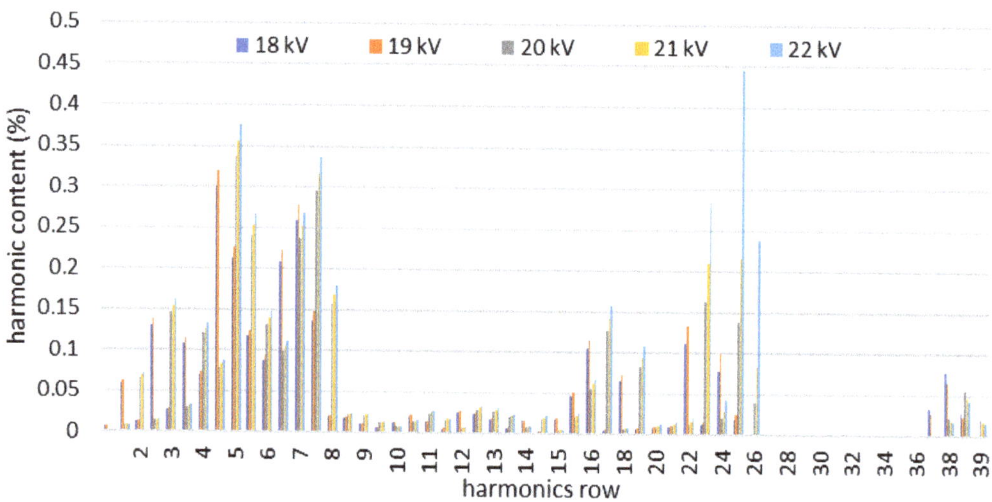

Figure 9. Impact of voltage level on voltage harmonics at PCC.

Further, the impact on the capacitor bank is analyzed and presented in Figure 10. As can be seen in the analyzed case, the resonance band is not shifted, but is dampened with the increase in capacitor power. One has to underline that it is also possible that the resonance band will be shifted depending on the power system's conditions.

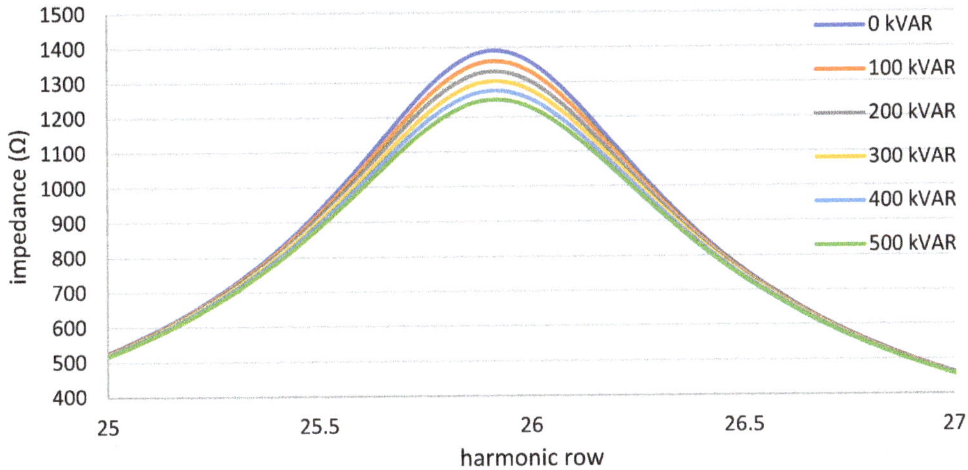

Figure 10. Impact of capacitor bank on impedance characteristic at point of coupling (PCC).

The next stage of the simulation is to indicate the impact of network expansion accompanying the connection of a photovoltaic power plant on the impedance characteristics of the power grid—Figure 11. The initial value of the network impedance is marked in blue. The addition of a power cable representing the new PV source power output system is marked with the following colors:

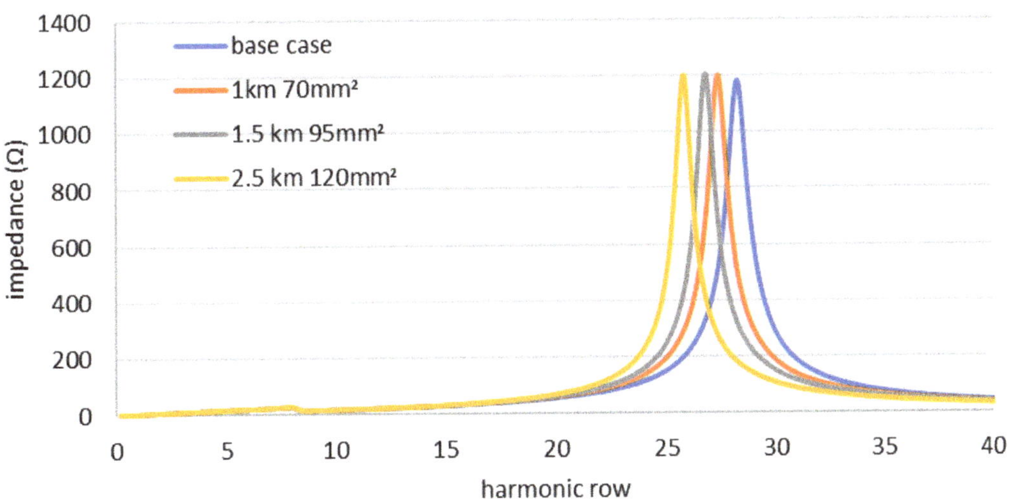

Figure 11. Impedance characteristics of the network as a function of PV plant cable marked in gray in Figure 7.

Orange—cable with a cross-section of 70 mm^2 and a length of 1 km;
Gray—cable with a cross-section of 95 mm^2 and a length of 1.5 km;
Yellow—cable with a cross-section of 120 mm^2 and a length of 2.5 km.

It was found that the addition of further cable lines led to a further shift in the resonance band toward lower harmonic orders, which in practice means that the risk of exceeding the permissible harmonic values in extensive networks with a large share of cable lines or networks operating with only one supply transformer (two in the normal mode) is greater. Theoretically, to minimize the shift in the resonance band toward lower harmonics, it is worth considering the use of overhead lines with lower capacitance, but this is rather unlikely, since cable lines are characterized by higher reliability, which is often a superior feature.

In the next stage, the impact of power plant transformers and the exemplary low-pass filter installed in 750 kW inverters is analyzed [48]. The filter consists of series-connected inductances with L1 = 170 mH, L2 = 80 mH, C = 466 μF, R = 0.1Ω (Figure 12). The gray color shows the impedance characteristics of the network expanded with a 1.5 km 95 mm^2 cable line. The orange color shows the network expanded with two 1.6 MVA transformers. The impact of transformers in the analyzed case is negligible, as evidenced by the overlapping characteristics. The impact of transformer capacity is also verified, but no significant impact on the results in the analyzed network is found [49]. The impact of four inverter filters is shown in blue (covered by yellow). Adding filters leads to a clear shift in the resonance band toward higher harmonics, which is usually beneficial, but at the same time, a deterioration in the conditions in the lower harmonics is observed (small resonance band around the eighth harmonic).

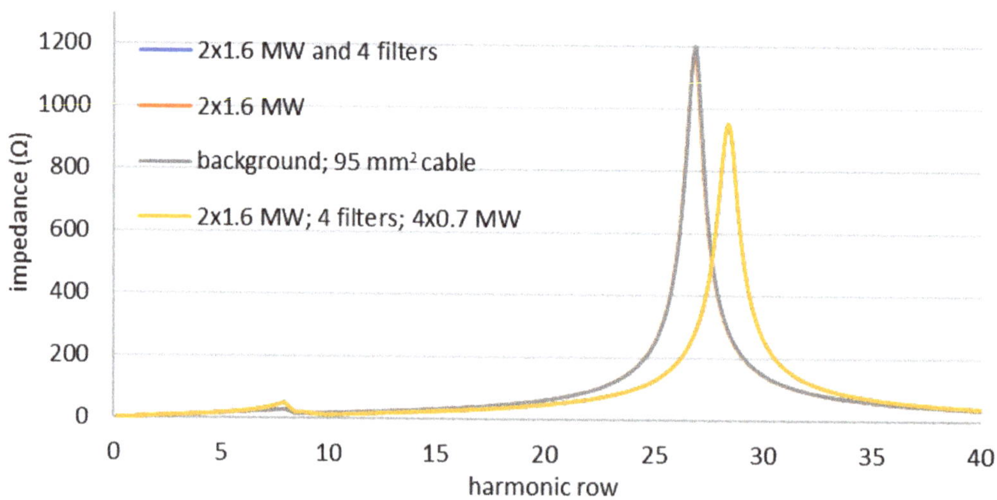

Figure 12. Impedance characteristics of the network.

The presented simulation shows that the resonance band may be significantly shifted when regulation activities within normal operating limits are performed. The modification of the power system due to the new power plant installation may also have a noticeable impact on impedance characteristics.

4. Harmonic Mitigation—Alternative Methods

There are different mitigation measures against harmonics—passive and active filters or sophisticated control algorithms; however, traditionally, three winding transformers are used as a basic harmonic mitigation measure. The mitigation methods need to be considered at the design stage or afterward, but additional devices are required e.g., filters, which increase the cost [50].

To minimize the risk of exceeding harmonic values, it is worth considering the selection of inverters that generate harmonics that do not coincide with the resonance band. An example is shown in Figure 13. As can be seen, the large variety of characteristics allows the selection of devices that do not exceed the permissible values. However, it should be emphasized that, regardless of the amplitudes of individual voltage harmonics, the relative value of THDu harmonics cannot be exceeded.

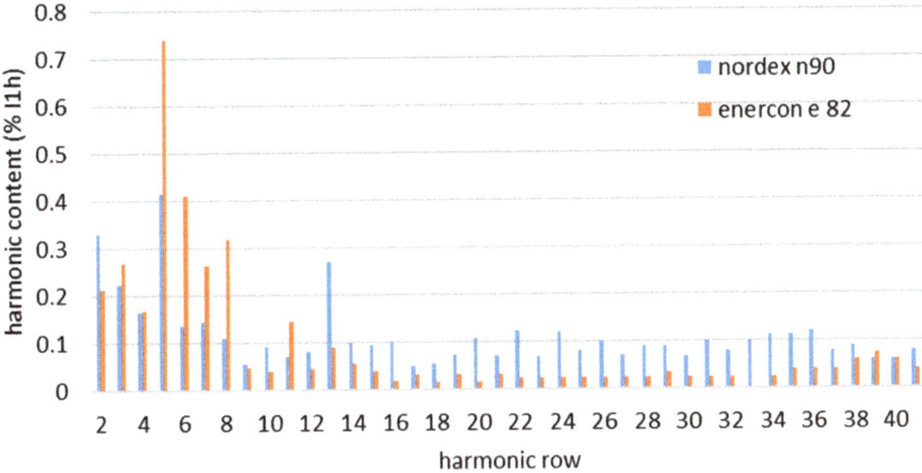

Figure 13. Harmonic spectrum emitted by 2 exemplary wind turbines; based on [51].

The simulation software allows for the analysis of the phase shifts of harmonics—however, for the phase correct option, the data required for such simulations are difficult to obtain as only harmonics amplitudes are reported by manufacturers.

The harmonics amplitude can be reduced if the harmonics sources are in antiphase. In the case of the most common fifth and seventh harmonics, the antiphase occurs when the same loads are supplied with a 30° phase shift, which is obtained by a properly designed three-winding transformer and proper load distribution [52].

Measurements allowed us to notice that the new generation of PV inverters generates a stable harmonics spectrum, not only in the context of harmonics amplitude, but also harmonics phase, which is the basis for the utilization of phase shift transformers for harmonics mitigation. Measurements (Figures 14 and 15) show that twin-type PV sources located nearby generate almost the same harmonics, including angles that allow for the utilization of different transformers' phase shifts to minimize harmonic distortion.

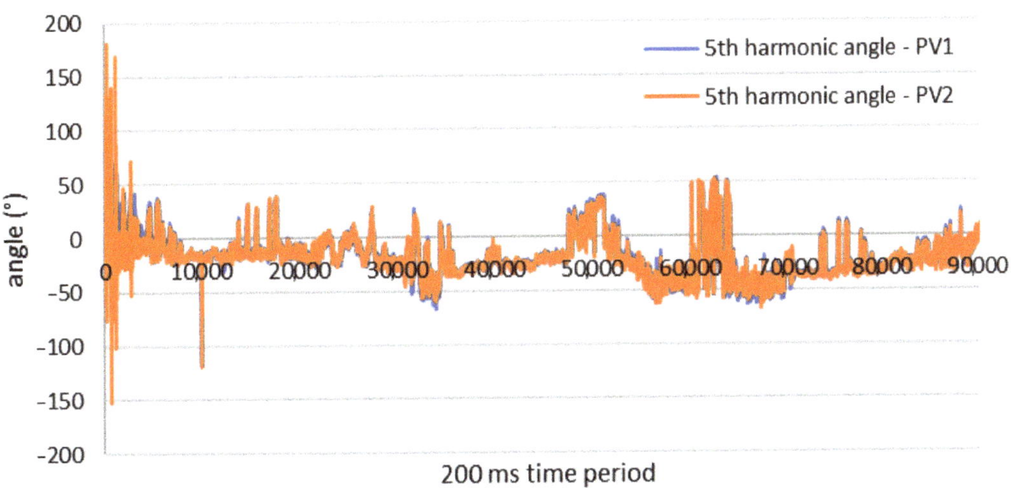

Figure 14. Angles of harmonic shifts of 5th harmonics of two PV power plants located in close proximity.

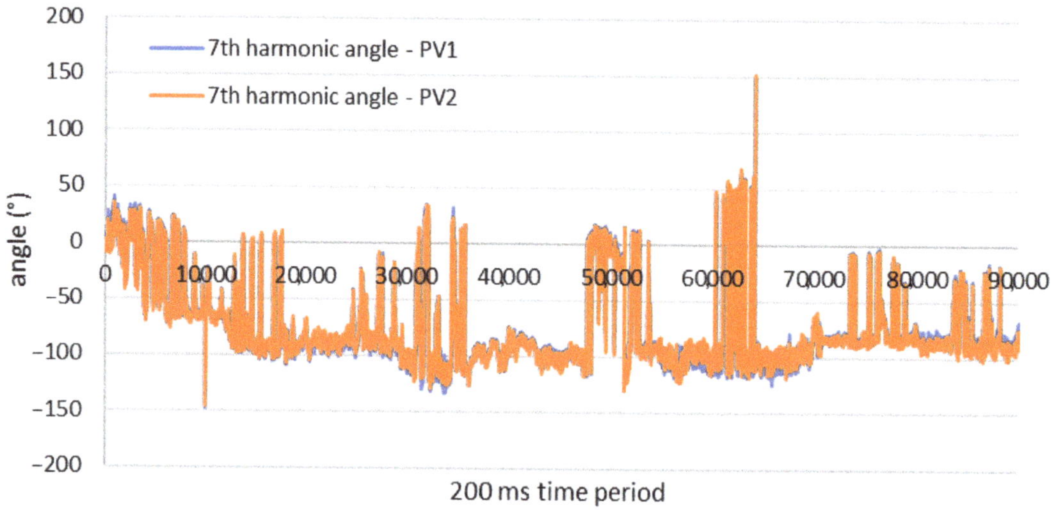

Figure 15. Angles of harmonic shifts of 7th harmonics of two PV power plants located nearby.

PV plants typically deliver the power using two winding transformers with the same phase shift. Two winding transformers with different phase shifts are rare but available on the market [53].

A change in reactive power causes a phase shift of the generated current and therefore a shift in the harmonics [54]. The effect of reactive power generation requires on-site verification; nevertheless, assuming that the harmonics shift ideally with the fundamental component of current, one may conclude that it is possible to cancel chosen harmonics by shifting the reactive power. To cancel the harmonic completely, one must create a signal in the opposite phase to the existing signal, which is obtained by a signal shift equal to half of the harmonic period (7). If the number of inverters taking part in harmonics cancellation is an odd number e.g., 3, then one needs to ensure phase shift equal to one-third of the harmonic period (8)

$$\varphi_h = \frac{1}{2} \cdot \frac{1}{50 \cdot h} \tag{7}$$

$$\varphi_h = \frac{1}{3} \cdot \frac{1}{50 \cdot h} \tag{8}$$

The methodology of using reactive power to mitigate harmonics under resonance conditions is presented in Figure 16. The initiation phase requires on-site measurement, which confirms the proper operation of the presented algorithm for given inverters. After successful verification, the algorithm continuously extracts information about harmonics to detect potential resonance bands. If resonance is detected, the harmonics limit of the harmonics under resonance is violated, the inverter generates resonating harmonics, and the next action is performed. The algorithm assesses the required phase shift (7) or (8) depending on the configuration; further, the phase angle is calculated (or measured) and the required reactive power is calculated based on the sum of the measured angle and required angle change according to (9). Finally, the reactive power change is added to the measured reactive power to obtain a new operating point and the power is changed according to the ramp function, so the best compensation point can be chosen. One needs to note that the detailed phase shift is impossible to calculate because harmonics may consist of different-frequency signals, not just the ideal frequency. Further, it may be required to set an active power threshold above which regulation can be activated because of unstable operation in low harmonics ranges.

$$Q_r = P{\cdot}tg(\varphi + \varphi_h) \tag{9}$$

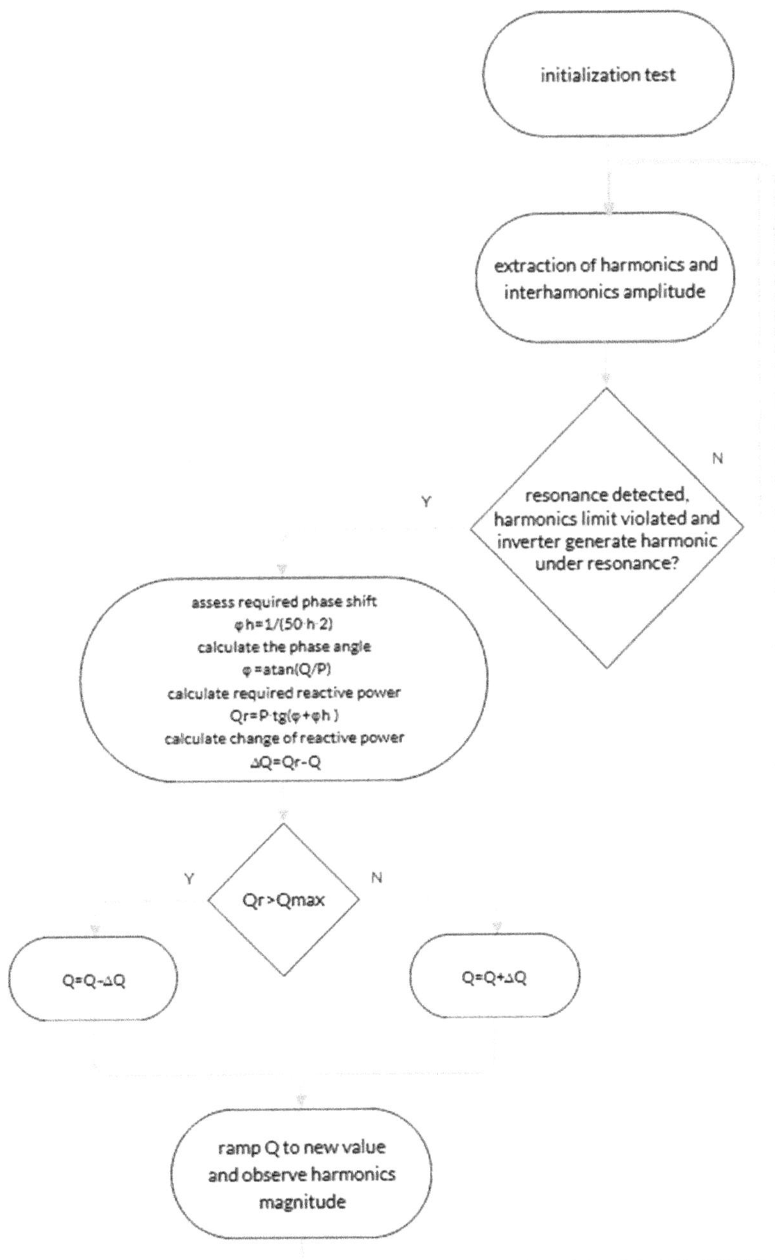

Figure 16. Utilization of reactive power to dampen harmonic resonance.

Calculation of the Q adjustment range is performed in order to avoid unnecessary changes in reactive power and their impact on voltage level. One should also consider that inverters need to change the reactive power according to the Q(U) characteristic (Figure 17), which could result in a change in harmonics flow. The typical V2 and V3 setting is ~±5% Vn. That of V1 and V2 is typically ~±8% Vn. The proposed algorithm can shift the reactive power within the calculated band to compensate for the potential negative effects. It is believed that the risk of the unwanted resonances is higher in generation stations, since typically the same inverters are used, whereas in distribution substations, the negative effect is often reduced by interactions between many different types of devices.

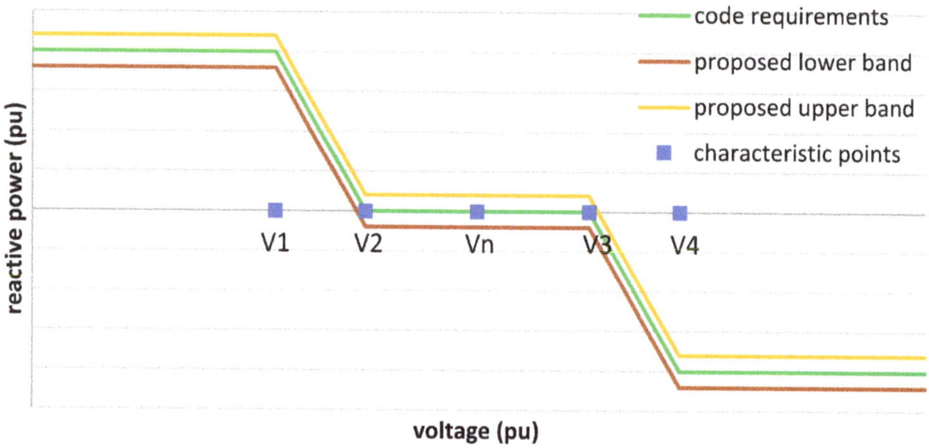

Figure 17. Typical volt VAR regulation characteristics and the proposed exemplary regulation band; where V1—max reactive power generation, V2—start of reactive power generation, Vn—nominal voltage, V3—start of reactive power consumption, and V4—max reactive power consumption.

A general diagram of controlled objects is presented in Figure 18. The voltage and current transformer provide signals to the PQ analyzer installed on-site. The PQ analyzer sends signals to the controller, which run the algorithm presented in Figure 16. The presented diagram does not consider PQ measurement on the low-voltage side of the power plant, because the whole operating range should be verified at the commissioning stage. In case one has any doubts, additional PQ measurements are recommended.

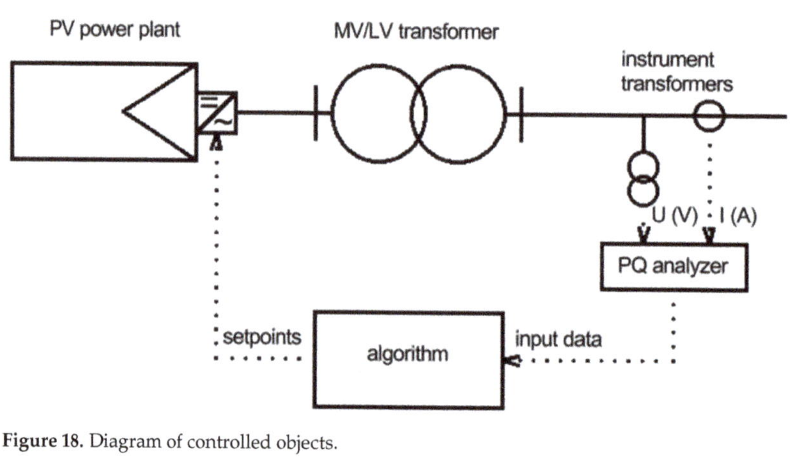

Figure 18. Diagram of controlled objects.

To ensure the proper operation of the proposed algorithm, one needs to consider the measuring errors. In order to ensure acceptable measurement errors, one needs to utilize high-accuracy equipment. In particular, one needs to utilize:

- An A-class power quality analyzer, which is characterized by measurement uncertainty equal to 0.1% of the declared input voltage;
- A high-accuracy instrument transformer, e.g., 0.2 class (0.2 amplitude error) [55].

One needs to make sure that the measurement range of the PQ analyzer and its current sensors are adapted to the current flowing at the secondary side of current transformer. Both current clamps or a Rogowski coil could be used. One needs to be aware that the instrument transformer can introduce additional measurement errors. The Rogowski coils can be installed directly on the cable lines and therefore the current transformer and its measurement errors are eliminated. Despite the amplitude measurement errors, one needs to consider the frequency characteristics of the instrument transformers. According to [56], inductive transformers are suitable for measurements of 50th-order harmonics, but only in 10 kV networks. In case of 20 kV networks, only harmonics up to the 20th order are measured within an acceptable error range. Taking into account numerous designs of instrument transformers, it is recommended to contact manufacturers or analyze the technical specifications in order to validate the harmonics measurement accuracy.

The power plant controller is required by distribution system operators and needs to fulfill global requirements e.g., the NCRFG grid code, and local requirements. There are different communication standards available e.g., DNP3.0, 61400-25, 61850-7, and 61850-5, which allow for receiving commands from the power system operator. The above-mentioned modifications of reactive power and ramping are state-of-the-art functionalities of controllers and inverters. The controller sets parameters according to the grid code requirements e.g., above-mentioned Q(U) characteristic, so the modifications to the reactive power under the U regulation band may be blocked. One needs to underline that the settings could be modified only with the consent of the distribution network operator.

Ongoing efforts connected to the optimization of inverters and other electronic devices allow for reducing the risk of harmonic-related issues. Typically, in order to reduce the harmonics content, passive or active filters are used. A general comparison of harmonics mitigation methods is presented in Table 3. Passive filters may be susceptible to overloads and may be ineffective in case of power system structure modification. Active filters are effective for different configurations of power systems and any harmonics order. Additionally, despite harmonic filtering and reactive power correction, active filters are able to perform phase-balancing. Active filters are rather expensive; therefore, they are typically used in case of low-power applications. In case of high-power applications, hybrid filters, which are a combination of passive and active filters, are used [57]. Filters are often used in industrial networks; however, the filters are relatively rarely used in distribution system networks. In case of harmonics issues in a distribution system, the harmonic source is often transferred to an individual power transformer in order to minimize the negative impact on the surrounding area. In case of railway substations, the three winding phase-shifting transformers and filters e.g., gamma filters, and 12 pulse converters are typically used in order to minimize harmonics pollution [58]. Furthermore, in cases of high-power distortion sources, the series reactor is used. Different filter types can be adapted to the situation in the network [59].

The proposed solution could reduce the resonance effect and reduce the risk of resulting failures. If problems are detected, it is recommended to install conventional filters. It is also possible to simultaneously use an active filter and the proposed solution if the filter power is not sufficient. Further research is planned in order to validate the proposed solutions.

Table 3. General comparison of harmonic mitigation methods.

	Passive Filters	Active Filters	Inverters' Built-in Algorithms	Proposed Solution
Cost	Big	Big	No extra cost	No extra cost
Experience	Big	Big	Depending on the solution	Low
Analysis required	Project needed	Sizing of device and limitations of sizes	No extra project	No extra project
Individual harmonics limitation	Harmonics designed to be damped	Yes, the full spectrum of problematic harmonics	Depending on the construction	Yes, but only the resonance band
THD reduction	Yes if sized properly	Yes	Yes	Yes, under resonance conditions
Losses	Extra losses	Extra losses	Switching losses could increase	Change of reactive power could have an impact on losses however in general the impact is minimal
Impact of power system reconfiguration	Big impact	No, until power limitations are met	Yes, could be reduced if adaptive algorithms are used	Effectiveness could drop in case of multi resonance point

One needs to be aware that the proposed alternative harmonic mitigation methods are not perfect.

- Different transformers' phase shifts may be used only in the case of twin PV designs, and it may be difficult to find transformers with different phase shifts in the market;
- The harmonic spectrum of energy sources is one of many parameters e.g., short circuit power, cost, warranty, etc.; therefore, it may not always be possible to find an optimal harmonic spectrum for the location. Moreover, it is often difficult to obtain the harmonic spectrum for analysis;
- Active changing the reactive power to mitigate harmonics is effective in case of high-frequency harmonics; however, in the case of low-frequency resonances, the method requires high reactive power, which could have an impact on voltage level or energy losses. The effectiveness of the proposed method may be different depending on the inverter type; therefore, a special test may be required to confirm that the solution is effective, e.g., the harmonic angle changes as a function of the generated power.

Future research is connected to the testing of the proposed method. Both laboratory and on-site tests are recommended at the initial stage. The laboratory tests allow for observing the exact impact of reactive power change on harmonics amplitudes and harmonic angles. The differences between theoretical and real values may be a result of apparent power change as a result of reactive power change (negligible for higher-order harmonics) and nonlinear dependency of harmonics as a function of the apparent power. Furthermore, additional errors are expected at the low power level. Typically, the current distortions is strong at low power levels, e.g., 10–20% THD for powers below 10% of the nominal value. Taking into account the great number of inverters available on the market, it is recommended to start the tests using the most popular inverters. Alternatively, it is possible to choose inverters that are installed in locations where harmonics resonances are detected. In that case, however, it is possible to run on-site tests and verify the harmonic mitigation methods under real conditions. The on-site tests should be preceded by developing a simulation model, which indicate the impedance characteristics of the network in order to observe the resonance points. This analysis is needed to verify if the mitigation of one harmonic resonance band would not cause a violation of another harmonic band, since, in some cases, a few resonance bands may be observed. Another solution to observe resonance bands is to observe harmonics histograms. If there is no harmonic resonance band, the harmonics amplitudes decrease with the increase in harmonics rows. In the case of resonance bands, one can observe peaks on the histogram. One must underline that the voltage harmonics peaks may be observed only for specific hours.

After the effectiveness of the proposed method is confirmed, however, the mitigation strength is too low, and one can use another inverter within the network. In order to avoid erroneous results of harmonics compensation, inverter regulation should be performed in a coordinated manner. One needs to develop the regulation plan and adapt time delays between inverters as a function of the harmonics row e.g., inverter A starts the regulation first in case of a specific harmonics violation and inverter B starts the regulation after the inverter A regulation cycle is finished. For different harmonics rows, inverter B starts the regulation first, and inverter A follows. The advanced coordination schemes could be performed by advanced systems, e.g., artificial intelligence.

5. Conclusions

This paper presents the authors' measurements and experiences with harmonics resonance, which is supplemented with examples from the literature. It is believed that the risk of resonance is particularly high in the case of generation stations with a high share of cable lines and uniform inverters; therefore, a simulation model presenting the interaction between volt/VAR/harmonics control is analyzed.

The contribution of this paper is:

- Pointing out the potential problem in the power system—the risk of resonance under various conditions like control actions in the power system and power system extension;
- Providing measurements that show examples of resonances;
- Development of simulation study cases that present the interaction between volt/VAR control and harmonics interactions;
- Recommendations about the need to monitor harmonics in case of volt/VAR regulation;
- Alternative passive and active harmonics mitigation methods are proposed:
    - Utilization of transformers with different phase shifts for mitigation of harmonics emitted by twin PV plants and proof of concept measurements are presented;
    - Active reactive power regulation is proposed in order to shift harmonics in the resonance band to reduce the risk of serious issues. The general methodology and further research are presented;
    - The adaptation of the design to mitigate harmonic issues is presented, e.g., utilization of different line types or sources that are the sources of harmonics outside the potential resonance band.

The proposed methods are compared with typical harmonic mitigation methods. It is believed that the proposed reactive power regulation method could reduce the risk of failure in distribution systems in the case of resonance band activation due to sudden changes in power system structure, e.g., during power system restoration or failures of the power system.

Author Contributions: Conceptualization, K.L.; methodology, K.L.; software, K.L.; validation, K.L., J.G. and M.U.; formal analysis, M.U. and Z.N.; investigation, K.L., J.G. and M.U.; resources, J.G.; data curation, K.L.; writing—original draft preparation, K.L., J.G. and M.U.; writing—review and editing, K.L.; visualization, K.L.; supervision, J.G. and Z.N.; project administration, J.G.; funding acquisition, K.L., J.G. and Z.N. All authors have read and agreed to the published version of the manuscript.

Funding: This research was funded by 0711/SBAD/4616.

Data Availability Statement: The data presented in this study are openly available according to references below.

Conflicts of Interest: The authors declare no conflicts of interest.

Appendix A

Table A1. Cable parameters.

	R (Ω)	X (Ω)	B (mS)
Cable Grid Connection	0.0204	0.033	23.12
Cable 1.1	0.422	0.244	159.59
Cable 1.2	0.168	0.098	63.92
Cable 2.1	0.633	0.367	239.39
Cable 2.2	0.232	0.134	87.78
Cable 2.3	0.232	0.134	87.78
Cable 2.4	0.317	0.183	119.69
PV 1 km 70 mm^2	0.446	0.137	61.57
PV 1.5 km 95 mm^2	0.501	0.196	101.79
PV 2.5 km 120 mm^2	0.641	0.317	197.07

Table A2. Transformer parameters.

Wind Turbine Transformer		PV Transformer
Voltage	20/0.69	20/0.8
Short circuit voltage	6%	6%
Cooper losses	20 kW	14 kW
No load loss	10 kW	2.4 kW
No load current	0.8%	0.15%

Table A3. Wind turbine harmonics and interharmonics (percentage of fundamental component).

Harmonic Row	Harmonic Content (%)	Interharmonic Row	Interharmonic Content
2	0.28	1.5	0.11
3	0.41	2.5	0.12
4	0.25	3.5	0.18
5	0.91	4.5	0.35
6	0.29	5.5	0.85
7	0.43	6.5	0.29
8	0.24	7.5	0.72
9	0.06	8.5	0.1
10	0.02	9.5	0.05
11	0.08	10.5	0.05
12	0.02	11.5	0.05
13	0.07	12.5	0.08
14	0.02	13.5	0.05
15	0.01	14.5	0.04
16	0.1	15.5	0.04
17	0.2	16.5	0.
18	0.01	17.5	0.
19	0.1	18.5	0.
20	0.01	19.5	0.
21	0.01	20.5	0.
22	0.01	21.5	0.
23	0.1	22.5	0.
24	0.01	23.5	0.
25	0.05	24.5	0.
26	0.01	25.5	0.
27	0	26.5	0.
28	0	27.5	0.
29	0	28.5	0.

Table A3. *Cont.*

Harmonic Row	Harmonic Content (%)	Interharmonic Row	Interharmonic Content
30	0	29.5	0.
31	0	30.5	0.
32	0	31.5	0.
33	0	32.5	0.
34	0	33.5	0.
35	0	34.5	0.
36	0	35.4	0.
37	0	36.5	0.
38	0.02	37.5	0.
39	0.02	38.5	0.05
		39.5	0

References

1. Miller, P.; Wancerz, M. Praca zrodeł fotowoltaicznych przy zmianach i zanikach napiecia w sieci nn. *Rynek Energii* **2017**, *1*, 56–61.
2. Fantana, N.L.; Lowczowski, K.; Javora, R.; Myrda, P.; Gegot, F.; Brun, P.; Carpenter, J.; Beaudry, J.; Kaiser, N.; Yusa, H.; et al. *Substation Servicing and Supervision Using Moblie Devices and Smart Sensing*; CIGRE: Paris, France, 2020; ISBN 978-2-85873-528-0.
3. Musial, E. Ocena jakosci energii elektrycznej w sieciach przemyslowych. *Autom. Zaklocenia* **2010**, *1*, 30–45.
4. Barros, J.; de Apráiz, M.; Diego, R.I. A review of international limits for rapid voltage changes in public distribution networks. *Renew. Sustain. Energy Rev.* **2021**, *144*, 110966. [CrossRef]
5. Lodetti, S.; Azcarate, I.; Gutierrez, J.J.; Leturiondo, L.A.; Redondo, K.; Saiz, P.; Melero, J.J.; Bruna, J. Flicker of Modern Lighting Technologies Due to Rapid Voltage Changes. *Energies* **2019**, *12*, 865. [CrossRef]
6. Kostic, M.; Nikolic, A. Negative Consequence of Motor Voltage Asymmetry and Its Influence to the Unefficient Energy Usage. *WSEAS Trans. Circuits Syst.* **2010**, *9*, 547–552.
7. Mariscotti, A.; Mingotti, A. The Effects of Supraharmonic Distortion in MV and LV AC Grids. *Sensors* **2024**, *24*, 2465. [CrossRef]
8. Grady, M. Understanding Power System Harmonics. *IEEE Power Eng. Rev.* **2012**, *21*, 8–11. [CrossRef]
9. Arranz-Gimon, A.; Zorita-Lamadrid, A.; Morinigo-Sotelo, D.; Duque-Perez, O. A Review of Total Harmonic Distortion Factors for the Measurement of Harmonic and Interharmonic Pollution in Modern Power Systems. *Energies* **2021**, *14*, 6467. [CrossRef]
10. Pinyol, R. *Harmonics: Causes, Effects and Minimization*; Salicru: Barcelona, Spain, 2015.
11. Landis Gyr. *Landis+Gyr S650 Smart Grid Terminal A Smart Grid Innovation*; Landis Gyr: Zug, Switzerland, 2013.
12. Tran, Q.-T.; Truong, A.V.; Le, P.M. Reduction of harmonics in grid-connected inverters using variable switching frequency. *Int. J. Electr. Power Energy Syst.* **2016**, *82*, 242–251. [CrossRef]
13. Prodanovic, M.; Green, T. Control of Power Quality in Inverter-Based Distributed Generation. In Proceedings of the 28th Annual Conference of the IEEE Industrial Electronics Society, Sevilla, Spain, 5–8 November 2002; Volume 2, pp. 1185–1189. [CrossRef]
14. Liu, W.; Chung, I.-Y.; Liu, L.; Leng, S.; Cartes, D.A. Real-time particle swarm optimization based current harmonic cancellation. *Eng. Appl. Artif. Intell.* **2011**, *24*, 132–141. [CrossRef]
15. Zhang, M.; Wang, J.; Zhang, S.; Gao, L.; Guo, X.; Chen, L.; Xu, Y. Harmonic Resonance Analysis and Impedance Remodeling Method of Multi-Inverter Grid-Connected System. *Electronics* **2023**, *12*, 3684. [CrossRef]
16. Shen, X.; Liu, J.; Liu, Z.; Gao, Y.; Leon, J.I.; Vazquez, S.; Wu, L.; Franquelo, L.G. Sliding-Mode Control of Neutral-Point-Clamped Power Converters with Gain Adaptation. In *IEEE Transactions on Power Electronics*; IEEE: New York, NY, USA, 2024; Volume 39, pp. 9189–9201. [CrossRef]
17. Kalair, A.; Abas, N.; Kalair, A.R.; Saleem, Z.; Khan, N. Review of harmonic analysis, modeling and mitigation techniques. *Renew. Sustain. Energy Rev.* **2017**, *78*, 1152–1187. [CrossRef]
18. Jha, K.; Shaik, A.G. A comprehensive review of power quality mitigation in the scenario of solar PV integration into utility grid, e-Prime—Advances in Electrical Engineering. *Electron. Energy* **2023**, *3*, 100103. [CrossRef]
19. Micallef, A. Review of the current challenges and methods to mitigate power quality issues in single-phase microgrids. *IET Gener. Transm. Distrib.* **2019**, *13*, 2044–2054. [CrossRef]
20. Michalec, Ł.; Jasinski, M.; Sikorski, T.; Leonowicz, Z.; Jasinski, L.; Suresh, V. Impact of Harmonic Currents of Nonlinear Loads on Power Quality of a Low Voltage Network–Review and Case Study. *Energies* **2021**, *14*, 3665. [CrossRef]
21. Zagrajek, K.; Paska, J.; Klos, M.; Pawlak, K.; Marchel, P.; Bartecka, M.; Michalski, L.; Terlikowski, P. Impact of Electric Bus Charging on Distribution Substation and Local Grid in Warsaw. *Energies* **2020**, *13*, 1210. [CrossRef]
22. Baraniak, J.; Starzynski, J. Modeling the Impact of Electric Vehicle Charging Systems on Electric Power Quality. *Energies* **2020**, *13*, 3951. [CrossRef]
23. Miller, P.; Wancerz, M. Wpływ sposobu wyznaczania parametrów linii 110 kV na dokładność obliczeń sieciowych. *Prz. Elektrotech.* **2014**, *4*, 187–190.
24. ABB. Technical Guide No. 6 Guide to Harmonics with AC Drives. Available online: https://library.e.abb.com/public/bc35ffb4386c4c039e3a8ec20cef89c5/Technical_guide_No_6_3AFE64292714_RevF_EN.pdf (accessed on 10 September 2024).

25. Dziennik Ustaw Rzeczypospolitej Polskiej, Rozporzadzenie Ministra Klimatu I Srodowiska z dnia 22 Marca 2023 r. w Sprawie Szczegolowych Warunkow Funkcjonowania Systemu Elektroenergetycznego. Available online: https://isap.sejm.gov.pl/isap.nsf/download.xsp/WDU20230000819/O/D20230819.pdf (accessed on 10 September 2024).
26. Energa Operator. Available online: https://energa-operator.pl/przylaczenie-do-sieci/definicje (accessed on 10 September 2024).
27. IEEE. *IEEE Recommended Practice and Requirements for Harmonic Control in Electric Power Systems*; IEEE: New York, NY, USA, 2014; Volume 519.
28. Villanueva, I.; Vazquez, N.; Vaquero, J.; Hernandez, C.; Lopez, H.; Osorio, R. L vs. LCL Filter for Photovoltaic Grid-Connected Inverter: A Reliability Study. *Int. J. Photoenergy* **2020**, *2020*, 7872916. [CrossRef]
29. EMC/EMI Filter for PV Inverters. Available online: https://www.schaffner.com/product/FN2200/Schaffner_datasheet_FN2200.pdf (accessed on 10 September 2024).
30. Dash, D.K.; Sadhu, P.K. A Review on the Use of Active Power Filter for Grid-Connected Renewable Energy Conversion Systems. *Processes* **2023**, *11*, 1467. [CrossRef]
31. Enea Operator, Instrukcja Ruchu I Eksploatacji Sieci Dytaustrybucyjnej. 2024. Available online: https://www.operator.enea.pl/bricks/instrukcja-ruchu (accessed on 15 August 2024).
32. Fiorina, J.N. *Inverters and Harmonics (Case Studies of Non-Linear Loads)*; Cahier Technique Merlin Gerin: Lyon, France, 1993.
33. Lowczowski, K. Wplyw odksztalcenia napiecia na przepiecia ziemnozwarciowe. *Prz. Elektrotech.* **2018**, *94*, 41–44. [CrossRef]
34. Zheng, R.; Bollen, M. *Harmonic Resonances Associated with Wind Farms*; Lules University of Technology: Skelleftes, Sweden, 2010.
35. Rozporządzenie Komisji (UE) 2016/631 z dnia 14 Kwietnia 2016 r. Ustanawiające Kodeks Sieci Dotyczący Wymogów w Zakresie Przyłączenia Jednostek Wytwórczych do Sieci. Available online: https://eur-lex.europa.eu/legal-content/PL/TXT/PDF/?uri=CELEX:32016R0631&from=SK (accessed on 10 September 2024).
36. Bollen, M.; Mousavi-Gargari, S.; Bahramirad, S. Harmonic resonances due to transmission-system cables. *Renew. Energy Power Qual. J.* **2014**, 712–716. [CrossRef]
37. *PN-EN 50160:2023-10*; Supply Voltage Parameters in Public Power Grids. Polish Committee for Standardization: Warsaw, Poland, 2023.
38. Tauron Dystrybucja, Kryteria Techniczne Oceny Możliwości Przyłączenia Jednostek Wytwórczych, Magazynów Energii Oraz OSDn do Sieci Dystrybucyjnej Średniego Napięcia, Krakow. 2014. Available online: https://www.tauron-dystrybucja.pl/przylaczenie-do-sieci/przylaczenie/zrodla-wytworcze (accessed on 15 August 2024).
39. Enea Operator, Kryteria Techniczne Oceny Mozliwosci Przylaczenia Jednostek Wytworczych, Magazynow Energii oraz OSDn do Sieci Dystrybucyjnej Sredniego Napiecia ENEA Operator sp. z o.o. August 2023. Available online: https://www.operator.enea.pl/uploads-ev2/Operator/ospolce/raporty/Kryteria%20przy%C5%82%C4%85czania%20obiekt%C3%B3w%20do%20sieci%20SNna%20stronk%C4%99.pdf (accessed on 15 August 2024).
40. Available online: https://forum.huawei.com/enterprise/en/Harmonic-spectrum-for-SUN2000-100-KTL-M1-inverter/thread/667249710882242560-667213868771979264 (accessed on 10 September 2024).
41. Available online: https://pgeenergetykakolejowa.pl/kolej/1-3-zielona-kolej (accessed on 10 September 2024).
42. PLK SA, Standardy Techniczne, Tom IV Urzadzenia Trakcji Elektrycznej/Elektroenergetyki Trakcyjnej. Available online: https://www.plk-sa.pl/files/public/user_upload/pdf/Akty_prawne_i_przepisy/Standardy_techniczne/11.04.2022/3_TOM_IV.pdf (accessed on 10 September 2024).
43. Rojek, A. *Jakosc Energii Elektrycznej w Kolejowym Systemie Elektroenergetycznym*; Z. 137/138; Problemy Kolejnictwa: Warszawa, Polnad, 2003.
44. Kuznetsov, V.; Hubskyi, P.; Rojek, A.; Udzik, M.; Lowczowski, K. Progress and Challenges Connected with the Integration of Renewable Energy Sources with Railway Distribution Networks. *Energies* **2024**, *17*, 489. [CrossRef]
45. Lowczowski, K.; Nadolny, Z. Voltage Fluctuations and Flicker in Prosumer PV Installation. *Energies* **2022**, *15*, 2075. [CrossRef]
46. Power Monitors. *White Paper: Key Issues with DG/DER Monitoring*; Power Monitors: Mt Crawford, VA, USA, 2016.
47. DigSilent. *Power Factory User Manual*; DigSilent: Gomaringen, Germany, 2023.
48. Han, G.; Sang, S.; Cai, X. Impedance analysis and stabilization control of the LCL-type wind power inverter under weak grid conditions. *J. Renew. Sustain. Energy* **2018**, *10*, 035301. [CrossRef]
49. Andrieu, C.; Dauphant, E.; Boss, D. A Frequency Dependent Model For a MV/LV Transformer. In Proceedings of the International Conference on Power System Transients, Budapest, Hungary, 20–24 June 1999.
50. Bula, D.; Grabowski, D.; Maciazek, M. A Review on Optimization of Active Power Filter Placement and Sizing Methods. *Energies* **2022**, *15*, 1175. [CrossRef]
51. Bollen, M.; Yang, K. Harmonics—Another Aspect of the Integration between Wind-Power Installations and the Grid. In Proceedings of the 22nd International Conference on Electricity Distribution, Stockholm, Sweden, 10–13 June 2013.
52. Qian, G.; Wang, Q.; He, S.; Dai, W.; Wei, N.; Zhou, N. Harmonic Modeling and Analysis for Parallel 12-Pulse Rectifier under Unbalanced Voltage Condition in Frequency-Domain. *Energies* **2022**, *15*, 3946. [CrossRef]
53. Parmar, J. Understanding Vector Group of Transformer. 2012. Available online: https://electrical-engineering-portal.com/understanding-vector-group-transformer-1 (accessed on 15 August 2024).
54. Zainuddin, M.; Putra Surusa, F.E.; Syafaruddin, S.; Manjang, S. Constant Power Factor Mode of Grid-Connected Photovoltaic Inverter for Harmonics Distortion Assessment. *Int. J. Renew. Energy Res.* **2020**, *10*, 1525–1535. [CrossRef]

55. ABB. Instrument Transformers, Technical Information and Application Guide. 2004. Available online: https://library.e.abb.com/public/e2462bd7f816437ac1256f9a007629cf/ITTechInfoAppGuide.pdf (accessed on 15 August 2024).
56. CIGRE/CIRED, Guidelines for Power Quality Monitoring WG C4.112 TECHNICAL BROCHURE 596. 2014. Available online: https://www.e-cigre.org/publications/detail/596-cigrecired-guidelines-for-power-quality-monitoring-measurement-locations-processing-and-presentation-of-data.html (accessed on 15 August 2024).
57. Lumbreras, D.; Gálvez, E.; Collado, A.; Zaragoza, J. Trends in Power Quality, Harmonic Mitigation and Standards for Light and Heavy Industries: A Review. *Energies* **2020**, *13*, 5792. [CrossRef]
58. Olczykowski, Z.; Starula, P. Power quality in the circuits of traction substation. *J. Civ. Eng. Transp.* **2023**, *5*, 29–52. [CrossRef]
59. Kazem, H.A. Harmonic Mitigation Techniques Applied to Power Distribution Networks. *Adv. Power Electron.* **2013**, *2013*, 591680. [CrossRef]

energies

MDPI

Article

Multi-Fault-Tolerant Operation of Grid-Interfaced Photovoltaic Inverters Using Twin Delayed Deep Deterministic Policy Gradient Agent

Shyamal S. Chand [1,*], Branislav Hredzak [1,*] and Maurizio Cirrincione [2,†]

[1] School of Electrical Engineering and Telecommunications (EET), University of New South Wales (UNSW), Sydney, NSW 2052, Australia

[2] Université Marie et Louis Pasteur, UTBM, CNRS, Institut FEMTO-ST, F-90010 Belfort, France; maurizio.cirrincione@usp.ac.fj

[*] Correspondence: shyamal.chand@unsw.edu.au (S.S.C.); b.hredzak@unsw.edu.au (B.H.)

[†] School of Information Technology, Engineering, Mathematics, and Physics (STEMP), University of the South Pacific (USP), Laucala Campus, Suva, Fiji.

Abstract: The elevated penetration of renewable energy has seen a significant increase in the integration of inverter-based resources (IBRs) into the electricity network. According to various industrial standards on interconnection and interoperability, IBRs should be able to withstand variability in grid conditions. Positive sequence voltage-oriented control (PSVOC) with a feed-forward decoupling approach is often adopted to ensure closed-loop control of inverters. However, the dynamic response of this control scheme deteriorates during fluctuations in the grid voltage due to the sensitivity of proportional–integral controllers, the presence of the direct- and quadrature-axis voltage terms in the cross-coupling, and predefined saturation limits. As such, a twin delayed deep deterministic policy gradient-based voltage-oriented control (TD3VOC) is formulated and trained to provide effective current control of inverter-based resources under various dynamic conditions of the grid through transfer learning. The actor–critic-based reinforcement learning agent is designed and trained using the model-free Markov decision process through interaction with a grid-connected photovoltaic inverter environment developed in MATLAB/Simulink® 2023b. Using the standard PSVOC method results in inverter input voltage overshoots of up to 2.50 p.u., with post-fault current restoration times of as high as 0.55 s during asymmetrical faults. The designed TD3VOC technique confines the DC link voltage overshoot to 1.05 p.u. and achieves a low current recovery duration of 0.01 s after fault clearance. In the event of a severe symmetric fault, the conventional control method is unable to restore the inverter operation, leading to integral-time absolute errors of 0.60 and 0.32 for the currents of the *d* and *q* axes, respectively. The newly proposed agent-based control strategy restricts cumulative errors to 0.03 and 0.09 for the *d* and *q* axes, respectively, thus improving inverter regulation. The results indicate the superior performance of the proposed control scheme in maintaining the stability of the inverter DC link bus voltage, reducing post-fault system recovery time, and limiting negative sequence currents during severe asymmetrical and symmetrical grid faults compared with the conventional PSVOC approach.

Keywords: reinforcement learning; twin delayed deep deterministic policy gradient; data-driven control; grid fault ride-through; voltage-oriented control; asymmetrical and symmetrical faults

 https://doi.org/10.3390/en18010044

1. Introduction

The revolution in renewable energy is evolving substantially with the advancement of energy policies, the development of infrastructure, and the integration of continually improving power technologies. Renewable energy sources (RES), such as photovoltaics (PVs), wind energy conversion systems (WECSs), hydroelectric power stations (HPSs), and fuel cells (FCs), combined with various forms of energy storage devices, such as batteries and flywheels, are becoming increasingly dominant components of the power network. Power generation using distributed energy resources (DER) requires optimal deployment of extraction and control techniques for large-scale integration and transmission. Improving the stability and increasing the operational reliability of these microgrids is a topic of significance among many researchers around the world today [1,2].

Microgrids (MGs) generally operate in two main modes of operation: islanded or standalone and grid-connected. In islanded mode, central or decentralized control can be used to provide set voltage and frequency points, such as droop control or master–slave methods [3]. However, when coupled to the upstream electricity network, the grid-connected inverter is programmed to follow the voltage and frequency of the grid using control methods such as voltage-oriented control [2]. Faults in power systems can generally be categorized as symmetrical and asymmetrical faults. An asymmetric fault occurs when the voltage phases of the grid are unequally affected, creating an imbalance in the system, resulting in different fault currents in each phase. Due to the unbalanced nature of asymmetrical faults, they are more challenging to model and analyze. However, a symmetric fault affects the three phases of the grid voltage equally, causing the fault current in all three phases to be identical in magnitude and phase angle. Symmetrical faults are typically the most severe type and generate the highest fault currents, but occur less frequently than asymmetrical faults.

Most MGs connected to the grid are not of special concern for the conventional power network, as they cannot influence its operating parameters (voltage and frequency). However, when several MGs are integrated into a low- or medium-voltage network, the stability and reliability of the larger grid can be compromised [4]. To ensure standard operation, virtual inertias are often incorporated into droop control loops, allowing the voltage and frequency magnitudes to be adjusted according to the active and reactive power generated by the inverter using low-bandwidth communication. This approach can effectively prevent voltage sagging while giving flexibility to transition between island-connected and grid-connected operations through a phase-locked loop (PLL). Droop control has a simple implementation; however, there are a few downfalls, such as a slow transient response, circulating current between converters, and an imprecise power distribution during grid faults [5]. Furthermore, most standalone industrial grid-interfaced power converters are now required to have multiple fault ride-through (FRT) capabilities [6]. For instance, IEEE Std. 2800-2022—"IEEE Standard for Interconnection and Interoperability of Inverter-Based Resources (IBRs) Interconnecting with Associated Transmission Electric Power Systems"—clearly specifies performance requirements for dynamic active and reactive power support under abnormal frequency or voltage, negative sequence current injection, and low voltage or high voltage ride-through operation. As such, the IBR plant must be able to switch between multiple modes of operation when required [7].

Consequently, in the event of voltage imbalance faults at a particular node, the effect is transmitted to power transformers, converters, and the larger microgrid, eventually affecting the synchronization of inverters and sensitive loads. The protection system of MGs, whether in grid-connected or islanded operation, must respond to abnormal grid conditions. In comparison, the fault current in the connected mode of the grid is 10–50 times the full load per unit current due to the low impedance of the utility network, but in the islanded mode it is only 1.2–5 times higher [8]. As an ancillary service, negative sequence currents need to be injected to compensate for any voltage imbalances at the node. A complex proportional resonant controller can be designed to achieve this by sharing negative sequence currents between participating inverters connected to an AC microgrid using a communication link that transmits at 10 Hz [9]. However, the communication link is prone to failure and low-bandwidth data transfer can increase communication delays, resulting in deteriorated robustness during faults. In [10], an improved inverter control technique based on a grid impedance estimation technique is proposed using a Newton–Raphson algorithm implemented in the stationary reference frame that takes into account positive sequence phasors as input. This method iteratively adjusts the virtual impedance and stabilizes the inverter under unbalanced and harmonic-inflicted grid voltages. Similarly, another method involves a positive–negative sequence synchroconverter capable of limiting output currents to reduce power oscillations when subjected to disturbances in the grid [11]. When the grid voltages abruptly differ from the nominal operating set points, these control approaches have poor transients and a long settling time.

Moreover, to leverage the operation of smart inverters, low-voltage ride-through (LVRT) capability is essential. An autonomous model predictive controller (AMPC), which is an improvement over the classical MPC, can be adopted for adjustment of the injected active and reactive powers of a grid-connected qausi-Z-source inverter during such scenarios. Compared with conventional MPC, the trial-and-error design stage of the weight factor is replaced with an auto-tuning objective normalization function. An additional benefit of this method is the provision for a transition between maximum power point tracking (MPPT) and FRT operation with ease based on the command of the grid operator or the grid condition [12]. Intrinsically, it can sometimes be difficult to reduce inverter currents during LVRT, as it adjusts the reactive current to support the voltage of the grid [13]. As such, [14] proposes a method to reduce limit fault currents using phase angle adjustment and achieve improved voltage support by embedding the network impedance in the current control loop. In other works, a bridge-type fault current limiter (BFCL) is connected between the distributed energy resources and the main grid at the point of common coupling (PCC) to limit fault currents and provide reactive power support through cooperative control with the voltage source converter [15]. For a switched resistance wind generator (SRWG), an optimal proportional–integral (PI) voltage controller tuning method using the elephant herd algorithm (EHA) is another way to provide LVRT capabilities according to the grid code [16]. Further analysis of other existing grid fault-ride-through techniques is presented in Table 1.

Table 1. Critical analysis of existing fault-ride-through methods.

Ref.	System Architecture	Type of Fault(s)	Effects	Adopted Mitigation Technique	Advantages	Remarks/Limitation
[17]	Grid-connected multi-modular converter	Voltage swell	Cause over-modulation of multi-modular converter leading to distortions in current waveforms and instability	Fundamental zero-sequence voltage injection	Gurrantees even modulation margins in all three phases. Improves multi-modular converter tolerance during high voltage ride-through	No provision for compensating other major types of fault in the network
[18]	Grid-connected three-phase four-wire inverter	Voltage sag (symmetrical and asymmetrical) and short-circuit ratio variation	Significant increase in current harmonics. Significant drop in inverter power at severe sags. Oscillation in DC-link voltage	Constant average active power/constant active current/constant peak current injection	Constant average active power injection can assure operated at rated conditions during faults	Does not offer smooth transition to pre-fault condition. Multiple current references must be specified for fault ride-through

Table 1. *Cont.*

Ref.	System Architecture	Type of Fault(s)	Effects	Adopted Mitigation Technique	Advantages	Remarks/Limitation
[19]	Three-phase energy storage converter	Unbalanced voltage sags	Causes drastic changes in power angle. Disruptions in power supplied to the grid.	Current droop control	Provide reactive power support. Ability to limit power angle deviation. Does not require power calculations in the outer control loop.	Modified control hinders commanding in terms of required power. The battery state of charge is not considered during fault ride-through.
[20]	Modified IEEE 9-bus network with solar, wind, and battery storage	Asymmetrical and symmetrical	Affects stability and reliability of the entire network.	Sliding mode linearization recurrent high-order neural controller trained using an extended Kalman filter	Allows mitigation of nonlinear dynamics due to parameter variation or faults. Reduced divergence in the DC link voltage and active and reactive power during abnormal operating conditions.	The approach does not consider nonlinear loading (inductive or capacitive). The use of a recurrent high-order neural network, as well as a sliding mode controller, complicates the design process.
[21]	Doubly-fed induction generator-based wind turbine systems	Voltage sags, asymmetrical and symmetrical	Increase in stator currents which leads to transient overshoots in rotor currents due to magnetic coupling.	Nonlinear sliding mode controller for bridge-type fault current limiter	To overcome the chattering phenomenon, an exponential reaching law is proposed for the sliding surface design. Current limiting is achieved during the fault period.	There is a lack of stability analysis as operating margins of the proposed control technique are not specified. Does not account for scalability issues that may arise for larger wind farms.
[22]	Standalone inverter	Asymmetrical and symmetrical	Can damage the semiconductor switches. May cause the inverter to shut down, depriving the power supply to critical loads.	Finite control set model predictive control	High-speed fault detection. Ability to limit total harmonic distortion of currents during faults. Fast recovery to nominal mode of operation.	Requires operation in dual modes, nominal and fault conditions. The objective functions need to be well defined to suit the system. Relies on accurate fault modeling and system parameters.
[23]	Doubly-fed induction generator-based wind turbine systems	Asymmetrical and symmetrical	Current surge in stator and armature winding. Comprises drivetrain integrity due to induced torque transients.	Event-triggered sliding mode control combined with supplementary energy storage support interfaced via an alternate active bridge converter	Effectively regulates the DC link voltage. Low use of communication channels.	The use of supercapacitor banks, disturbance observer, and additional dual active bridge converters complicates the design process. Requires supplementary energy sources.

Recent advances in computational power density have led to the rise of numerous applications of machine learning (ML) or artificial intelligence (AI) in demand-side energy management, distribution networks, and improving the stability of power converters in MGs [24]. In an effort to advance the application of the Markovian jump stochastic neural network, a framework is particularly necessary to model abrupt variation of the system architecture and parameters in the context of stability and immunity to input disturbances [25]. The analysis of the Cohen–Grossberg bidirectional associative memory neural network comprising time delay can be reconfigured to inverter control under uncertainties [26]. In addition, synchronization of control systems techniques with sampled data prior to additive delay has also been investigated in detail [27]. Such approaches can be applied to improve the reliability and synchronization of multiple grid-connected inverters for optimal power distribution. In addition, fast and accurate fault detection often relies on appropriate data selection and feature extraction methods. High-quality hybrid feature selection based on correlated items applied in the context of classification can be used to identify critical aspects of inverter performance by detecting voltage variation or unpredictable changes in load [28–30]. Through careful evaluation and reduction of the feature space with a focus on the most relevant data points, the computational cost for training and deploying sophisticated algorithms such as the adaptive neural fuzzy interference system for strategic decision-making in the energy sector has also been progressively advanced due to improved overall reliability in dynamic environments [31, 32].

In hierarchical control, machine learning-based approaches have induced significant improvements in power system performance at all levels (primary, secondary, and tertiary) and have been shown to be effective in load forecasting, generation prediction, and fault diagnosis [33]. In a low-voltage grid, LVRT and constant reactive current injection during voltage sags can be achieved through online detection of grid abnormalities based on wavelet transforms and shallow neural networks and the resultant adjustments in the power references. The fault identification technique reported a classification precision of 98.3% with PCC voltage and frequency as input data [34]. Terminal voltage regulation is crucial during transitions between grid-connected and standalone operations. A DG inverter can be controlled using a deep neural network (DNN) trained offline in conjunction with a proportional integral derivative (PID) controller to ensure a smooth transition. Additionally, a feedforward loop can be incorporated to mitigate harmonics for nonlinear loads [35]. Battery energy storage systems (BESS) in MGs are commonly used for voltage support and fault compensation. To address the issue of short circuit-induced voltage sags in power lines using BESS in hybrid MG, a recurrent wavelet petri fuzzy neural network (RWPFNN) can be used in the control loop for LVRT and fast voltage restoration [36]. Other studies have employed neural networks with an extended Kalman filter to handle the control of doubly fed induction generators during sensor failure on controlled variables of back-to-back power converters [37].

In addition, reinforcement learning (RL) techniques have been extensively utilized to design better secondary control (power and frequency) of power inverters. For unstructured MGs with tidal power units and vehicle-to-grid connections, load frequency control is extremely important to restore stability. A fractional gradient descent based on a fuzzy logic controller can be implemented as the main frequency regulator with a deep deterministic policy gradient (DDPG) composed of an actor–critic architecture to produce additional control signals for frequency stabilization [38]. In [39], a quantum neural network (QNN) is formulated to cooperatively control the frequency of an isolated MG. It combines DRL and quantum machine learning to minimize the number of parameters and minimize training requirements for a 13-bus network. A data set containing the frequency response of unoptimized linear controllers is initially required to train the network through supervised learning. In [40], a frequency regulation approach is proposed using deep reinforcement learning (DRL), considering both the frequency performance and the economic operational limits of the MG. The twin delayed deep-deterministic policy gradient (TD3PG) algorithm is implemented based on historical data for adaptive selection of the best commands to ensure frequency stability.

Moreover, a virtual inertia emulation technique is established in [41] using the twin delayed deep deterministic policy gradient (TD3PG) reinforcement learning algorithm for frequency regulation in weak grids with reduced inertia and damping capability. Verification shows the ability of such controllers to better stabilize frequency in grid-connected and standalone MGs. In another study, a Kuramoto-based consensus algorithm is used to implement multi-agent reinforcement learning to handle frequency deviations. In [42], an adaptive controller based on TD3PG is proposed to replace virtual synchronous generators

(VSG) in multilevel modular converter control. This approach reported better system strength and higher resistance to disturbances than conventional VSGs. Reinforcement learning algorithms can also be used in designing parameters of existing controllers such as proportional–integral, proportional–resonant, and fractional-order proportional–integral regulators, etc. In [43], a deep Q network is adopted to adjust the parameters of the proportional–integral controllers for a dual active bridge converter according to the modulation phase shifted angles. This allows real-time mapping of phase-shift angles to the necessary control parameters for stability.

Furthermore, reinforcement learning is applied to design Q-learning and TD3PG current controllers in variable-speed doubly fed induction generators for the rotor and grid-side inverters [44]. The performance of both agents is evaluated under dynamic conditions, leading to the conclusion that Q-learning agents may not perform well under significant disturbances compared to TD3 agents. In another study, the TD3PG agent is utilized again to attain superior performance of permanent magnet synchronous machines (PMSM) governed using the classical field-oriented control scheme (FOC). The agent works in tandem with ESO-type speed observers, which allows sensorless FOC control through online estimation of the machine speed. Multiple neural networks are used to obtain the load torque with the TD3 agent performing a real-time estimated speed correction. The results indicate a better reference tracking capability with the use of the RL TD3PG agent [45]. DDPG agents have also been implemented in a similar way and have been trained to work in parallel with the outer loop speed controller designed using sliding mode control (SMC) to improve the load torque rejection capability in PMSM [46]. In this work, it is shown that RL agents have the ability to directly manipulate control signals to obtain the desired performance without a high computational or mathematical burden. Although there are extensive applications of RL agents and other intelligent control methods in machine control [47], there is limited literature that focuses on designing robust and adaptive artificial-based controllers for inverters to withstand adverse grid disturbances.

With the increasing penetration of renewable energy into the electrical network and the increasing demand for resilient inverter-based technology, there is a growing need to develop advanced control methods to stabilize MGs under various dynamic conditions. The expanding nature of the incorporation of DERs with utility and standard ancillary practices makes the operation and control of voltage, frequency, and FRT mechanisms more challenging. AI has the capability to bring about a new frontier in power engineering with promising advances in processing density and improving deep learning algorithms [48]. If implemented appropriately, a self-governing, highly intelligent controller capable of withstanding various grid abnormalities is definitely a possibility in MG applications. Although the above-mentioned model-free control methods deliver promising results, most depend on the type of data used during the training process. As such, considering deep learning methods that do not require an exact MG model for efficient FRT and have provision for transfer learning is desirable [49]. Evidently, during grid faults, such as voltage swells, dips, unbalances, frequency disturbances, harmonics, asymmetrical, and symmetrical faults, the current control loop on IBRs is important in regulating power quality and ensuring stability. Most of the methods available in the literature often target compensating for one or a few types of fault using a particular approach, while some techniques require switching between multiple control modes depending on the induced fault. Some classical FRT controllers perform additional computations to generate negative sequence components for fault compensation; however, this requires exact plant modeling, which makes the design process monotonous.

In addition, the current loop with cross-coupling terms in conventional voltage-oriented control in inverter-based MGs causes instabilities during abnormal grid conditions, which are attributed to poor proportional–integral controller dynamics and saturation limits. Based on the aforementioned constraints and inspired by reinforcement learning techniques, a novel twin delayed deep deterministic policy gradient-based reinforcement learning control approach is proposed to provide effective FRT operation and ensure optimal current injection from RES. The main contributions of this research article are as follows.

- A novel twin delayed deep-deterministic policy gradient agent is formulated and trained to generate the voltages of the direct and quadrature axes in real time for resilient operation of the inverter considering abnormal conditions of the grid based on a set of observed states from the environment developed in MATLAB/Simulink®.
- A framework for training and deploying fully AI-based current controllers has been established to achieve the optimal response of grid-interfaced inverters using a model-free Markov decision process guided through a suitable reward function, thus eliminating the need to design controllers based on uncertain operating parameters and an accurate system model.
- Initially, the agent is trained for a single asymmetric phase-to-phase fault and then applied to improve the inverter response for other asymmetric and symmetric grid faults using transfer learning. This feature allows the controller to continuously adapt to variations in operating conditions and be resilient despite deviations in grid voltages without requiring complex deep neural network architectures and minimizes computational or training costs.

Such control approaches have not been studied in depth in the field of inverter fault-ride operation. Therefore, this article focuses on developing the actor–critic-based data-driven control strategy, which has been thoroughly verified under asymmetric faults (L-L and L-L-G) and symmetric faults (L-L-L). The proposed fault ride-through approach is also compared with the standard positive sequence voltage-oriented control technique.

The organization of this paper is as follows. Section 1 provides a detailed background and an extensive review of the literature on the subject. Section 2 covers the description of the power conversion system, while Section 3 describes the traditional positive sequence voltage-oriented control method. The novel design process for the twin delayed deep deterministic policy gradient agent, intended for grid fault ride-through, is detailed in Section 4. This includes specifics on the architecture of the deep neural network, the formulation of the reward function, and the training procedure. Section 5 presents a comparative evaluation of the proposed fault ride-through approach against the conventional control method. The research considers three distinct fault scenarios, with the DC link voltage response, post-fault recovery time, and cumulative integral-time error analysis serving as primary evaluation metrics. Section 6 discusses the limitations of the study and future research opportunities, and the article is concluded in Section 7.

2. System Description

The IEEE 13-bus network is a benchmark test case system used for research to implement new concepts in electricity generation, transmission, and distribution. For modeling purposes, single AC generators are represented as the swing voltage source with a fixed X/R ratio of 10. The classical 13 bus system has multiple predetermined single- and three-phase loads connected to all buses. For this study, an inverter is connected to an existing 4160/415 V transformer between buses 633 and 634 to facilitate the integration of a 100 kW solar system as shown in Figure 1.

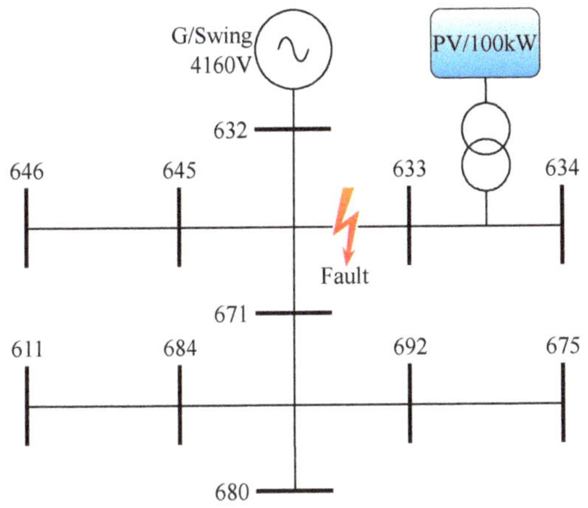

Figure 1. Single line diagram of the IEEE 13 bus network with an interconnected solar system.

In the primary power conversion circuit (Figure 2), a three-phase two-level inverter is connected to the IEEE 13 bus network between buses 633 and 634 with an LCL-type current harmonic filter. A DC/DC boost converter is used to increase the voltage of the PV panels to charge the DC link capacitors. The scheme follows a centralized inverter topology in which a single inverter encompasses the entire power conversion circuit and controls the power using a single DC link [50]. The incremental conductance (INC) maximum power point tracking algorithm is used for optimum power extraction. INC requires measurement of the array voltage and current to calculate the instantaneous conductance and update the reference duty cycle accordingly. It has the benefit of faster adaptation to rapidly changing climate conditions and fewer oscillations when attaining the MPP. The size of the perturbation voltage step is set to 1×10^{-3} V to minimize tracking error after reaching the desired operating point. Other parameters of the power conversion circuit are stated in Table 2.

Table 2. Power conversion circuit and controller specifications.

Parameter	Symbol	Value	Unit
Grid frequency	ω	60	Hz
Rated PV power	P_n	100	kW
Utility line voltage	E_{LL}	415	V
Rated DC bus voltage	V_{dc}	700	V
Inverter switching frequency	F_{sw}	15	kHz
Input capacitor	C_{in}	10	μF
Boost Inductance	L	2.4	mH
DC link capacitor	C_{dc}	2000	μF
Boost switching frequency	F_b	5	kHz
Damping resistor	R_d	0.4	Ω
Filter capacitance	C_f	77	μF
Internal resistance of L_g	R_g	18	mΩ
Internal resistance of L_i	R_i	28	mΩ
Grid side filter inductance	L_g	177	μH
Inverter side filter inductance	L_i	280	μH
Current controller gains		$K_p = 0.6$ $K_i = 243$	
DC link controller gains		$K_p = 1.35$ $K_i = 395$	

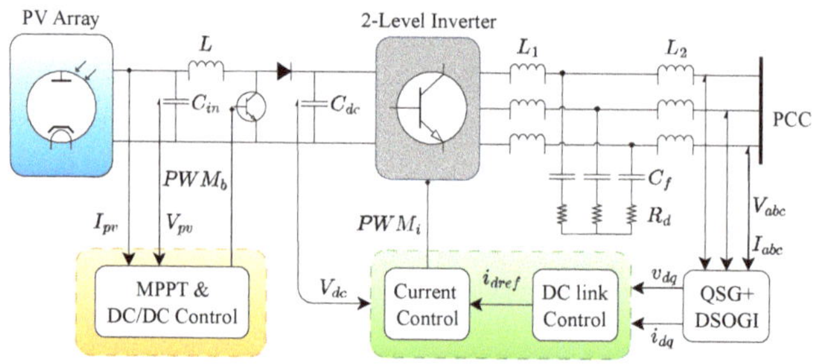

Figure 2. Dual-stage photovoltaic power conversion circuit.

For this study, as a prerequisite, voltages and currents are normalized to per unit values. The use of per-unit (p.u.) measurements significantly enhances scalability, making it easier to design, analyze, and operate systems of varying sizes and complexities with various control techniques. Using a base apparent power (S_{base}) of 100 kVA and a nominal AC phase-to-phase voltage (V_{AC}) of 415 V, the voltage and current bases can be specified as

$$V_{AC(base)} = \sqrt{\frac{2}{3}} \times V_{AC} \qquad I_{AC(base)} = \sqrt{\frac{2}{3}} \times \frac{S_{base}}{V_{AC}} \tag{1}$$

Thus, the instantaneous voltages and currents of the grid at each phase measured at PCC are independently converted to their equivalents per unit using Equation (2). The transformed voltages and currents are then fed to a phased-locked loop to extract their d and q counterparts in normalized form.

$$V_{abc}(p.u.) = \frac{V_{PCC}}{V_{AC(base)}} \qquad I_{abc}(p.u.) = \frac{I_{PCC}}{I_{AC(base)}} \tag{2}$$

Grid synchronization is crucial to maintaining quality power injection into the grid despite the volatile nature of the grid. As such, the double second-order generalized integrator phase-locked loop (DSOGI-PLL) is often used to synchronize the inverter voltages with the main grid. DSOGI-PLL has established superior performance in stabilizing the inverter under numerous disturbances of the grid [51,52]. It is based on an instantaneous symmetric component method with two inbuilt quadrature signal generators (QSG) to filter out additional harmonics present in the grid voltages in the first stage. Each QSG produces two voltages on the quadratic axis (90° apart) that can be recombined to form positive and negative stationary reference frame grid components using Clarke's transformation of the grid voltage at PCC (v_α and v_β). Since in the case of a grid, the frequency is constant under normal operating conditions, the filter bandwidth is essentially dependent on the damping factor (K_γ), which is generally chosen as $\sqrt{2}$. This makes it also suitable for variable-frequency applications [53].

Taking $v_\alpha{}'$, $v_\beta{}'$, $qv_\alpha{}'$, and $qv_\beta{}'$ as QSG output, the decoupled positive and negative sequence components of the grid voltage can be obtained by using the sequence calculation in Equation (3). Thus, the corresponding positive and negative voltages on the d and q axes can be easily obtained by Park's transformation of the respective stationary reference frame voltages with ωt as the estimated grid angle according to Equations (4) and (5). v_d^+ gives the amplitude of the input voltage, while v_q^+ gives information on the phase error. A PI controller is used as a loop filter to maintain v_d^+ at zero and send it to the voltage-controlled oscillator represented by the integrator, thus allowing the phase angle of the input signal to be estimated [54]. Transformation of three-phase voltage and currents from their natural states (abc) into (dq) synchronous rotating frame DC quantities allows decoupled control of grid-connected inverters in both positive and negative domains.

$$\begin{aligned} v_\alpha^+ &= \tfrac{1}{2}(v_\alpha{}' - qv_\beta{}') \quad \& \quad v_\beta^+ = \tfrac{1}{2}(v_\beta{}' + qv_\alpha{}') \\ v_\alpha^- &= \tfrac{1}{2}(v_\alpha{}' + qv_\beta{}') \quad \& \quad v_\beta^- = \tfrac{1}{2}(v_\beta{}' - qv_\alpha{}') \end{aligned} \tag{3}$$

$$v_d^+ = v_\alpha^+ \sin(\omega t) - v_\beta^+ \cos(\omega t) \quad \& \quad v_q^+ = v_\alpha^+ \cos(\omega t) + v_\beta^+ \sin(\omega t) \tag{4}$$

$$v_d^- = v_\alpha^+ \sin(-\omega t) - v_\beta^+ \cos(-\omega t) \quad \& \quad v_q^- = v_\alpha^+ \cos(-\omega t) + v_\beta^+ \sin(-\omega t) \tag{5}$$

The approach can also be applied to the measured grid currents at PCC to obtain its positive and negative sequence components. With i_α and i_β as inputs to the dual QSG, the corresponding outputs $i_\alpha{}'$, $i_\beta{}'$, $qi_\alpha{}'$, and $qi_\beta{}'$ can be recombined and produce the positive and negative current sequence components required for control purposes.

$$\begin{aligned} i_\alpha^+ &= \tfrac{1}{2}(i_\alpha{}' - qi_\beta{}') \quad \& \quad i_\beta^+ = \tfrac{1}{2}(i_\beta{}' + qi_\alpha{}') \\ i_\alpha^- &= \tfrac{1}{2}(i_\alpha{}' + qi_\beta{}') \quad \& \quad i_\beta^- = \tfrac{1}{2}(i_\beta{}' - qi_\alpha{}') \end{aligned} \tag{6}$$

$$i_d^+ = i_\alpha^+ \sin(\omega t) - i_\beta^+ \cos(\omega t) \quad \& \quad i_q^+ = i_\alpha^+ \cos(\omega t) + i_\beta^+ \sin(\omega t) \tag{7}$$

$$i_d^- = i_\alpha^+ \sin(-\omega t) - i_\beta^+ \cos(-\omega t) \quad \& \quad i_q^- = i_\alpha^+ \cos(-\omega t) + i_\beta^+ \sin(-\omega t) \tag{8}$$

3. Conventional Voltage-Oriented Control

The positive sequence voltage-oriented control (PSVOC) is most famously used to interface photovoltaic inverters with the grid. It has two cascaded control loops; the inner loop regulates the inverter currents on the d and q axes, and the outer loop ensures the stability of the DC link voltage, as shown in Figure 3. Normally, the DC link voltage reference is kept constant, and a PI controller is used to generate the direct axis current reference that corresponds to the net active power to be injected by the inverter into the grid [55]. Transient conditions and variations in the DC bus voltage are regulated by the capacitor's charge-and-discharge process. In a grid-connected situation, the voltage can fluctuate due to changes in solar irradiation or temperature levels of the photovoltaic array, as well as oscillations in the AC power caused by grid imbalances.

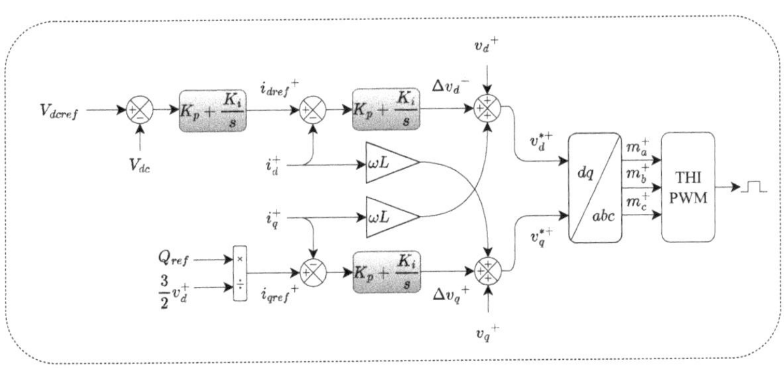

Figure 3. Conventional positive sequence voltage oriented control (PSVOC).

Injection of active power into the grid can be indicated by an increase in three-phase currents at the PCC, assuming that the grid voltage remains steady during normal operation. The change in DC link voltage is determined by the power balance between the solar panels and the currents injected into the grid. The reactive power supply can be commanded directly by specifying the required amount and calculating the current on the quadrature axis. The error for both currents can be processed by inner-loop dual PI controllers independently. Typically, the interconnecting filter transfer function is used to tune the proportional (K_p) and integral (K_i) controller gains. The final inverter reference voltage is the combined output of the current PI controllers, the grid voltages, and the cross-coupling terms [56].

$$\begin{aligned} v_d^{*+} &= \left(K_p + \tfrac{K_i}{s}\right)\left(i_{dref}^{+} - i_d^{+}\right) + \omega L i_q^{+} + v_d^{+} \\ v_q^{*+} &= \left(K_p + \tfrac{K_i}{s}\right)\left(i_{qref}^{+} - i_q^{+}\right) + \omega L i_d^{+} + v_q^{+} \end{aligned} \tag{9}$$

Decoupling allows independent control of active and reactive powers for grid-connected inverters. According to instantaneous power theory, the active and reactive powers on the AC side per unit can be evaluated using $P^{+} = (v_d^{+} i_d^{+} + v_q^{+} i_q^{+})$ and $Q^{+} = (v_q^{+} i_d^{+} - v_d^{+} i_q^{+})$, respectively. It is evident that the active and reactive power produced by the inverter is directly proportional to the direct- and quadrature-axis currents in the synchronous reference frame. Therefore, for the unity power factor, the reference reactive power (Q_{ref}) can be kept zero. The PI controllers employ a fixed frequency free-running carrier signal, resulting in a well-bounded current total harmonic distortion (THD) with constant frequency inverter operation. This aids in instantaneous takeover of current control and has the advantage of eliminating the effect of DC side ripple on the phase currents of the inverter.

The third harmonic injection pulse width modulation (THIPWM) strategy is highly desirable for producing IGBT gate pulses, as it can help improve inverter performance under conditions of low DC link voltage. In THIPWM, a sine wave with a higher oscillating frequency is added to the modulating signals, leading to a reduction in the peak of the resulting modulating signal (m_a^{+}, m_b^{+}, and m_c^{+}). With v_a^{+}, v_b^{+}, and v_c^{+} as the amplitude of the modulating signals of each phase and v_3 as the amplitude of the third harmonic, the positive sequence inverter modulating signals with the embedded third harmonic are given in Equation (10). v_3 is the amplitude of the third harmonic component, which is usually between 0.15 and 0.2 [57]. The resulting modulating signals are compared with a higher-frequency triangular wave generator to produce logic pulses for semiconductor devices.

$$\begin{aligned} m_a^{+} &= v_a^{+} \sin(\omega t) + v_3 \sin(3\omega t) \\ m_b^{+} &= v_b^{+} \sin(\omega t - \tfrac{2\pi}{3}) + v_3 \sin(3\omega t) \\ m_c^{+} &= v_c^{+} \sin(\omega t - \tfrac{4\pi}{3}) + v_3 \sin(3\omega t) \end{aligned} \tag{10}$$

The gains of the PI controller in such control approaches directly affect the stability of the inverter and the response to abrupt changes in operating conditions. Although there are many methods for designing these controller parameters, such as Ziegler–Nichols and pole placement, many cannot guarantee the best performance during grid disturbances and nonlinear loading conditions. For a more robust operation, metaheuristic-based algorithms inspired by nature have gained popularity in inverter control applications [58–61]. For this article, the widely used genetic algorithm (GA) optimization algorithm is implemented to iteratively determine the best set of PI values for the regulation of DC link voltage and inverter current. GA has been adopted because it guarantees convergence given certain constraint bands and does not require knowledge of derivatives, as the algorithm can process with input–output mapping [62]. For the inner current loop, the PI values depend on the transfer function of the LCL filter. With a cost function derived based on the integral time absolute error (ITAE), a population size of 10, and lower/upper limits for PI values as [0 500], 52 generations are sufficient for convergence to optimal values. Similarly, the DC link transfer function established in [63] is utilized for optimization with the established parameters. In this case, a total of 107 generations are required to achieve convergence to the optimum values. Thus, the PI values of the best fit for the current controller are determined as $K_p = 0.60$ and $K_i = 243$, while for the DC link PI controller, $K_p = 1.35$ and $K_i = 395$. All controllers are equipped with anti-windups to reduce the effect of saturation and prevent large overshoots or oscillations.

4. Reinforcement Learning Agent Design for Grid Fault Ride-Through

Fault ride-through requirements state that all IBRs must remain connected to the grid for a certain duration and wait for clearance to confirm that the fault is temporary. From the inverter's perspective, low voltage is simply a dip/sag in the voltage profile at PCC causing a drop in the computed d and q axes components. Depending on the percentage of dip and duration, the currents injected from the inverter may increase considerably. During faults, conventional PI controllers suffer from poor performance and are unable to track the designated current references in the synchronous reference frames. Prolonged symmetric or asymmetric faults can cause DC link spikes, overcurrents, and deviations of active and reactive powers. The dynamic response of the feedforward decoupled control approach also deteriorates during fluctuations in grid voltages due to the presence of voltage terms of the d and q axes in cross-coupling.

Although controller gains can be optimized, obtaining knowledge of the grid condition and incorporating it into the process is difficult due to the grid's dynamic nature. Additionally, PI controllers are insensitive to changes in grid operating conditions and pose a difficulty in maintaining inverter stability during abnormal grid conditions due to predefined saturation limits. Due to the presence of the grid voltage in the cross-coupling in the inner control loop, the PI controller tends to increase the inverter modulating voltages, causing an increase in the supplied currents. Prolonged divergence of the inverter currents results in the output of the PI controller reaching maximum thresholds, and thus the inverter goes into overmodulation mode.

Most other types of controller, such as proportional resonant, deadbeat, and fractional-order PI controllers, are inherently designed to provide a steady-state response and may not respond rapidly to sudden variations during grid faults. Nonlinearities are often introduced in the system due to disturbances in grid voltage and frequency and cause responses that standard controllers struggle to handle effectively. Other advanced control techniques, such as sliding mode and model predictive control, also encounter issues during grid transients due to the chattering phenomenon, sensitivity to uncertain parameter variation, tuning complexities, and dependency on accurate system models. As such, a twin delayed deep-determined policy gradient (TD3PG) current controller is formulated and trained to provide effective current control of inverter-based resources under various severities of grid disturbances. The improved voltage-oriented control based on reinforcement learning for effective fault management of grid-connected inverters is depicted in Figure 4.

Reinforcement learning is an effective tool that can greatly improve the design process of power converter controllers. It provides various benefits, particularly in intricate and dynamic environments where conventional control methods may face difficulties. RL algorithms are capable of continually learning and adapting to changing operational conditions, such as load variations or fluctuations in input voltage. This is particularly advantageous in the control of grid-feeding inverters, where the power network can be highly unstable. Inverters frequently demonstrate non-linear behavior, especially under fault conditions. RL agents are anticipated to perform well in handling such non-linearity without requiring explicit modeling, enabling more robust and flexible control mechanisms. Traditional control techniques typically depend on precise mathematical models of the inverter and its operational context. This model-free approach eliminates the need for an explicit system model. Instead, it derives the control strategy directly through interaction with the environment, making it ideal for systems where precise models are hard to develop, enhancing fault resilience.

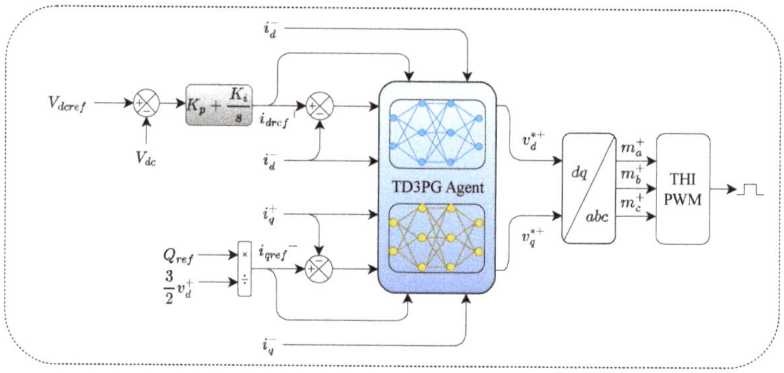

Figure 4. Twin delayed deep deterministic voltage-oriented control (TD3VOC) for inverter.

Twin delayed deep deterministic policy gradient is an advanced reinforcement learning algorithm that serves as an upgrade to the deep deterministic policy gradient algorithm. RL can be applied to control systems to train deep neural networks by interacting with an environment, enabling the achievement of tasks that are difficult to achieve using traditional control methods. The actor–critic agent is widely used in these applications, where the actor neural network determines the actions to be taken and the critic network assesses the effectiveness of those actions based on the rewards received from the environment (Figure 5).

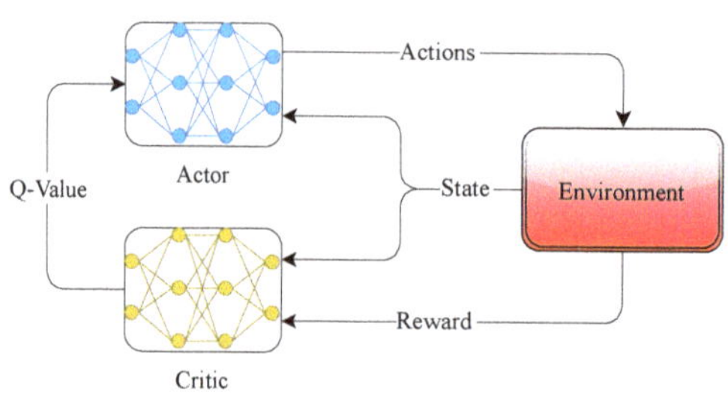

Figure 5. General reinforcement learning workflow.

Markov decision process (MDP) is a mathematical modeling tool that is used to address decision-making problems involving an RL agent and its interaction with the environment. It provides a framework for modeling sequential decision processes or a continuous action space in uncertain states. Optimizing an agent involves finding the best policy that maximizes the expected cumulative rewards for the agent using a suitable algorithm such as Q-learning or policy gradient. In MDP, states (s) represent all the information an agent receives from the environment, actions (a) are the array of numeric input (continuous or discrete) applied in the environment determined by the actor, and (r) is the cumulative reward obtained for the applied action. During training, the actor maps the states and actions, eventually converging to a deterministic policy (π) after numerous iterations [64].

4.1. Deep Neural Networks

To denote the primary indicators for selecting various actions in the subspace, $[i^{+}_{dref}, i^{+}_{qref}, (i^{+}_{dref} - i^{+}_{d}), (i^{+}_{qref} - i^{+}_{q})]$ are relevant states that are chosen as observations for the agent, while the modulation voltages of the inverter $[v_d^{*+}, v_q^{*+}]$ are two actions required from the actor network to provide enhanced stability to the inverter. The critic network has two feature input layers (FILs): one for the states and the other for the corresponding actions. Information measured via sensors is passed through two fully connected layers (FCLs), and the output is summed together, forming a single feedforward path. The rectified linear unit (ReLu) activation function performs a threshold operation before the data is inputted to the next FCL, where all negative values are set to zero. The final FCL outputs the required Q-value at each time step.

In a similar manner, the states are fed to the actor network using a FIL and passed through an FCL with 64 outputs, which are given two separate but identical paths with two FCLs. The outputs are summed and propagated through two more FCLs, and the final modulation voltages are produced. In the actor network, the hyperbolic tangent activation function (tanh) is employed between FCL due to its high performance and naturally normalizes inputs in the range $[-1, 1]$, which act as saturation limits for direct and quadrature voltages. The deep neural networks of critics and actors are shown in Figure 6. The 'Adam' learning algorithm is used to train dual critic and actor networks, each with a total of 3100 and 5800 learnable parameters, respectively. In general, both networks follow a similar pattern in which the number of outputs in the FCL is progressively reduced up to the output layer. This prevents overfitting and ensures a more stable training convergence.

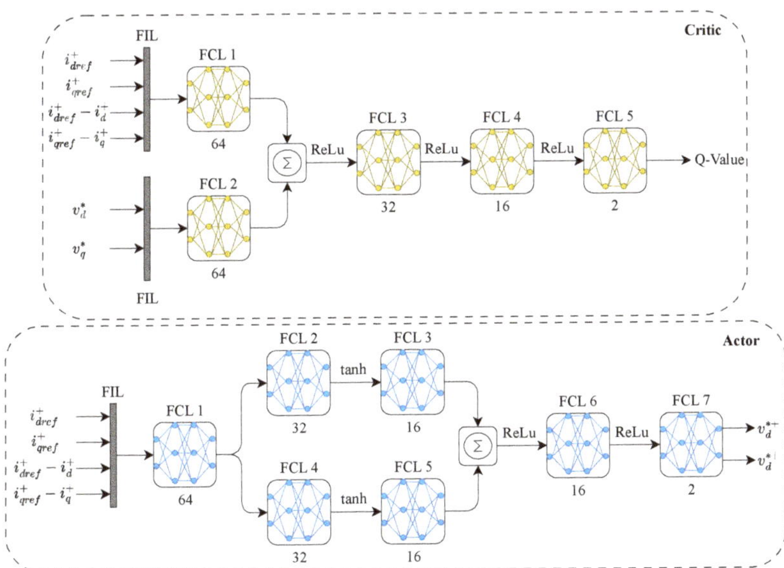

Figure 6. Critic and actor deep neural network architecture.

4.2. Reward Function

It is necessary to use a reward function that will allow the agent to learn the best possible policy by encapsulating a function that minimizes the inverter current deviation from the nominal set points at each discrete time step during abrupt disturbances on the grid side. The reward function is derived on the premise of effectively controlling the direct- and quadrature-axes currents based on the observed states. It is a combination of the measured d and q current errors with respect to the reference, the integral square error, and the negative sequence currents of the inverter, which must be minimized with any imbalance induced in the grid voltage. The total episode reward is the sum of all instantaneous rewards at each time step until the execution is terminated, given as

$$\Re = \sum_{t=0}^{T} \begin{pmatrix} M\left(\int (i^{+}_{dref} - i^{+}_{d})^2 dt + \int (i^{+}_{qref} - i^{+}_{q})^2 dt\right) + \\ N\left((i^{+}_{dref} - i^{+}_{d})^2 + (i^{+}_{qref} - i^{+}_{q})^2\right) + \\ O\left((i^{-}_{d})^2 + (i^{-}_{q})^2\right) \end{pmatrix} \quad (11)$$

The reward function assigns different weights to different aspects of the reward, represented by the arbitrary gain values M, N, and O. In this case, N and O are set to -1 and M to -0.5. The weights are selected to guarantee quick convergence during training. The adjustable gains permit customization of the significance assigned to each part of the reward function. For example, the squared errors of the positive and negative sequence currents are assigned the same gains so that both errors are equally minimized as the injected inverter currents deviate during faults. In contrast, the cumulative error increase over time is assigned a lower weight to ensure that it does not affect the real-time inverter regulation under the usual grid conditions. In addition, an average reward is calculated based on the length of the score averaging window. To simulate the agent's adaptation, a short-circuit fault is introduced between phases B and C at bus 633. The agent is expected to adjust its behavior based on the defined reward function to minimize current deviations in the inverter and improve the resilience of the inverter. All measurements are taken as per unit values.

4.3. TD3PG Training Process

The agent training process involves systematically updating the parameters of the actor–critic network. In TD3PG, two critic networks are present that help mitigate the issue of overestimation bias of the DDPG agent and also incorporate delayed updates of the target networks to stabilize convergence during training. The output of the critic networks, known as the Q-value, is an important component of the training process and acts as an indicator for the estimated cumulative reward given a set of observations and actions data influenced by certain policies. The complete agent training process is shown in Figure 7, and the general agent training pseudocode is presented as Algorithm 1.

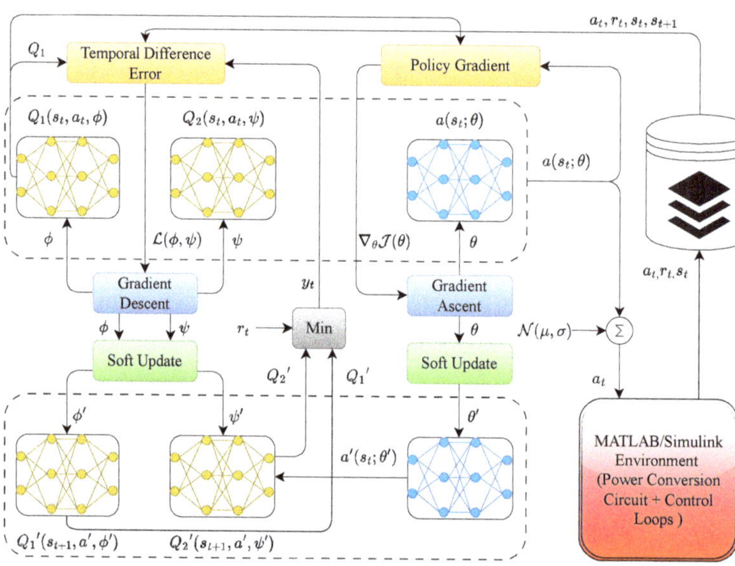

Figure 7. Training process of TD3PG RL agent incorporating data flow and environment interactions.

Algorithm 1: Pseudocode for training the TD3PG agent for inverter regulation

Setup the agent in the MATLAB/Simulink environment containing the grid-following inverter topology with appropriate grid fault models
Initialize critic networks Q_1, Q_2 with random parameters ϕ, ψ
Initialize actor network μ with random parameters θ
Initialize target networks Q_1', Q_2', μ' with ϕ' and ψ' and θ'
Set replay buffer $\mathcal{R}$ and exploration noise $\mathcal{N}$
Define hyperparameters: discount factor, learning rates, target update rate, regularisation, etc
for *each episode* **do**
  Initialize environment and get initial state (s_t)
  for *each time step (t) in the episode* **do**
    Select action + exploration noise: $a_t' = a(s_t;\theta) + \mathcal{N}(\mu,\sigma)$
    Execute action a in the environment and observe reward r and next state s_{t+1}
    Store transition (s_t, a_t, r_t, s_t') in replay buffer $\mathcal{R}$
    if *time to update* **then**
      for *each update step* **do**
        Sample a random mini-batch of $\mathcal{M}$ transitions (s_i, a_i, r_i, s_{t+1}) from $\mathcal{R}$
        Compute target Q-values using target networks and clipped double Q-learning:
        Compute target Q-values: $y_t = r_t + \gamma \min(Q_1'(s_{t+1}, a'(s_{t+1};\theta');\phi'), (Q_2'(s_{t+1}, a'(s_{t+1};\theta');\psi'))$
        Update the critic networks by minimising the loss:
        $\mathcal{L}(\phi) = \frac{1}{\mathcal{M}}\sum_{i=1}^{n}(Q_1(s_t, a_t, \phi) - y_t)^2$ and $\mathcal{L}(\psi) = \frac{1}{\mathcal{M}}\sum_{i=1}^{n}(Q_1(s_t, a_t, \psi) - y_t)^2$
        Update critic parameters ϕ and ψ using gradient descent
        $\phi \rightarrow \phi + \alpha_{critic}\nabla_{\phi}\mathcal{J}(\phi,\psi)$ and $\psi \rightarrow \psi + \alpha_{critic}\nabla_{\psi}\mathcal{L}(\phi,\psi)$
        if *update step policy delay == 0* **then**
          Update actor network by maximizing the Q-value:
          $\nabla J(\theta) \approx E_{s_t \sim \mathcal{R}}[\nabla_{\theta} Q_1(s_t, a_t;\phi)|_{a_t = a(s_t;\theta)} \nabla_{\theta} a(s_t;\theta)]$
          Update actor parameters θ using gradient ascent: $\theta \rightarrow \theta + \alpha_{actor}\nabla_{\theta} J(\theta)$
          Update target networks:
          $\phi' \rightarrow \phi'(1-\tau) + \tau\phi$, $\psi' \rightarrow \psi'(1-\tau) + \tau\psi$ and $\theta' \rightarrow \theta'(1-\tau) + \tau\theta$
        end
      end
    end
    Update the state: $s_{t+1} \rightarrow s_t$
    if *the episode is done* **then**
      Break the loop
    end
  end
end
return *The trained actor network or the final optimised policy* (π)

Denoting the parameters of the actor network as θ, the weight of the first critic network (Q_1) as ϕ, the weights of the second critic network (Q_2) as ψ, and the parameters of the corresponding target network as θ', ϕ', and ψ'. The actor is expected to maximize the expected cumulative long-term reward by minimizing the negative Q-value function given as

$$\mathcal{J}(\theta) = -E_{s_t,a_t \sim \mathcal{R}}[Q_1(s_t, a_t; \phi)] \tag{12}$$

E represents the expectation operator that is associated with the stochastic property of the transition probability and policy, π. Assuming $a(s_t;\theta)$ is the output of the actor network given a state in a time interval (s_t) and specific weights and $\nabla_\theta Q_1(s_t, a_t; \phi)$ is the calculated gradient of the first critic network with respect to the action, the gradient ascent can be used to update the parameters of the actor network using the chain rule applied to the expected return objective according to the actor parameters [65].

$$\nabla \mathcal{J}(\theta) \approx E_{s_t \sim \mathbf{R}}[\nabla_\theta Q_1(s_t, a_t; \phi)|_{a_t = a(s_t;\theta)} \nabla_\theta a(s_t; \theta)] \tag{13}$$

Therefore, the new weights of the actor network can be calculated using the return objective and the actor learning rate (α_{actor}) as follows.

$$\theta \rightarrow \theta + \alpha_{actor} \nabla_\theta \mathcal{J}(\theta) \tag{14}$$

Furthermore, critic networks are trained to reduce the Q-value output of critic networks (Q_1 and Q_2) and target critic networks Q_1' and Q_2'. The targeted Q-value (y_t) is obtained by evaluating the minimum of the two Q-values of the target networks, which also include the discounted reward (r_t) and the predicted Q-values in the next state (s_{t+1}), commonly calculated using the Bellman equation. Minimum operation ensures reduced overestimation bias. A discount factor (γ), generally between 0 and 1, determines the weight given to future rewards. High γ values give more importance to compounded rewards, and a small value prioritizes immediate rewards, creating a trade-off between past and instant rewards.

$$y_t = r_t + \gamma \min(Q_1'(s_{t+1}, a'(s_{t+1}; \theta'); \phi'), (Q_2'(s_{t+1}, a'(s_{t+1}; \theta'); \psi')) \tag{15}$$

The critic tends to reduce the difference between the target Q-value (from the Bellman equation) and the estimated Q-values, known as the temporal difference error (TDE), using a formulated loss function stated in Equation (8). Given M as the size of the mini-batch and the set of state-action data from the stored experiences in the memory at each time step, the critic networks obtain a better approximation for the true Q-value via minimization of this loss function.

$$\mathcal{L}(\phi, \psi) = \frac{1}{M} \sum_{i=1}^{N} (Q_1(s_t, a_t, (\phi, \psi)) - y_t)^2 \tag{16}$$

Eventually, both parameters of the critic network are updated using gradient descent considering a reasonable learning rate (α_{critic}).

$$\phi \rightarrow \phi + \alpha_{critic} \nabla_\phi \mathcal{J}(\phi, \psi) \tag{17}$$

$$\psi \rightarrow \psi + \alpha_{critic} \nabla_\psi \mathcal{L}(\phi, \psi) \tag{18}$$

The target networks (Q_1', Q_2', and a') are updated through a soft update process that involves slowly evolving the network parameters with the current policy network parameters. A predefined target smooth factor (τ) is used to provide a more stable learning process.

$$\theta' \rightarrow \theta'(1-\tau) + \tau\theta \tag{19}$$

$$\phi' \rightarrow \phi'(1-\tau) + \tau\phi \tag{20}$$

$$\psi' \rightarrow \psi'(1-\tau) + \tau\psi \tag{21}$$

During the training process, noise exploration data are often added to actions to allow for more robust policy development. Sampled Gaussian noise $\mathcal{N}$ characterized by a normal distribution with reasonable mean (μ) and standard deviation (σ) can be used to modify the selected actions before being applied.

$$a'_t = a(s_t;\theta) + \mathcal{N}(\mu,\sigma) \tag{22}$$

In this way, the agent can explore various regions of the action space via controlled randomness. It prevents it from obtaining a suboptimal policy, which is especially crucial in discovering effective continuous-deterministic actions. However, high values of μ and σ lead to more exploration but may cause overfitting, resulting in less deterministic behavior. Therefore, it is important to maintain balance without compromising training or policy stability. In some cases, a standard deviation decay rate (ε) is used, allowing controlled reduction in the amount of noise that is added as agents take progressive steps and helping to shift from exploration to exploitation.

Furthermore, the TDP3PG agent offers several notable advantages that make it highly suitable for improving the effectiveness of such control tasks. It is specifically designed to handle problems with continuous action spaces, such as the regulation of power converters. This design choice eliminates the need for sampling from a probability distribution, as the agent can directly output continuous actions. Consequently, the complexity of decision making is reduced. Additionally, the use of the minimum operation in the Bellman equation combined with clipped double Q-learning critic networks ensures a more stable training process by mitigating issues related to overestimated Q-values and training divergence. Moreover, delayed updates of critic networks, in comparison to the actor, result in decorrelated target Q-values. This contributes to reducing variance and facilitates efficient training [41].

5. Results and Discussion

Training is performed using the high-performance shared computational cluster "Katana" with 64 GB of memory using a single CPU core. It takes approximately 22 h for the training to be completed, with 3000 steps per episode. All training hyperparameter specifications are provided in Table 3.

As a secondary training criterion, all agents that achieve an average of −500 reward points or more are saved for further evaluation, resulting in a total of 833 being saved. Training progress with the episode reward obtained, the average reward, and the Q-value is depicted in Figure 8. Initially, the agent starts with a low reward (−15,000), but slowly progresses to higher values as the actor neural network parameters become accustomed to the objective based on the updated weights. It is observed that after 840 episodes there is a sharp drop in the episode reward, after which a decline is noted in the training behavior. This can be an indication of overfitting, as the actor may already have reached its optimum policy. It may also be an indication that the actor is stuck in a local suboptimal policy that causes a temporary decrease in reward due to too much or insufficient exploration.

Table 3. TD3 agent training parameters.

Hyper Parameter	Symbol	Value
Sample time	T_s	100 µs
Mini batch size	$\mathcal{M}$	64
Discount factor	γ	0.99
Maximum episodes	$\mathcal{E}_{max}$	1000
Actor learning rate	α_{actor}	1×10^{-3}
Critic learning rate	α_{critic}	1×10^{-3}
Target smooth factor	τ	1×10^{-3}
Policy update frequency	ξ	1
L2 regularisation (actor)	ρ_{actor}	1×10^{-4}
L2 regularisation (critic)	ρ_{critic}	1×10^{-4}
Experience buffer length	$\mathcal{R}$	1×10^{6}
Number of steps to look ahead	n	1
Score averaging window length	$\mathcal{K}$	15
Standard deviation	σ	0.1
Minimum standard deviation	σ_{min}	0.025
Standard deviation decay rate	σ_{decay}	1×10^{-6}

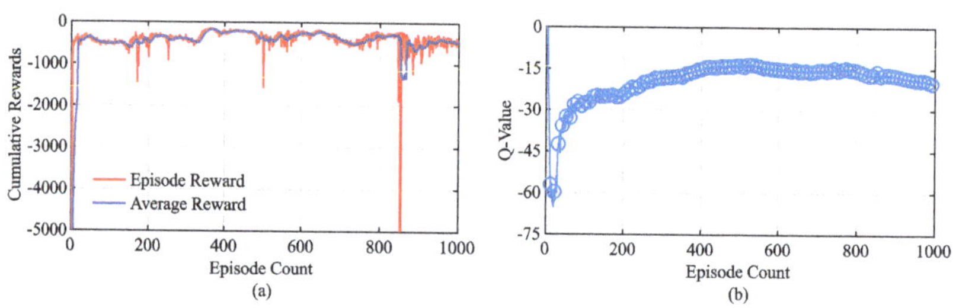

Figure 8. (**a**) Cumulative reward obtained and (**b**) Q-value progression at each successive iteration.

Since the actor–critic agent is deployed in a detailed time-varying environment containing feedback signals with inverter switching noise, observations and action pairs in the experience buffer length can also cause ambiguity, thus reducing training effectiveness and reward accumulation. If this pattern continues, the final agent at the end of maximum episodes may not perform well even if the learning rate is slow with a small regularization parameter. Since the approach deals with inverter current control at the primary level, to select the finest possible policy, the agent that attains the highest average reward is chosen as the best contender for controlling the inverter during faults. Using this criterion, the agent in episode 364 is adopted, which attains the highest reward of −151 with a Q-value of −16.70.

Under all test conditions, the DC link reference is 1 p.u. while the reactive power reference is kept at 0 p.u. It should be noted that the TD3PG agent is trained only for a single asymmetrical fault between phases B and C across a 0.25 Ω resistor. However, its effectiveness in mitigating other types of symmetrical and asymmetrical faults through transfer learning has been verified. Using leverage of the pre-trained model significantly reduces the training time, especially for various types of grid faults. Training the agent for only one fault also allows the actor neural network to learn more generalized actions and leads to better accuracy. In addition, applying transfer learning reduces the need for computational resources, making the approach more efficient. The performance of the positive sequence voltage-oriented control and the designed twin delayed deep deterministic policy gradient voltage-oriented control is evaluated under both asymmetric and symmetric grid faults based on the adaptation to faults and operation restoration time.

5.1. Asymmetrical Fault: Short Circuit Between Phases B and C

In this test case, a phase-to-phase fault is induced between phases B and C (Figure 9a) over a fault resistance of 75 mΩ between time 0.20 and 0.25 s on bus 633. In particular, the trained agent has significantly reduced the post-fault recovery time of the inverter currents with immediate settling after fault clearance. With PSVOC, the current waveform settles at around 0.60 s. However, the proposed control approach takes only 0.06 s for the current to return to nominal values, and the waveform is only distorted for the duration of the fault without any significant increase in peak values, as shown in Figure 9b,c. This also helps to maintain the integrity of the grid voltages, as prolonged injection of distorted currents adds unwanted harmonics and may cause the inverter to isolate from the grid.

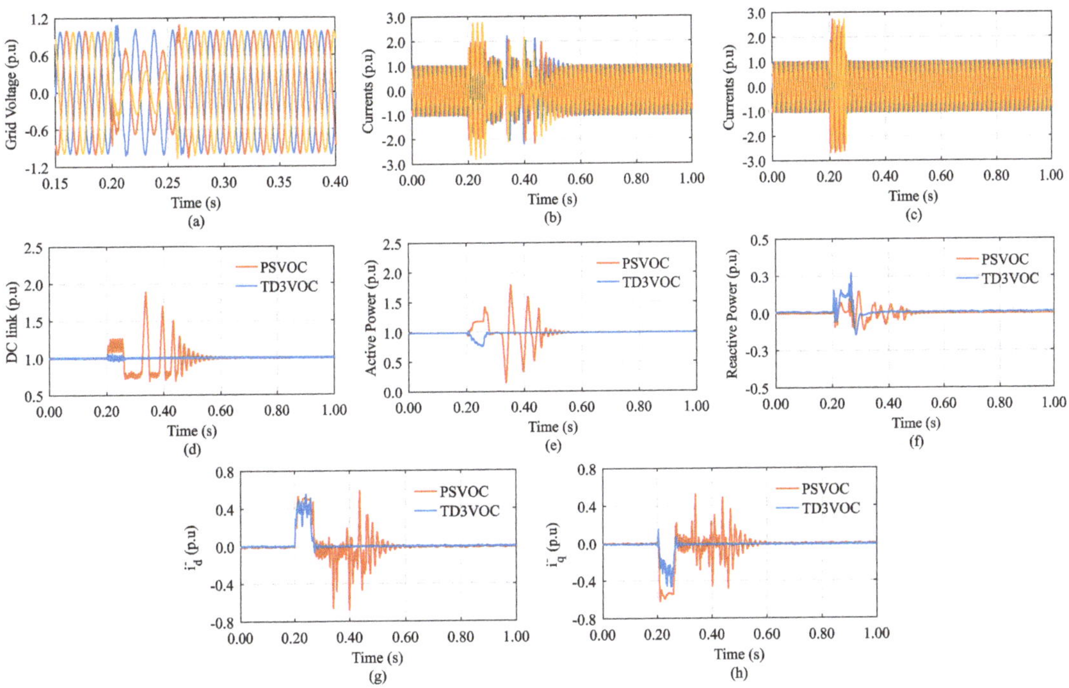

Figure 9. (**a**) Grid voltages, three-phase current waveform with (**b**) PSVOC and (**c**) TD3VOC, (**d**) DC link voltage, (**e**) active power, (**f**) reactive power, and negative sequence current on (**g**) direct and (**h**) quadrature axis response during asymmetrical short circuit fault between phases B and C.

As seen in Figure 9d, with PSVOC, an overshoot in the DC link voltage is observed during the fault. As the fault clears, an undershoot is seen as the maximum power point, and the DC-link PI controller attempts to recover from the disturbance. During the recovery period, the inverter consistently undergoes over- and undermodulation as a result of a destabilized DC link reaching a maximum voltage of 1.85 p.u. Following a period of 0.35 s after the fault has cleared, the nominal operating point is finally reached. However, the trained agent is far better at stabilizing the DC link voltage and shows immediate post-fault recovery. Consequent fluctuations in active and reactive powers (Figure 9e,f) are detected with large overshoots using the conventional controller, and although a slight drop in injected active power is observed with the TD3PG agent, the nominal operating point is quickly restored without any overshoot, showing a more robust response despite deviations during the fault. In addition, the TD3 agent performs well in minimizing negative sequence direct and quadrature axis currents during the fault and immediate retention after the termination of the fault, as presented in Figure 9g,h.

5.2. Asymmetrical Fault: Phases A and B to Ground Fault

To further investigate the performance of the neural-based current controller, a double-phase-to-ground fault is applied at the PCC voltage for 50 ms (Figure 10a). The fault and ground resistance are assumed to be 50 mΩ. The three-phase currents experience oscillations and take about 0.55 s to settle after fault clearance when PI-based VOC is used. High current peaks are momentarily seen as the inverter attempts to recover from the fault, as shown in Figure 10b. This effect is completely eliminated with the RL agent. Although the current increases during the fault period, the inverter is quick to recover and attains stability without any post-fault transients, as depicted in Figure 10c. The superior response of the neural network is attributed to its fast input-to-output mapping. A similar effect is seen in the negative sequence currents on the direct and quadrature axes (Figure 10g,h). While the inverter is in overmodulation, the current oscillates and has several overshoots and undershoots during recovery with the conventional control approach. This problem is completely eliminated with the trained actor–critic neural networks ensuring a robust response.

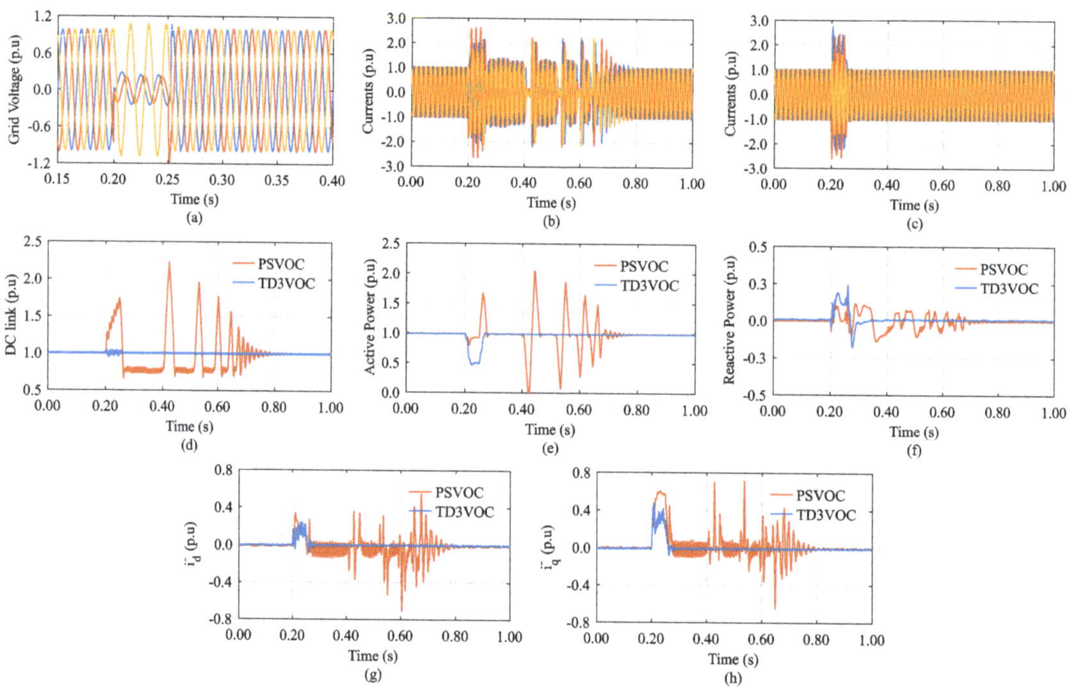

Figure 10. (**a**) Grid voltages, three-phase current waveform with (**b**) PSVOC and (**c**) TD3VOC, (**d**) DC link voltage, (**e**) active power, (**f**) reactive power, and negative sequence current on (**g**) direct and (**h**) quadrature axis response during phases A and B to ground fault.

During this period, a drastic response is observed for the DC link voltage compared with the previous case, reaching a peak of 2.25 p.u. It undergoes transients and settles at 0.80 s, as seen in Figure 10d when the inverter is controlled using PSVOC. With the TD3VOC, during the fault period, the voltage experiences slight oscillations but does not diverge from the reference value. When using cascaded PI controllers, inverter stability depends on the accurate functioning of each loop. As such, a disturbance may cause the entire system to respond in an unpredictable manner, even if the controllers are properly tuned. This scenario is also seen in the active and reactive power response in Figure 10e,f, where the active and reactive powers diverge considerably due to the unstable DC link. Nevertheless, with the TD3 agent, immediate recovery is observed, as nominal operating points are restored without transients.

5.3. Symmetrical Fault: Phases A, B, and C Short Circuit Fault

A severe symmetric three-phase fault is applied between phases A, B, and C on bus 633 across a fault resistance of 35 mΩ (Figure 11a). The conventional PSVOC approach does not restore the inverter current waveform shown in Figure 11b due to the reaching of controller saturation limits. As a result, high oscillations and distortions in the currents are observed, even though they remain in phase with the grid voltage. Prolonged injection of highly distorted currents would eventually trigger protection devices, causing the inverter to disconnect from the grid, disrupting power transfer. In contrast, the RL agent can restore the inverter currents within 0.05 s after the fault is cleared, as shown in Figure 11c. The voltage diverges and reaches more than 2.5 times the nominal operating range, which will eventually activate the circuit tripping mechanisms. In contrast, the difference is significant when TD3VOC is used instead, leading to a much smoother response of the DC link voltage without much divergence (Figure 11d).

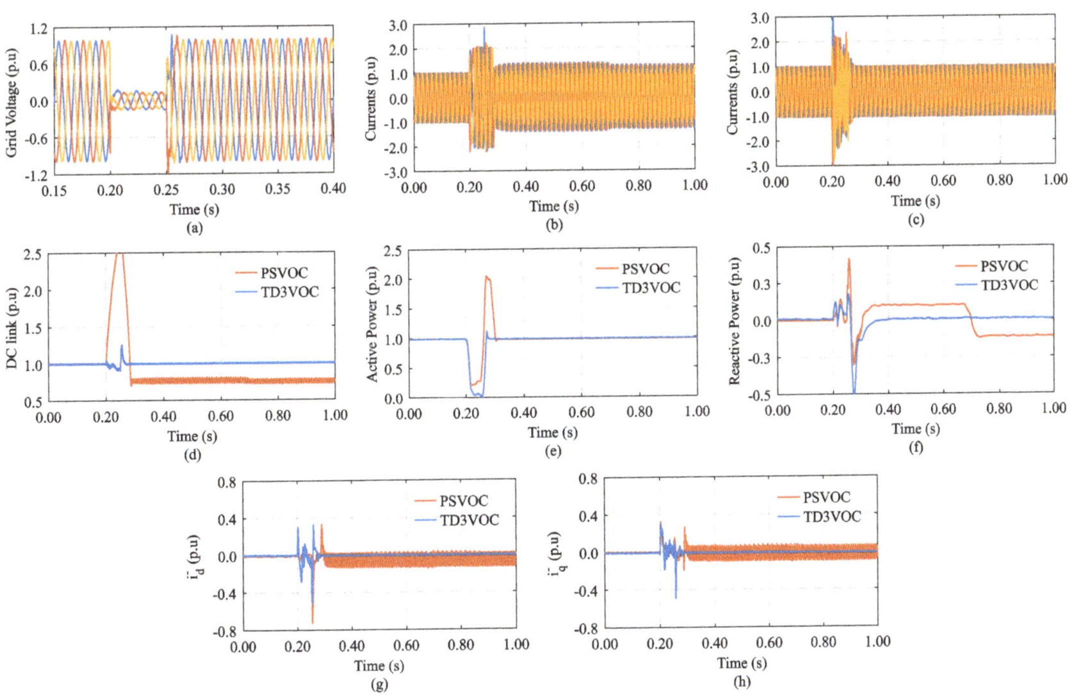

Figure 11. (**a**) Grid voltages, three-phase current waveform with (**b**) PSVOC and (**c**) TD3VOC, (**d**) DC link voltage, (**e**) active power, (**f**) reactive power, and negative sequence current on (**g**) direct and (**h**) quadrature axis response to a symmetrical short-circuit fault between phases A, B, and C.

Likewise, active power experiences disruptions during a fault, and a large overshoot occurs as the inverter tries to restore power post-fault. This recovery time is reduced with the proposed TD3PG current controller, as shown in Figure 11e,f. Although the active power is restored to nominal values, it is at the cost of a lower DC link voltage and an overmodulating inverter. Due to not recovering from saturation and continuous operation with a lower DC link voltage, the negative sequence currents suffer from high oscillations after fault clearance when the inverter is controlled with PSVOC. However, the TD3VOC scheme performs quite well in restoring and reducing negative sequence currents even after severe disturbances (Figure 11g,h). It should be noted that although the agent was not specifically trained for symmetrical faults, it still demonstrates a good response to such severe disturbances, thus proving its adaptability through transfer learning.

5.4. Error Analysis

To accurately assess the performance of the proposed fault-tolerant agent-based control method, calculating the integral-time absolute error (ITAE) is crucial. The ITAE metric multiplies the absolute error by time, thereby giving greater importance to errors that occur later in the system compared to those at the start. Minimizing ITAE is important to achieve optimal performance during faults in inverter control applications. The error data for the d and q axes currents obtained from the evaluation models have been analyzed under the three types of faults, and the results are shown in Figure 12. With the conventional controller, ITAE on the d and q axes increases significantly as faults become more severe. During the phase-to-phase asymmetric fault between phases B and C, a maximum of 0.20 and 0.15 is reached for the d and q axes, respectively. However, this increases to 0.32 and 0.60 after a symmetric fault is induced in the system. With the proposed TD3PG current controller, it is observed that during all test cases, the error remains below 0.10 and 0.03 for the d and q axes, respectively. This metric authenticates the viability of the formulated agent in providing a reasonably better inverter response compared with the standard control approach.

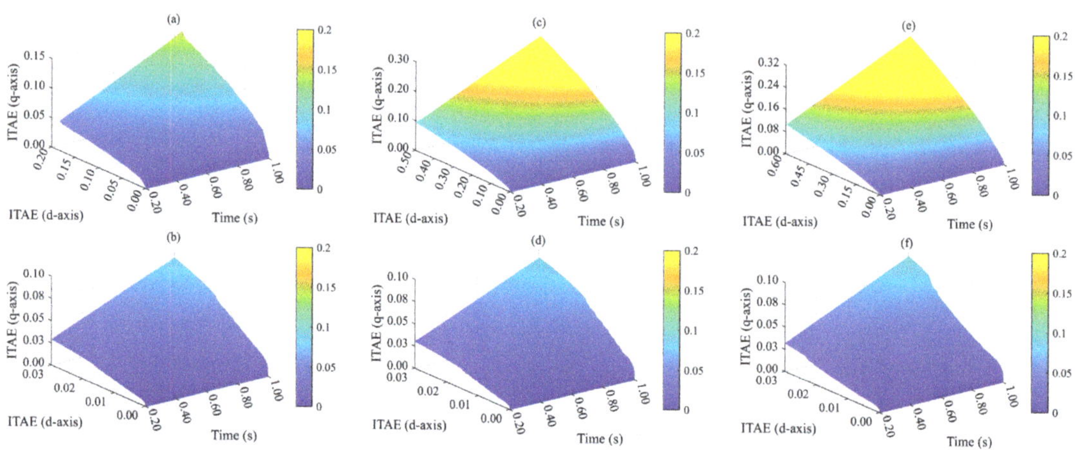

Figure 12. Integral-time error visualization for d and q axes currents using (**a**,**c**,**e**) PSVOC and (**b**,**d**,**f**) TD3VOC for (**a**,**b**) asymmetric fault between phases B and C, (**c**,**d**) asymmetric phases A and B to ground fault, and (**e**,**f**) symmetric short-circuit fault between phases A, B and C.

5.5. Summary

In summary, the key results obtained in this research for the existing voltage-oriented control scheme and the redesigned TD3PG voltage-oriented control architecture are stated in Tables 4 and 5. Key performance indicators were selected as the peak DC link voltage magnitude during the fault period with a nominal value of 1 p.u., the post-fault recovery time after fault clearance (PFRT), and ITAE on the d and q axes. In every scenario, the suggested TD3VOC exhibits a robust fault-ride-through capability in all evaluations by the use of transfer learning. It exhibits minimized DC link voltage peaks during faults and reduced post-fault system recovery time. A significant attribute that has directly affected the performance of the TD3PG agent is its structural design. Bellman's minimum equation paired with clipped double Q-learning mitigates the overestimation bias found in other continuous-time agents. The algorithm's lower sensitivity to hyperparameter selection is due to postponed policy updates and target smoothing. These sophisticated features, along with its adaptability to unpredictable inverter operating conditions, make TD3PG an exceptionally effective option to improve the control of grid-feeding photovoltaic systems during grid disruptions. In addition, the per-unit system is beneficial in minimizing the alterations needed to apply this method to larger power systems. It ensures that the neural network generates modulation voltages in a consistent manner, regardless of the differing power ratings in various systems.

Table 4. Performance of conventional positive-sequence voltage-oriented control.

Type of Fault	Voltage (p.u.)	PFRT (s)	ITAE (d-axis)	ITAE (q-axis)
Phase-to-phase (B to C)	1.85	0.35	0.20	0.15
Phase-to-ground (A, B to G)	2.25	0.55	0.50	0.28
Phases A, B and C short circuit	>2.50	>0.80	0.60	0.32

Table 5. Performance of proposed twin delayed deep deterministic voltage-oriented control.

Type of Fault	Voltage (p.u.)	PFRT (s)	ITAE (d-axis)	ITAE (q-axis)
Phase-to-phase (B to C)	1.05	0.01	0.03	0.09
Phase-to-ground (A, B to G)	1.05	0.01	0.03	0.09
Phases A, B and C short circuit	1.25	0.02	0.03	0.09

6. Limitations and Future Works

Despite the impressive fault-resistant capabilities of the TD3VOC, it should be noted that achieving the optimal policy for effective real-time inverter control still requires approximately 20–25 h of episodic iteration. The duration of training is predominantly influenced by the processor power, the number of CPU cores, and the available memory. In addition, the environment plays a significant role in determining the required computational power. In this study, we developed a highly detailed grid-interfaced photovoltaic inverter environment in MATLAB/Simulink® with which the RL agent interacted. An average model could also be used to train the agent; however, it is uncertain whether the same agent would perform well when faced with a more elaborate model that includes switching noise and variability in feedback signals.

The results presented in this article are obtained using the twin delayed deep-deterministic policy gradient agent; however, there is the possibility of exploring other continuous-time agents such as soft actor–critic, trust region policy optimization, model-based policy optimization, etc. A comparative study can be conducted to determine the best candidate for developing data-driven controllers based on training time, number of hyperparameter selections, and adaptation of various grid conditions. It will also be interesting to compare the formulated fault ride-through approach with other established methods such as sliding mode and model reference adaptive controllers.

The proposed agent-based current controller can be trained and implemented for various other grid-feeding inverter control applications in weak microgrids that require a robust fault ride-through mechanism, such as wind energy conversion systems, grid-forming battery energy storage systems, fuel cell systems, and other inverter control schemes featuring an inner current loop. This direct design method can also be applied to develop robust data-driven controllers for electrical machines that are controlled using cascaded field-oriented control or the direct torque control scheme to provide better torque handling capabilities, smoother acceleration, and increased energy efficiency.

In addition, the experimental validation of such advanced control techniques is still a subject of particular interest to numerous researchers today. Implementing RL agents for optimal regulation of the inverter during dynamic grid conditions will require an embedded system that involves a combination of advanced hardware and software features. It will be necessary to utilize high-performance microcontroller units (MCUs), digital signal processors (DSPs), or systems on chips (SoCs) capable of handling both RL algorithms and real-time inverter control with appropriate communication interfaces. It is suggested to use the Xilinx field-programmable gate array board, OPAL-RT technology, or DSPACE Microlabbox due to compatibility with MATLAB/Simulink®, customized software options, and expandable hardware. There is immense potential for RL-based current controllers to enhance the performance and reliability of inverter systems in various domains, particularly in the context of developing modern and intelligent energy systems.

7. Conclusions

In this paper, a robust data-driven grid fault ride-through approach is proposed for photovoltaic inverters. A twin delayed deep-deterministic policy gradient agent is embedded in the conventional voltage-oriented control scheme as an upgrade to the PI controllers of the d and q axes and cross-coupling terms. Classical current controllers are designed assuming a balanced and stable grid, which is not always the case due to the presence of unpredictable fault conditions, and also suffer from dependence on controller gains and predefined saturation limits. To address these issues, the reinforcement learning workflow is adopted to train actor–critic deep neural networks to improve the inverter current response under various grid conditions. A dual stage inverter environment with a suitable reward function was set up to guide the training process. Initially, the agent is trained for a single asymmetric fault between phases B and C. However, it is possible to use the same agent to improve inverter stability across a wide spectrum of grid faults via transfer learning. When using the PSVOC method, the DC link overshoots to 1.85 p.u. and 2.25 p.u. while a current post-fault restoration time of 0.35 s and 0.55 s for phase-to-phase and phase-to-ground faults, respectively. With the developed TD3VOC technique, voltage overshoot is limited to 1.05 p.u. with a current recovery time of 0.01 after fault clearance in both cases. For a severe symmetric fault, the conventional control approach cannot restore inverter operation with a voltage overshoot of more than 2.50 p.u. with integral-time absolute errors of 0.60 and 0.32 on the d and q axes, respectively. The proposed agent-based control method limits the voltage overshoot to 1.25 p.u. and achieves a cumulative error of 0.03 and 0.09 on the d and q axes, respectively. The proposed control approach is implemented using per-unit measurements to allow for scalability and can be extended to grid-connected wind turbine generators, grid-forming inverters, and other inverter-based resources that require robust fault ride through operation.

Author Contributions: Conceptualization, S.S.C.; methodology, S.S.C.; software, S.S.C.; validation, S.S.C.; formal analysis, S.S.C.; investigation, S.S.C.; resources, B.H.; writing—original draft preparation, S.S.C. and B.H.; writing—review and editing, S.S.C. and B.H.; supervision, B.H. and M.C.; project administration, B.H. and M.C. All authors have read and agreed to the published version of the manuscript.

Funding: This research received no external funding.

Data Availability Statement: The original contributions presented in this study are included in the article. Further inquiries can be directed to the corresponding authors.

Conflicts of Interest: The authors declare no conflicts of interest.

List of Acronyms

MG	Microgrid
IBR	Inverter Based Resources
INC	Incremental Conductance
FRT	Fault-Ride Through
MDP	Markov Decision Process
PFRT	Post-fault Recovery Time
TD3PG	Twin Delay Deep Deterministic Policy Gradient
PSVOC	Positive Sequence Voltage-Oriented Control
TD3VOC	Twin Delay Deep Deterministic Voltage-Oriented Control
PV	Photovoltaic
DER	Distributed Energy Resource
QSG	Quadrature Signal Generator
PCC	Point of Common Coupling
DRL	Deep Reinforcement Learning
ITAE	Integral-time Absolute Error
DDPG	Deep Deterministic Policy Gradient
DSOGI	Double Second Order Generalized Integrator
THIPWM	Third Harmonic Injection Pulse Width Modulation

References

1. Panda, A.; Dauda, A.K.; Chua, H.; Tan, R.R.; Aviso, K.B. Recent advances in the integration of renewable energy sources and storage facilities with hybrid power systems. *Clean. Eng. Technol.* **2023**, *12*, 100598. [CrossRef]
2. Shuai, Z.; Sun, Y.; Shen, Z.J.; Tian, W.; Tu, C.; Li, Y.; Yin, X. Microgrid stability: Classification and a review. *Renew. Sustain. Energy Rev.* **2016**, *58*, 167–179. [CrossRef]
3. Rajesh, K.S.; Dash, S.S.; Rajagopal, R.; Sridhar, R. A review on control of ac microgrid. *Renew. Sustain. Energy Rev.* **2017**, *71*, 814–819. [CrossRef]
4. Ahmed, M.; Meegahapola, L.; Vahidnia, A.; Datta, M. Stability and Control Aspects of Microgrid Architectures-A Comprehensive Review. *IEEE Access* **2020**, *8*, 144730–144766. [CrossRef]
5. Nawaz, F.; Pashajavid, E.; Fan, Y.; Batool, M. A Comprehensive Review of the State-of-the-Art of Secondary Control Strategies for Microgrids. *IEEE Access* **2023**. [CrossRef]
6. Qais, M.H.; Hasanien, H.M.; Alghuwainem, S. Whale optimization algorithm-based Sugeno fuzzy logic controller for fault ride-through improvement of grid-connected variable speed wind generators. *Eng. Appl. Artif. Intell.* **2020**, *87*, 103328. [CrossRef]
7. *IEEE Std 2800-2022*; IEEE Standard for Interconnection and Interoperability of Inverter-Based Resources (IBRs) Interconnecting with Associated Transmission Electric Power Systems. IEEE: Piscataway, NJ, USA, 2012; pp. 1–180.
8. Singh, M.; Basak, P. Q and frequency component-based fault and nature detection for hybrid microgrid. *Sustain. Energy Grids Netw.* **2021**, *28*, 100552. [CrossRef]
9. Borrell, A.; Velasco, M.; Miret, J.; Camacho, A.; Marti, P.; Castilla, M. Collaborative Voltage Unbalance Elimination in Grid-Connected AC Microgrids with Grid-Feeding Inverters. *IEEE Trans. Power Electron.* **2021**, *36*, 7189–7201. [CrossRef]
10. Gomes, H.M.; Carralero, L.L.; Suárez, J.H.; Tahim, A.N.; Pinheiro, J.R.; Costa, F.F. A Grid Impedance Estimation Method Robust Against Grid Voltages Unbalance and Harmonic Distortions. *J. Control Autom. Electr. Syst.* **2023**, *34*, 289–301. [CrossRef]
11. Ferreira, R.V.; Silva, S.M.; Brandao, D.I. Positive–Negative Sequence Synchronverter for Unbalanced Voltage in AC Grids. *J. Control Autom. Electr. Syst.* **2021**, *32*, 711–720. [CrossRef]
12. Easley, M.; Jain, S.; Shadmand, M.; Abu-Rub, H. Autonomous Model Predictive Controlled Smart Inverter with Proactive Grid Fault Ride-Through Capability. *IEEE Trans. Energy Convers.* **2020**, *35*, 1825–1836. [CrossRef]
13. Nguyen, T.K.; Chang, G.W.; Liu, Y.J.; Li, G.Y. Improving Voltage Ride-Through Capability of Grid-Tied Microgrid With Harmonics Mitigation. *IEEE Trans. Power Deliv.* **2023**, *38*, 738–741. [CrossRef]
14. Liu, X.; Chen, X.; Shahidehpour, M.; Li, C.; Wu, Q.; Wu, Y.; Wen, J. Active Fault Current Limitation for Low-Voltage Ride-Through of Networked Microgrids. *IEEE Trans. Power Deliv.* **2022**, *37*, 980–992. [CrossRef]
15. Bahramian-Habil, H.; Abyaneh, H.A.; Gharehpetian, G.B. Improving LVRT capability of microgrid by using bridge-type fault current limiter. *Electr. Power Syst. Res.* **2021**, *191*, 106872. [CrossRef]

16. El-Naggar, M.F.; Mosaad, M.I.; Hasanien, H.M.; AbdulFattah, T.A.; Bendary, A.F. Elephant herding algorithm-based optimal PI controller for LVRT enhancement of wind energy conversion systems. *Ain Shams Eng. J.* **2021**, *12*, 599–608. [CrossRef]
17. Xiao, Q.; Jin, Y.; Chen, L.; Yu, X.; Jia, H.; Liu, H.; Teodorescu, R.; Blaabjerg, F. A Simple Operation Approach for Modular Multilevel Converter under Grid Voltage Swell in Medium-Voltage Microgrids. *IEEE Access* **2019**, *7*, 147280–147291. [CrossRef]
18. Brandao, D.I.; Mendes, F.E.; Ferreira, R.V.; Silva, S.M.; Pires, I.A. Active and reactive power injection strategies for three-phase four-wire inverters during symmetrical/asymmetrical voltage sags. *IEEE Trans. Ind. Appl.* **2019**, *55*, 2347–2355. [CrossRef]
19. Wang, H.; Zhang, Q.; Wu, D.; Zhang, J. Advanced Current-Droop Control for Storage Converters for Fault Ride-Through Enhancement. *IEEE J. Emerg. Sel. Top. Power Electron.* **2020**, *8*, 2461–2474. [CrossRef]
20. Djilali, L.; Sanchez, E.N.; Ornelas-Tellez, F.; Avalos, A.; Belkheiri, M. Improving Microgrid Low-Voltage Ride-Through Capacity Using Neural Control. *IEEE Syst. J.* **2020**, *14*, 2825–2836. [CrossRef]
21. Firouzi, M.; Nasiri, M.; Mobayen, S.; Gharehpetian, G.B. Sliding Mode Controller-Based BFCL for Fault Ride-Through Performance Enhancement of DFIG-Based Wind Turbines. *Complexity* **2020**, *2020*, 1259539. [CrossRef]
22. Aboelsaud, R.; Ibrahim, A.; Aleksandrov, I.V.; Ali, Z.M. Model predictive control algorithm for fault ride-through of stand-alone microgrid inverter. *Int. J. Electr. Power Energy Syst.* **2022**, *135*, 107485. [CrossRef]
23. Musarrat, M.N.; Fekih, A.; Rahman, M.A.; Islam, M.R.; Muttaqi, K.M. An Event Triggered Sliding Mode Control-Based Fault Ride-Through Scheme to Improve the Transient Stability of Wind Energy Systems. *IEEE Trans. Ind. Appl.* **2024**, *60*, 876–886. [CrossRef]
24. Jia, H.; Hu, Y.; Jiang, T.; Li, G.; Tightiz, L.; Yoo, J. A Review on a Data-Driven Microgrid Management System Integrating an Active Distribution Network: Challenges, Issues, and New Trends. *Energies* **2022**, *15*, 8739. [CrossRef]
25. Cao, Y.; Chandrasekar, A.; Radhika, T.; Vijayakumar, V. Input-to-state stability of stochastic Markovian jump genetic regulatory networks. *Math. Comput. Simul.* **2024**, *222*, 174–187. [CrossRef]
26. Radhika, T.; Chandrasekar, A.; Vijayakumar, V.; Zhu, Q. Analysis of Markovian Jump Stochastic Cohen–Grossberg BAM Neural Networks with Time Delays for Exponential Input-to-State Stability. *Neural Process. Lett.* **2023**, *55*, 11055–11072. [CrossRef]
27. Tamil Thendral, M.; Ganesh Babu, T.R.; Chandrasekar, A.; Cao, Y. Synchronization of Markovian jump neural networks for sampled data control systems with additive delay components: Analysis of image encryption technique. *Math. Methods Appl. Sci.* **2022**. [CrossRef]
28. Mamdouh Farghaly, H.; Abd El-Hafeez, T. A new feature selection method based on frequent and associated itemsets for text classification. *Concurr. Comput. Pract. Exp.* **2022**, *34*, e7258. [CrossRef]
29. Mamdouh Farghaly, H.; Abd El-Hafeez, T. A high-quality feature selection method based on frequent and correlated items for text classification. *Soft Comput.* **2023**, *27*, 11259–11274. [CrossRef]
30. Farghaly, H.M.; Ali, A.A.; El-Hafeez, T.A. Developing an Efficient Method for Automatic Threshold Detection Based on Hybrid Feature Selection Approach. *Adv. Intell. Syst. Comput.* **2020**, *1225*, 56–72. [CrossRef]
31. Mostafa, G.; Mahmoud, H.; Abd El-Hafeez, T.; ElAraby, M.E. Feature reduction for hepatocellular carcinoma prediction using machine learning algorithms. *J. Big Data* **2024**, *11*, 1–27. [CrossRef]
32. Eliwa, E.H.I.; El Koshiry, A.M.; Abd El-Hafeez, T.; Omar, A. Optimal Gasoline Price Predictions: Leveraging the ANFIS Regression Model. *Int. J. Intell. Syst.* **2024**, *2024*, 8462056. [CrossRef]
33. Li, S.; Oshnoei, A.; Blaabjerg, F.; Anvari-Moghaddam, A. Hierarchical Control for Microgrids: A Survey on Classical and Machine Learning-Based Methods. *Sustainability* **2023**, *15*, 8952. [CrossRef]
34. Bai, K.; Sindhu, V.; Haque, A. Fault Ride Through approach for Grid-Connected Photovoltaic System. *e-Prime-Adv. Electr. Eng. Electron. Energy* **2023**, *5*, 100232. [CrossRef]
35. Benbouzid, M.; Elghali, B.; El, A.; Bouzid, M.; Zerrougui, M.; El-Ebiary, A.H.; Attia, M.A.; Marei, M.I.; Sameh, M.A. An Integrated Seamless Control Strategy for Distributed Generators Based on a Deep Learning Artificial Neural Network. *Sustainability* **2022**, *14*, 13506. [CrossRef]
36. Lin, F.J.; Liao, J.C.; Chen, C.I.; Chen, P.R.; Zhang, Y.M. Voltage Restoration Control for Microgrid with Recurrent Wavelet Petri Fuzzy Neural Network. *IEEE Access* **2022**, *10*, 12510–12529. [CrossRef]
37. Conchas, R.F.; Sanchez, E.N.; Ricalde, L.J.; Alvarez, J.G.; Alanis, A.Y. Sensor fault-tolerant control for a doubly fed induction generator in a smart grid. *Eng. Appl. Artif. Intell.* **2023**, *117*, 105527. [CrossRef]
38. Khooban, M.H.; Gheisarnejad, M. A Novel Deep Reinforcement Learning Controller Based Type-II Fuzzy System: Frequency Regulation in Microgrids. *IEEE Trans. Emerg. Top. Comput. Intell.* **2021**, *5*, 689–699. [CrossRef]
39. Yan, R.; Wang, Y.; Xu, Y.; Dai, J. A Multiagent Quantum Deep Reinforcement Learning Method for Distributed Frequency Control of Islanded Microgrids. *IEEE Trans. Control. Netw. Syst.* **2022**, *9*, 1622–1632. [CrossRef]
40. He, X.; Ge, S.; Liu, H.; Xu, Z.; Mi, Y.; Wang, C. Frequency regulation of multi-microgrid with shared energy storage based on deep reinforcement learning. *Electr. Power Syst. Res.* **2023**, *214*, 108962. [CrossRef]
41. Egbomwan, O.E.; Liu, S.; Chaoui, H. Twin Delayed Deep Deterministic Policy Gradient (TD3) Based Virtual Inertia Control for Inverter-Interfacing DGs in Microgrids. *IEEE Syst. J.* **2023**, *17*, 2122–2132. [CrossRef]

42. Yang, M.; Wu, X.; Loveth, M.C. A Deep Reinforcement Learning Design for Virtual Synchronous Generators Accommodating Modular Multilevel Converters. *Appl. Sci.* **2023**, *13*, 5879. [CrossRef]
43. You, W.; Yang, G.; Chu, J.; Ju, C. Deep reinforcement learning-based proportional–integral control for dual-active-bridge converter. *Neural Comput. Appl.* **2023**, *35*, 17953–17966. [CrossRef]
44. Kosuru, R.; Liu, S.; Shi, W. Deep Reinforcement Learning for Stability Enhancement of a Variable Wind Speed DFIG System. *Actuators* **2022**, *11*, 203. [CrossRef]
45. Nicola, M.; Nicola, C.I.; Ionete, C.; Şendrescu, D.; Roman, M. Improved Performance for PMSM Sensorless Control Based on Robust-Type Controller, ESO-Type Observer, Multiple Neural Networks, and RL-TD3 Agent. *Sensors* **2023**, *23*, 5799. [CrossRef]
46. Wang, H.; Vukosavic, S.N.; Katic, V.; Nicola, M.; Nicola, C.I.; Selis, D. Improvement of PMSM Sensorless Control Based on Synergetic and Sliding Mode Controllers Using a Reinforcement Learning Deep Deterministic Policy Gradient Agent. *Energies* **2022**, *15*, 2208. [CrossRef]
47. Prasad, R.; Kumar, D.; Chand, S.; Fagiolini, A.; Mudaliar, H.; Benedetto, M.; Cirrincione, M. Enhancing Speed Loop PI Controllers with Adaptive Feed-forward Neural Networks: Application to Induction Motor Drives. In Proceedings of the 2022 International Conference on Electrical Machines and Systems (ICEMS 2022), Chiang Mai, Thailand, 29 November–2 December 2022. [CrossRef]
48. Hu, J.; Shan, Y.; Cheng, K.W.; Islam, S. Overview of Power Converter Control in Microgrids - Challenges, Advances, and Future Trends. *IEEE Trans. Power Electron.* **2022**, *37*, 9907–9922. [CrossRef]
49. Mohammadi, F.; Mohammadi-Ivatloo, B.; Gharehpetian, G.B.; Ali, M.H.; Wei, W.; Erdinc, O.; Shirkhani, M. Robust Control Strategies for Microgrids: A Review. *IEEE Syst. J.* **2022**, *16*, 2401–2412. [CrossRef]
50. Chand, S.; Kumar, R.; Prasad, R.; Cirrincione, M.; Raj, K. Open Circuit (OC) and Short Circuit (SC) IGBT Switch Fault Detection in Three-Phase Standalone Photovoltaic Inverters Using Shallow Neural Networks. In Proceedings of the 2022 International Conference on Electrical Machines and Systems (ICEMS 2022), Chiang Mai, Thailand, 29 November–2 December 2022. [CrossRef]
51. Xu, J.; Qian, H.; Hu, Y.; Bian, S.; Xie, S. Overview of SOGI-Based Single-Phase Phase-Locked Loops for Grid Synchronization under Complex Grid Conditions. *IEEE Access* **2021**, *9*, 39275–39291. [CrossRef]
52. Li, K.; Bo, A.; Zheng, H.; Sun, N. Performance Analysis of Three-Phase Phase-Locked Loops for Distorted and Unbalanced Grids. *J. Power Electron.* **2017**, *17*, 262–271. [CrossRef]
53. Zhang, R.; Chand, S.S.; Hredzak, B.; Bie, Z. A Novel Synchronization Strategy for Distributed Energy Resources in Weak Grids Using Remote Strong Grid Sensing. *Energies* **2024**, *17*, 6135. [CrossRef]
54. Chand, S.; Prasad, R.; Mudaliar, H.; Kumar, D.; Fagiolini, A.; Benedetto, M.; Cirrincione, M. Improving Power Delivery of Grid-Connected Induction Machine Based Wind Generators under Dynamic Conditions Using Feedforward Linear Neural Networks. *IEEE Access* **2023**, *11*, 63550–63564. [CrossRef]
55. Khan, M.Y.A.; Liu, H.; Yang, Z.; Yuan, X. A Comprehensive Review on Grid Connected Photovoltaic Inverters, Their Modulation Techniques, and Control Strategies. *Energies* **2020**, *13*, 4185. [CrossRef]
56. Chand, S.; Prasad, R.; Mudaliar, H.; Kumar, D.; Fagiolini, A.; Di Benedetto, M.; Cirrincione, M. Enhanced Current Loop PI Controllers with Adaptive Feed-Forward Neural Network via Estimation of Grid Impedance: Application to Three-Phase Grid-Tied PV Inverters. In Proceedings of the 2022 IEEE Energy Conversion Congress and Exposition (ECCE 2022), Detroit, MI, USA, 9–13 October 2022. [CrossRef]
57. Subsingha, W. A Comparative Study of Sinusoidal PWM and Third Harmonic Injected PWM Reference Signal on Five Level Diode Clamp Inverter. *Energy Procedia* **2016**, *89*, 137–148. [CrossRef]
58. Tarequzzaman, M.; Ahmed, S.; Moznuzzaman, M. Performance Improvement of Direct Torque Control Induction Motor Drive using Genetic Algorithm Optimized PI Controller. In Proceedings of the 2018 International Conference on Advancement in Electrical and Electronic Engineering (ICAEEE 2018), Gazipur, Bangladesh, 22–24 November 2018. [CrossRef]
59. Yang, Q.; Dai, N. Genetic Algorithm for PI Controller Design of Grid-connected Inverter based on Multilayer Perceptron Model. In Proceedings of the 2021 IEEE Sustainable Power and Energy Conference: Energy Transition for Carbon Neutrality (iSPEC 2021), Nanjing, China, 23–25 December 2021; pp. 2770–2775. [CrossRef]
60. Debdouche, N.; Zarour, L.; Chebabhi, A.; Bessous, N.; Benbouhenni, H.; Colak, I. Genetic algorithm-super-twisting technique for grid-connected PV system associate with filter. *Energy Rep.* **2023**, *10*, 4231–4252. [CrossRef]
61. de Oliveira Evald, P.J.D.; Hollweg, G.V.; Borin, L.C.; Mattos, E.; Tambara, R.V.; Montagner, V.F.; Gründling, H.A. An optimal initialisation for robust model reference adaptive PI controller for grid-tied power systems under unbalanced grid conditions. *Eng. Appl. Artif. Intell.* **2023**, *124*, 106589. [CrossRef]
62. Ali, M.; Tariq, M.; Upadhyay, D.; Khan, S.A.; Satpathi, K.; Alamri, B.; Alahmadi, A.A. Genetic Algorithm Based PI Control with 12-Band Hysteresis Current Control of an Asymmetrical 13-Level Inverter. *Energies* **2021**, *14*, 6663. [CrossRef]
63. Yang, Y.; Kim, K.A.; Blaabjerg, F.; Sangwongwanich, A. Control of PV systems under normal grid conditions. In *Advances in Grid-Connected Photovoltaic Power Conversion Systems*; Woodhead Publishing Ltd.: Cambridge, UK, 2019; pp. 75–112. [CrossRef]

64. Neal, C.; Dagdougui, H.; Lodi, A.; Fernandez, J.M. Reinforcement Learning Based Penetration Testing of a Microgrid Control Algorithm. In Proceedings of the 2021 IEEE 11th Annual Computing and Communication Workshop and Conference (CCWC 2021), Virtual, 27–30 January 2021; pp. 38–44. [CrossRef]
65. Fujimoto, S.; Van Hoof, H.; Meger, D. Addressing Function Approximation Error in Actor-Critic Methods. In Proceedings of the 35th International Conference on Machine Learning (ICML 2018), Stockholm Sweden, 10–15 July 2018; Volume 4, pp. 2587–2601.

energies

MDPI

Article

Planning for Medium- and Heavy-Duty Electric Vehicle Charging Infrastructure in Distribution Networks to Support Long-Range Electric Trucks

Joshua Then, Ashish P. Agalgaonkar * and Kashem M. Muttaqi

School of Electrical, Computer and Telecommunications Engineering, University of Wollongong, Wollongong, NSW 2500, Australia; jt381@uowmail.edu.au (J.T.); kashem@uow.edu.au (K.M.M.)
* Correspondence: ashish@uow.edu.au

Abstract: Electrification of the transport sector introduces operational issues in the electricity distribution network, such as excessive voltage deviation, substation overloading, and adverse power quality impacts on other network loads. These concerns are expected to grow as electrification expands to incorporate heavy vehicles such as trucks and buses due to their greater energy requirements and higher charging loads. Two strategies are proposed to support medium- and heavy-duty chargers which address their high power demand and mitigate power quality disturbances and the overloading of substations. The first is a dedicated feeder connected at the secondary of the substation directly to the charging station which aims to reduce the impact of high load on other customers. The second is the addition of a dedicated substation that solely provides power for charging stations in major corridors, alleviating stress on existing zone substations. Hosting capacity is measured using a voltage deviation index, describing the deviation in line voltage, which should experience a sag of no more than 6% of the nominal voltage, and a substation charging capacity index, describing the available capacity of each zone substation as a ratio of its total power capacity. Verification of the proposed strategies was performed on an MV-LV distribution network representative of an industrial Australian town with heavy-vehicle charging. Results showed that the network could handle ten 250 kW chargers, which was tripled to 35 with a dedicated feeder. The dedicated feeder alternatively allowed up to 10 megawatt-scale chargers, which was again tripled when a dedicated substation was introduced.

Keywords: electric vehicles; demand-side management; distributed generation; distribution system reconfiguration

1. Introduction

In Australia, the transport sector accounts for 19% of the national greenhouse gas emissions. Trucks and buses make up 21% of this, while light commercial vehicles make up 17%, leading to a combined 38% of transport emissions coming from road freight [1]. The Australian government is aiming to achieve a net-zero greenhouse gas emission target by 2050, with the aim of producing 43% less than 2005 levels by 2030. Heavy-duty vehicles account for 4% of road vehicles in Australia, with 500,000 rigid and 100,000 articulated trucks, but consume 23% of the fuel used in the transport sector [2]. Studies performed in the US found that electric trucks could reduce emissions by around 50%, with a similar reduction in energy consumption when they were used in urban driving with frequent stops [3,4].

 https://doi.org/10.3390/en18040785

The electrification of trucks is relatively new due to improvements in batteries that have enabled trucks to carry onboard batteries without becoming excessively heavy. This was an issue which impeded the development of electric trucks despite the ongoing electrification of the light-vehicle market. Improved battery technology is necessary to enable trucks to maintain a suitable driving range while minimising vehicle tare weight, which, in turn, maximises vehicle payload. Electric trucks are increasing in popularity due to decreasing battery prices and the increasing number of electric vehicle (EV) chargers that are being installed. They have been found to be 55% more energy efficient than the equivalent diesel truck, having an average energy consumption of 1.6 kWh/km [5].

The suitability of electric trucks to replace the conventional diesel fleet is dependent largely on driving range and payload. Electric trucks have a higher capital cost but lower operating costs due to the relatively lower price of electricity compared to diesel and simpler maintenance due to a lower number of moving parts. Diesel makes up 20% of short-haul costs and 35% of long-haul costs. A 22-tonne truck driving 300 km would require up to 85 L of diesel for AUD 116.34, while the equivalent electric truck would require 280 kWh of electricity for AUD 42 [2]. However, the requirement for heavy batteries can reduce the maximum payload that can be carried, which, along with routing constraints due to charging needs, may result in a larger fleet being used to meet the same demands as a diesel fleet. This diminishes the benefits gained from the improved drivetrain efficiency and reduced energy costs of running electric trucks. This makes the electrification of short-haul trucks more viable due to their shorter driving distances, low-speed driving, frequent stopping due to loading and unloading, predictable routes with destinations at depots and warehouses, and primarily day-time driving [6,7]. However, these cases only represent a fraction of the current usage of trucks. A study of Switzerland and Finland found that, with current technologies, electric trucks can only replace around 50% of trips [8]. Utilisation of electric trucks for long-haul trips will require further improvements in battery technology and an increase in the number of charging stations en route. The combination of fast charging and smaller batteries improves EV competitiveness as the lighter batteries allow for an increased payload but requires fast charging at intervals of less than 300 km [9]. The electrification of larger trucks also increases competitiveness as the payoff from a greater payload outweighs the increased energy consumption and vehicle costs [9]. Vehicle-to-grid and grid-to-vehicle operations can be performed to improve the utilisation of renewable energy by charging and discharging electric trucks according to the generated power and load demand [10].

Truck charging strategies can be grouped into three categories. The first is overnight slow charging using a DC charger, which provides between 50 and 100 kW over the course of 6 to 8 h. A suitable charging management strategy can be utilised to perform peak shaving and load balancing which alleviates the stress on the electricity network and reduces company costs, especially if variable tariffs are in place. The second category is destination charging, which is necessary if vehicles have smaller batteries than those needed to meet the daily travel distance. Vehicles such as the Navistar E-Star and Smith Newton have a driving range of less than 200 km, which is sufficient for a single delivery but not for a full day of driving [6]. Destination charging is performed at warehouses, logistic centres, and delivery hubs using fast chargers in the range of 150 to 400 kW. Charging is performed in the approximately hour-long time window that naturally occurs during vehicle unloading and loading. Highway charging is performed at urban nodes and highway charging stations using high-power chargers of up to 1.2 MW. Charging takes place during the half-hour mandatory driving breaks, minimising delays. This form of charging is used for long-haul transport which may not stop in any urban areas or delivery sites over the course of a day. Charging stations are being considered for highways to accommodate the longer road-trips of light-duty vehicles, with charging capacities of up to 300 kW [11]. Wireless charging has also been proposed and studied as an alternate means of charging. Its usage for electric trucks would reduce the battery size required while removing concerns over driving range, in addition to reducing downtime due to charging [12]. However, it presents significant costs due to installation and maintenance. Leakage flux from the large air gap reduces power transfer efficiency, and poses health and safety concerns. Additionally, the location of wireless chargers is restricted to areas with access to the distribution network.

A study from the US found that around 70% of heavy-duty vehicles drive less than 100 miles a day, while around 10% drive over 500 miles a day. However, the energy consumption of trucks is skewed, with 40% coming from trucks driving at least 500 miles [7]. A study on the long-haul industry in Europe found that, based on driving distances, stop location and duration, and energy requirements, a charging station should have four to five times more overnight chargers than high-power chargers. An average of eight overnight chargers (50–100 kW) and two megawatt-scale chargers (0.7–1.2 MW) was considered sufficient [13]. A road freight model for the greater Sydney metropolitan area in Australia found that there were approximately 36 million truck movements annually, which, combined, travelled over 6 billion kilometres and emitted almost 1000 tonnes of CO_2 in 2021 [14]. A study of short-haul trucks in south-east Queensland in Australia compared the coverage of electric trucks with the number of charging stations in the region, with eight sites being sufficient for driving distances of up to 100 km [15].

The high charging rate required to make heavy-duty EVs competitive presents an issue for the electricity network. The modern distribution network can accommodate light-duty EVs but will require significant upgrades to support the megawatt-scale chargers demanded by long-haul EVs. A study of Texas found that the simultaneous charging of 11% of heavy-duty vehicles, 30,000 vehicles in total, would lead to voltage violations and reliability issues, with only a few dozen EVs needed to have adverse effects on a local distribution grid [16]. The total annual charging for the US truck fleet with complete electrification would require 184 TWh, which is equivalent to 5% of the total US energy consumption [5]. Smart charging can be utilised for load shaping by modulating EV charging rates to smooth the load curve and distribute demand throughout the day [17]. However, this is limited by the available charging options and charging demand, along with the requirement for a centralised controller.

This paper aims to examine the impact of the destination charging of electric trucks on the urban electricity network. The hosting capacity of the electricity network when presented with the requirements of truck charging will be calculated to determine if the current network is sufficient. Indices are used to quantify the hosting capacity based on the bus voltages throughout the network and the transformer loading at each substation. The main contributions of this paper include:

- Utilisation of two indices to measure the hosting capacity of a distribution network for electric vehicle charging.
- Proposal of the introduction of a dedicated feeder connected at the secondary of the zone substation transformers to provide power for fast-charging stations while mitigating power quality issues in parallel feeders.
- Proposal of the usage of a dedicated zone substation connected to the transmission substation to provide power for a large fast-charging station with megawatt-scale chargers while minimising the impact on the distribution grid.

2. Proposed Methodology

The transition from internal combustion engines to electrified vehicles will require an increase in the number of electric charging stations. Coincidentally, this will also result in a reduction in the number of fuel station required. As such, the replacement of fuel stations with EV charging stations presents a suitable solution to address both concerns. This practice is currently being performed by some fuel companies in Australia on a small scale, an example being Ampol, which has 23 EV charging stations. This solution removes the need for locations to be traded between owners and reduces the construction required as some infrastructure, such as the convenience store and garage, will be retained. Figure 1 shows a futuristic concept for an en-route charging station, often located along motorways and other major corridors. It includes fast and slow charging for both light-duty passenger EVs and heavy-duty commercial EVs, along with rest-stop amenities and food services. Truck charging is estimated to have a capacity of 1 MW, but development of the megawatt-scale charging system could lead to a charging rate of up to 3.75 MW [18]. This type of charging station would either replace or be used in conjunction with the existing petrol stations and rest-stops currently present on major road corridors and motorways.

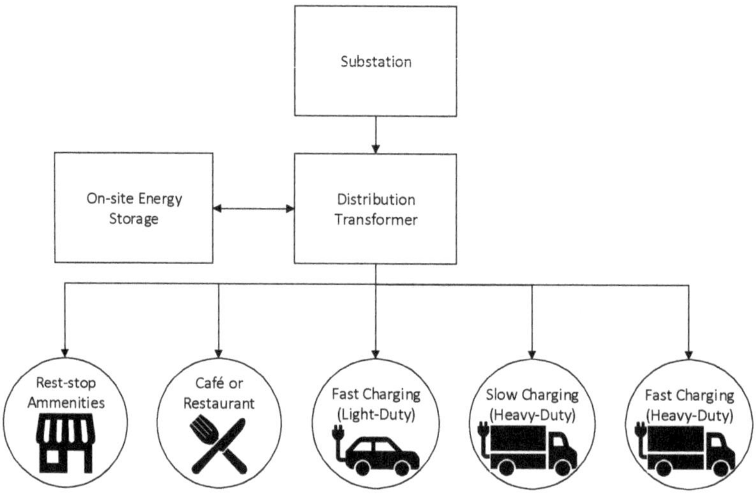

Figure 1. Futuristic concept for motorway rest-stop with light-duty and heavy-duty charging stations.

The charging of EVs is constrained by the state of charge (SOC) of the battery and the maximum charging rate that the onboard electronics allow. The maximum SOC of the battery is simply determined by the capacity of the battery, while the minimum SOC is a value greater than zero required to maintain battery life, typically between 10% and 20% of the maximum SOC. For this study, a 15% minimum SOC is used. The maximum charging rate is limited by onboard electronics, rather than the charger, and may result in a charging rate lower than what the charger can provide. However, for the proposed study, it is assumed that the heavy-duty EVs have electronics capable of charging at the maximum power output of the charging equipment. These are shown in (1) and (2) below.

$$0.15\ SOC_{max} \leq SOC \leq SOC_{max} \quad (1)$$

$$0 \leq P \leq P_{max} \quad (2)$$

where SOC and SOC_{max} are the vehicle state-of-charge and maximum state-of-charge, respectively, and P and P_{max} are the instantaneous and maximum charging rates of the vehicle, respectively. Other constraints to charging capacity are the maximum feeder load and the mitigation of reverse power flow if PV units are installed. Mitigation of feeder losses requires the charging station to be located close to the substation, limiting access for larger vehicles. Two indices are proposed to determine the hosting capacity of the network for medium- and heavy-duty electric vehicle charging, based on those presented in [19].

2.1. Substation Charging Capacity

The first index proposed determines the size of the charging station based on the available capacity of the substation. The substation charging capacity (SCC) describes the size and number of chargers that can be used at any time. This is given in (3). The SCC is the remaining capacity of the substation as a ratio of its maximum capacity when loaded. As such, a value of 1 indicates that the substation is completely unloaded and has its full capacity available, while a value of 0 indicates that the substation is fully loaded and cannot provide extra power for extended periods of time. Scheduling is performed by EV owners according to this value and charging demand to ensure that overloading of the substation does not occur. This can be performed by either the truck owner or their respective company according to delivery destinations and loading times, leading to a predictable timeline that, in some cases, can be planned several days or weeks in advance.

$$SCC_{feed} = 1 - \frac{|V_{zs}^{a} I_{zs}^{a}| + |V_{zs}^{b} I_{zs}^{b}| + |V_{zs}^{c} I_{zs}^{c}|}{S_{zs}} \tag{3}$$

where V_{zs}^{a}, V_{zs}^{b}, and V_{zs}^{c} are the voltages at the secondary of each phase of the substation, I_{zs}^{a}, I_{zs}^{b}, and I_{zs}^{c} are the current deliveries for each phase of the substation, and S_{zs} is the $N-1$ rating of the zone substation. The voltages and currents are multiplied to determine the overall power flow as power flow cannot be directly measured. A positive value indicates that there is available capacity at the substation, with a value of 1 indicating that the total capacity is available. A negative value indicates that the substation is overloaded.

The SCC has limits of 0 and 1, given in (4), to represent when loading has met substation capacity and when the substation is completely unloaded, respectively. A value greater than the upper limit would indicate reverse power flow through the zone substation from generators located downstream of the substation, such as distributed rooftop PV. Reverse power flow may cause maloperation of tap changers and voltage regulators, and increased voltage on the primary side of the transformer. Reverse power flow will not be examined for the proposed methodology as the additional load from truck charging is expected to far outweigh the power generated by distributed PV units. The addition of PV units could potentially increase the hosting capacity of the network for EV chargers during daylight hours.

$$0 < SCC \leq 1 \tag{4}$$

2.2. Voltage Deviation Index

The voltage deviation in the network is used to calculate a further index. Voltage sag occurs due to voltage losses in feeders due to line impedances and occurs in all areas of the network with load current. Excessive low voltage may lead to malfunctioning load equipment and blackouts in extreme cases. According to Australian Standard AS 60038, the limits for 230 V line voltages are +10% and −6% of the nominal voltage [20,21]. A voltage sag greater than this is unacceptable and indicates that the network is overloaded. The high current draw required for fast EV charging is expected to cause notable voltage sag within the feeder, especially when multiple EVs charge simultaneously at a local charging station. A voltage deviation index (VDI) is used to describe the sag in the network at various nodes, shown in (5). This is calculated from the ratio of the feeder voltage to its nominal voltage.

A value of 0 shows that the feeder is operating at its nominal voltage, with positive values indicating voltage drop and negative values indicated voltage rise. The 6% limit for voltage sag corresponds to a VDI of 0.06.

$$VDI(k) = 1 - \frac{V(k)}{V_{nom}} \tag{5}$$

where V_{nom} is the nominal voltage of the feeder and $V(k)$ is the feeder voltage at the k-th time step. The VDI is calculated for the feeder with the charging stations to observe the voltage deviation that arises from EV charging. The voltage deviation in parallel feeders connected to other customers will also be studied to examine the impact of EV charging on other parts of the network. Changes in network topology will be proposed to reduce the impact EV charging has on other customers in the distribution network to reduce disruptions to normal loads.

The VDI has limits of -0.1 and 0.06, given in (6) below. The upper limit corresponds to the 6% voltage sag discussed earlier. The lower limit represents a maximum 10% voltage rise in line voltages. Voltage rise may be caused by reverse power flow from distributed generation. Voltage rise can lead to increased line currents and overheating, along with maloperation of tap changers and voltage regulators. Voltage rise will not be extensively examined in this paper as the large load presented by truck chargers is expected to significantly outweigh any generated power from distributed PV.

$$-0.1 \leq VDI \leq 0.06 \tag{6}$$

2.3. Practical Implementation

In a typical distribution network, EV chargers are connected to existing feeders at the location of the charging station. This leads to limitations in charging capacity due to a limited combined line current and voltage sag. Substation capacity will also affect the maximum charging capacity. The adverse effects of excessive charging and voltage sag will impact other customers on the feeder and may be visible to customers on parallel feeders.

A typical LV distribution network can be summarised as shown in Figure 2a, in which the parts highlighted in blue indicate normal infrastructure and normal loads and the charging infrastructure for heavy-duty commercial vehicles is highlighted yellow. In this network, the transmission substation connects to multiple zone substations which each feed into their own local loads. These may be some combination of MV industrial loads and LV commercial and residential loads fed via a distribution transformer. EV fast charging will be present at the same level as the LV commercial and residential loads, and utilise the same feeders. Two separate distribution system topology upgrades are proposed and tested for this distribution network. The first is the introduction of a dedicated feeder, while the second is the introduction of a dedicated substation. These two were briefly described in [7] from an economical perspective. The introduction of a feeder was estimated to cost USD 400,000, while the substation would cost at least ten times this. However, the greater capacity of the substation allowed it to pay itself off within 4 years, while the feeder would require up to a year.

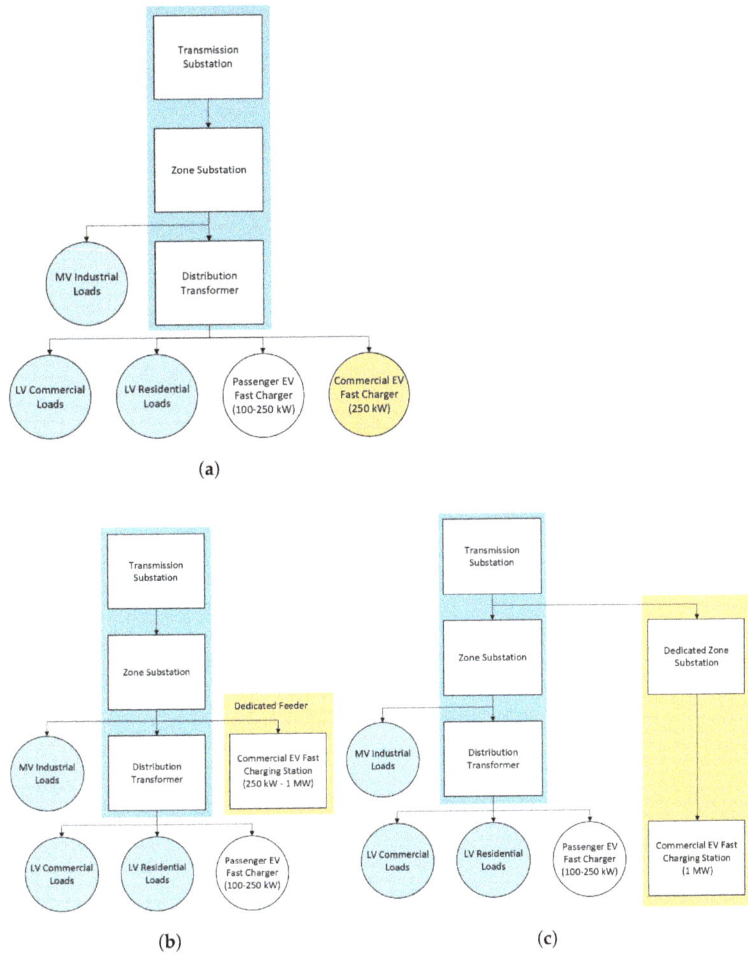

Figure 2. Typical network configuration and proposed network reconfigurations. (**a**) Simplification of a typical network topology. (**b**) Distribution network with a dedicated feeder. (**c**) Distribution network with a dedicated substation for commercial EV charging.

2.3.1. Dedicated Feeder Strategy

The first strategy introduces a dedicated feeder for charging stations which will connect directly to the substation. This will reduce impacts on the normal substation load connected to parallel feeders. This method is the simpler and cheaper of the two proposed but is limited as the capacity of chargers that can be used depends on the normal available capacity of the substation. Charging is controlled to ensure the substation is not overloaded during operation. The shift of the commercial EV charging station from shared LV feeders to a dedicated feeder is shown in Figure 2b, with the same highlighting as that of the previous figure.

2.3.2. Dedicated Substation Strategy

The second topology upgrade is the introduction of a dedicated substation as a means of meeting the electricity demand of the charging station without impacting the power quality provided to other customers within the network. Similar systems exist in countries with high EV adoption such as Sweden [22]. The dedicated substation reduces line losses due to a combination of reduced line currents and shorter feeders as the transformers can be located close to the charging station. Additionally, the substation can be sized such that it can meet the expected charging demand, reducing the need to schedule charging and accommodate normal customer loads. The network layout with the dedicated substation can be represented as shown in Figure 2c. The transmission substation and generator capacities are still constraints that must be considered. Locating the substation and charging station will require a large open area. However, greater flexibility is offered as they can be located near a main road, motorway, or industrial area, providing easier access for long-haul trucks. If the substation is still insufficient, or if multiple heavy-duty charging stations are required, then multiple dedicated substations can be introduced and located in separated areas of the distribution network.

The process for the proposed methodology is shown in Figure 3, where blue blocks are the start and finish, green blocks are actions and, orange blocks are decision points. It begins with the modelling and simulating of the network without truck charging to set up the base case. This verifies the test network as suitable for hosting heavy-duty charging. Distributed truck charging is then added to the LV network to determine the hosting capacity without any of the proposed changes to network configuration. Repeated simulations are performed, with the number of chargers being increased until one of the hosting capacity indices is reached. The two proposed network configurations are then introduced. Firstly, the dedicated feeder, followed by the dedicated substation. For both strategies, the network is simulated, and the resulting data are used to calculate the hosting capacity indices. If either index exceeds its allowable limits, then the hosting capacity is reached.

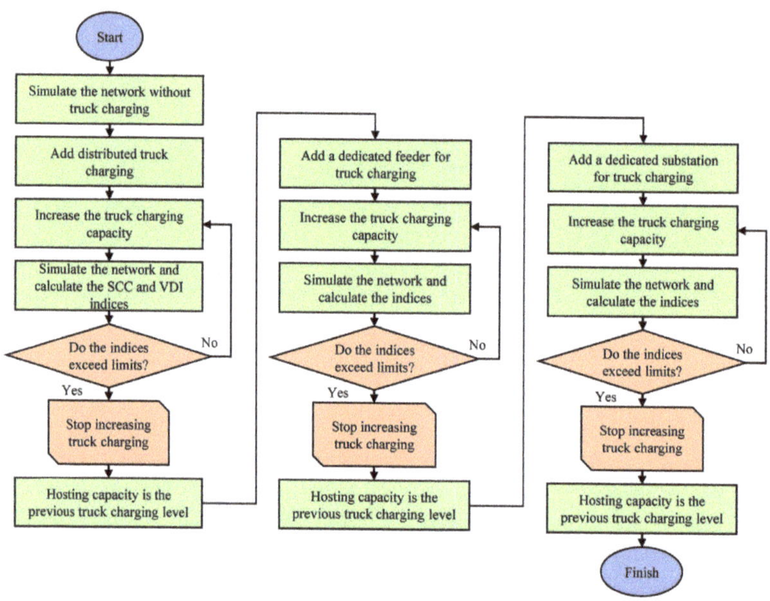

Figure 3. Proposed methodology.

3. Case Study

Verification of the proposed methodology for facilitating electric truck charging was performed using a modified test network. A 33 kV transmission substation with a downstream network of ten 11 kV zone substations was used to simulate the substation network of an Australian industrial city, shown in Figure 4a. Mesh connections exist between some zone substations due to their physical distance from the transmission substation. The substation labelled ZS7 contains purely industrial MV loads and, while simulated, will not be modified with any EV charging. An MV-LV distribution network was made based on the IEEE 13-bus network [23] with modifications to suit Australian standards with a 400 V, three-phase residential network based on a semi-rural Australian town. These modifications include the removal of capacitor banks and adjustments to line and transformer parameters to suit the different voltage used. A realistic representation of an LV distribution network in a semi-rural Australian town, based on the network data from [24], was added to the network model at node no. 634, as shown in Figure 4b. The LV network includes two feeders with four nodes each. Each residence has an EV comparable to a Nissan Leaf, as it is presumed that light-duty EVs will be adopted before heavy-duty EVs. These each have an 62 kWh battery connected to a 6.6 kW charger from the mains 240 V AC supply according to SAE J1772 [25]. Each residence also has a 5 kW PV unit, a typical size for Australian rooftop solar panels, under the assumption that rooftop solar power will be adopted concurrently with EV adoption. A power curve, given in [26], was used to estimate the power output of the PV units throughout the day. These PV systems are not intended to assist in the charging of heavy-duty vehicles due to their small size.

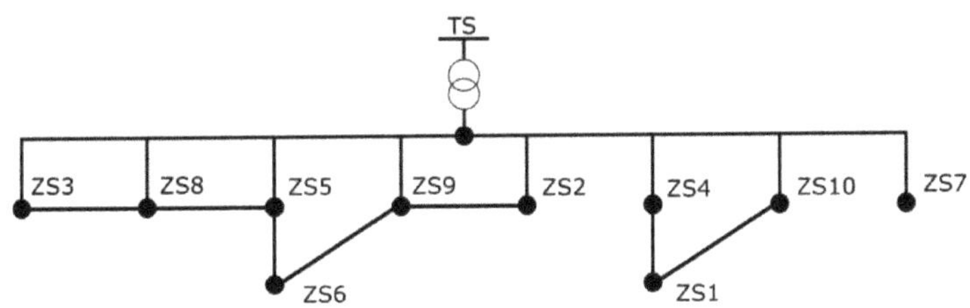

(**a**) Transmission and zone substations.

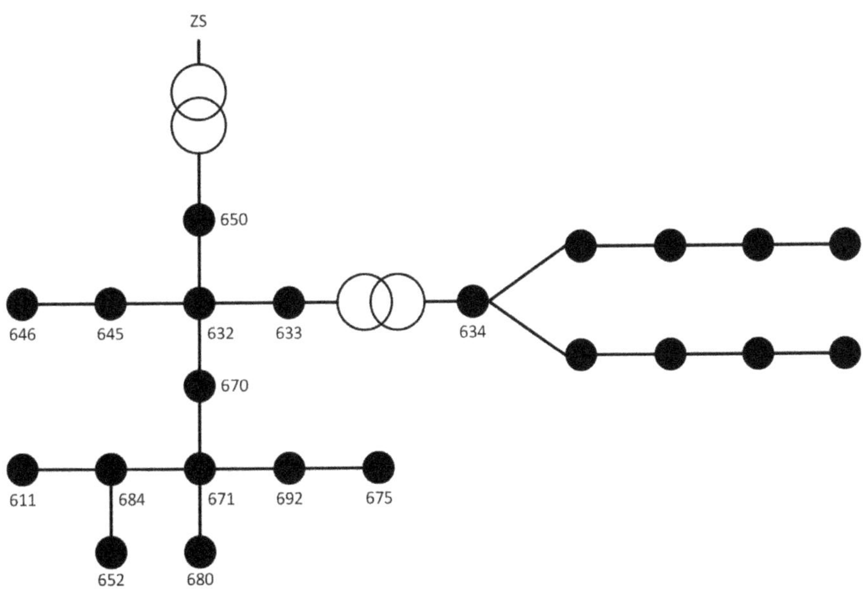

(**b**) Modified IEEE 13-bus network.

Figure 4. Test network layout.

Verification of the proposed strategies are performed through simulations of the distribution network in OpenDSS (9.8.0.1.). Each run simulates a 24-h period, recording nodal voltages, line currents and power flows. Data is sent to MATLAB (2023b) directly for analysis which determines the hosting capacity indices of the network and decides whether the network is capable to handling additional load. Network modifications are made in MATLAB which are returned to OpenDSS for the next run. Control strategies can be performed via MATLAB and Simulink (2323b), but this is outside the scope of this paper.

Three scenarios are used to examine the performance of the network with the proposed strategies, in line with the process described in Figure 3. The procedure undertaken to perform a single simulation run in MATLAB and OpenDSS is explained below:

1. Determine scenario and charging rate to be tested. The scenarios will be described later.
2. Define light-duty EV battery capacities and charging rates, and rooftop PV power output.
3. If truck charging is to be performed then initialise the network with the minimum number of truck chargers.
4. Generate distribution networks for OpenDSS with transmission and zone substations, and with or without truck charging.
5. Run simulation in OpenDSS with 24 time steps of one hour, exporting nodal voltages, line currents, power flows and line losses to MATLAB.
6. Calculate hosting capacity indices in MATLAB using data extracted from the OpenDSS results.
7. Determine if hosting capacity limits have been exceeded. If not exceeded, return to step 4 with an increased number of chargers, otherwise end the simualation.

The base case simulates the network model with residential EV car charging without the presence of any truck charging. This establishes the performance of the network before the addition of high-power DC chargers. This scenario represents the current network, which has very few truck chargers on existing MV feeders in the distribution network. The chargers are 250 kW DC chargers, comparable to modern chargers provided by EVSE and ABB. Fast charging is used during the day, with trucks charging at the maximum rate provided by the chargers, then they switch to slow charging overnight at 10% of the maximum charging rate. The energy capacities of heavy-duty trucks from makers such as Kenworth, Peterbilt, and Freightliner range between 300 and 500 kWh. It is assumed that there will always be a truck connected to each charger and drawing power. Truck charging is equally distributed across ten MV nodes in the network in parallel to previously existing MV loads. The number of chargers is initially set to a low number and is gradually increased until the hosting capacity indices exceed their limit. At this point, the network is at its maximum hosting capacity for truck charging without introducing modifications to the distribution network.

The next scenario is a near-future prediction in which truck charging stations are built in urban centres and near logistic centres and warehouses. A new dedicated feeder is located close to each substation for truck charging, with a length of 50 m. This represents a feeder connected directly from the secondary of the substation to the charging station and will minimise disruptions and power quality issues for other customers connected to the substation. There will initially be one 1 MW truck charging station per zone substation, each with four 250 kVA chargers. Larger chargers, rated at 1 MW each, will also be tested, representative of the heavy-duty charging required further in the future. The number of charging stations is increased until the hosting capacity indices are exceeded. The hosting capacity indices will be monitored for both the dedicated feeder and parallel feeders in the network to monitor the impact of truck charging on other customers.

The final scenario represents a more futuristic view, with a greater number of electric trucks and highway charging. A dedicated substation is added to the network solely for the charging of electric vehicles to simulate a large charging station located on the motorway for en-route charging. The charging station consists of 1 MW chargers to allow for faster charging when en route to delivery destinations. A dedicated MV feeder is used between the transmission substation and the newly introduced substation to minimise the power quality impacts on other customers in the grid and reduce line losses. The capacity of the substation is 50 MVA, which is comparable to the other zone substations in the network.

Trucks will charge at the maximum charging rate offered by the chargers during working hours, to reflect fast charging during stops at destinations such as warehouses or logistics centres. Outside of working hours, slow charging is used to preserve battery life and reduce the demand on the grid.

4. Results

The network was first run without truck charging. The end-of-feeder voltage deviation index is shown in Figure 5a. Phase unbalance in both MV and LV loads tends to cause a variation in the phase voltages; however, voltage deviation never exceeds the maximum allowable amount of 6%. Power flow through the zone substations is shown in Figure 6a. None of the substations is excessively loaded at any time. These results verify that the network obeys operational limits during normal loading without the presence of heavy-duty EV charging.

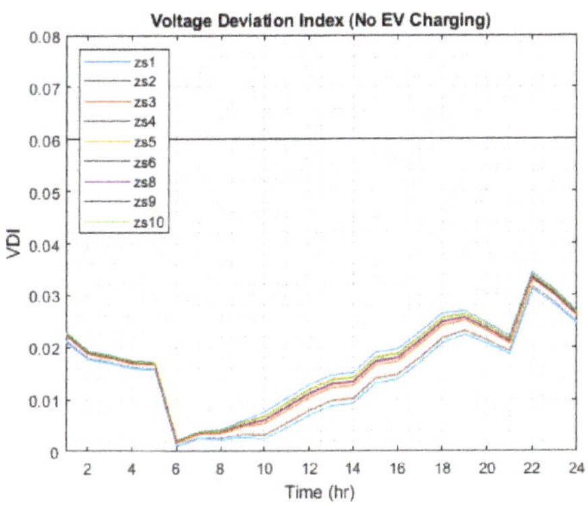

(**a**) Base case voltage deviation.

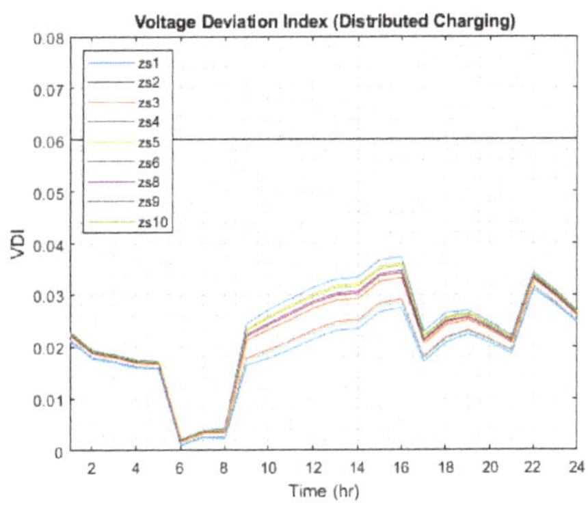

(**b**) Distributed charging voltage deviation.

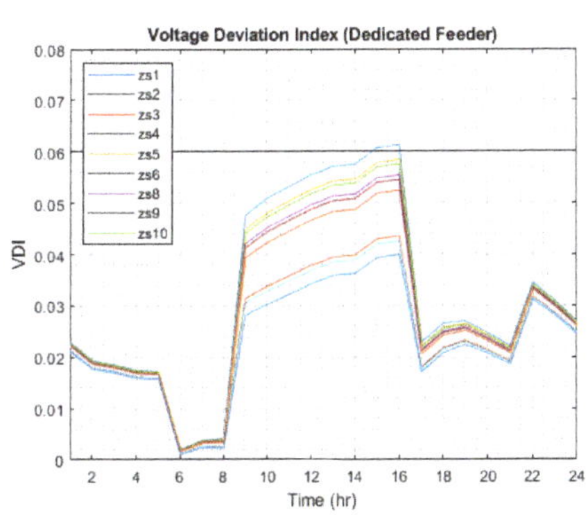

(**c**) Dedicated feeder voltage deviation.

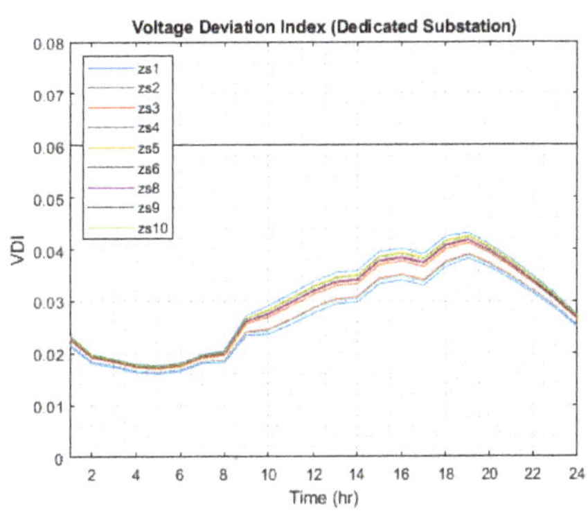

(**d**) Dedicated substation voltage deviation.

Figure 5. Voltage deviation index in different scenarios.

The first scenario with truck charging connects chargers to existing nodes within the distribution network. This is representative of the typical method currently used, which does not include any network upgrades to accommodate for the high load demand of truck chargers. Each zone substation is given ten 250 kW chargers, distributed to various MV nodes within the network. At ten heavy-duty EV chargers, the VDI of the network, shown in Figure 5b, is elevated at midday for all substations due to day-time charging, but does not exceed the 6% voltage drop allowed. However, substation power draw for zone substation 'zs5' exceeds its capacity during the middle of the day for five hours, as seen in Figure 6b, with another substation also experiencing a significant increase in loading during these times. As such, the hosting capacity of the network to the distributed 250 kW EV chargers is concluded to be 10.

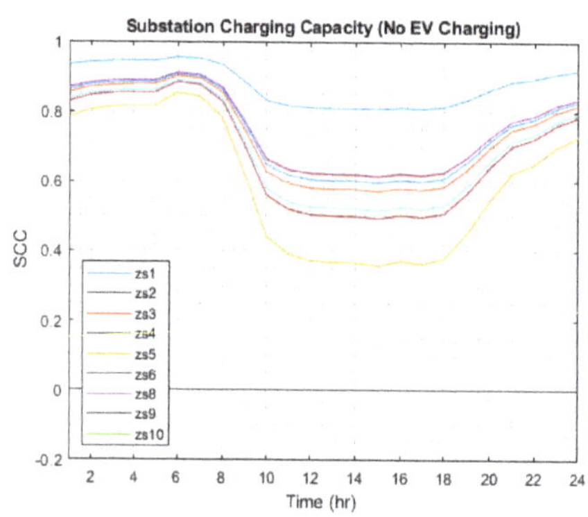

(**a**) Base case substation power delivery.

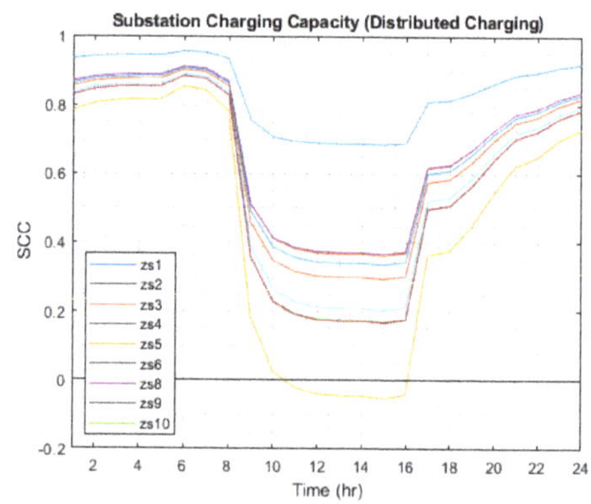

(**b**) Distributed charging substation power delivery.

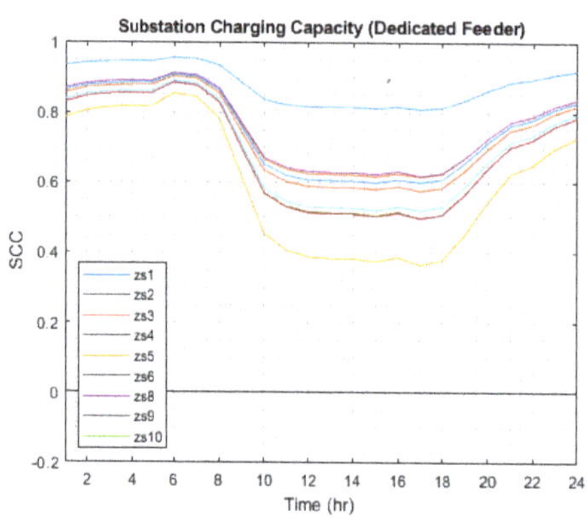

(**c**) Dedicated feeder substation power delivery.

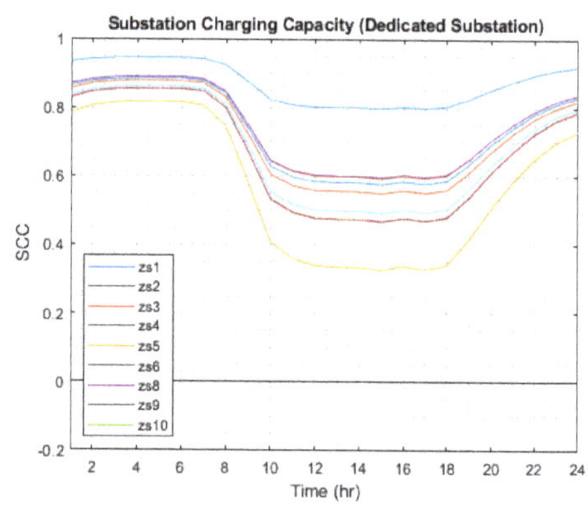

(**d**) Dedicated substation substation power delivery.

Figure 6. Substation charging capacity index in different scenarios.

To reduce line losses and substation power draw, a dedicated feeder is added at the secondary to the substation. Heavy-duty EV chargers are repositioned from their distributed locations to this dedicated feeder. With the current 250 kW chargers, the hosting capacity is increased to 35 chargers, at which point, customers at the feeder ends of one substation experience a voltage deviation greater than the allowable 6% drop, as shown in Figure 5c. Substation loading is reduced, shown in Figure 6c, and is not as severely affected as in the previous scenario with distributed charging. The reconfigured network increases the hosting capacity by three times and shifts the limit from the substation capacity to the voltage drop at the feeders. Increasing the individual charger size from 250 kW to 1 MW leads to similar results occurring when ten chargers are installed.

The final scenario is the introduction of a dedicated substation for truck charging. The voltage profiles and deviation index are shown in Figure 5d and substation power flow in Figure 6d. Shifting to the dedicated substation separates the normal customers and heavy-duty EV chargers. This results in voltages and currents downstream of the normal substation being almost unchanged from the base case scenario as the substation is no longer serving the load presented by truck chargers. The hosting capacity limit for the network is found to be 30 of the larger 1 MW truck chargers, after which, the voltage deviation observed exceeds the allowable 6% limit. This is seen in Figure 7, which shows the voltage profile of the three phases and deviation index for the new dedicated substation, where the horizontal lines show the nominal voltage and the allowed voltage sag of 6%. The substation does not experience any overloading and can be sized according to the maximum number of chargers that can be hosted. Variations in substation capacity can be applied as electric truck adoption increases since the substation is only required to serve one load. Additionally, excessive voltage deviation can be mitigated using step changers.

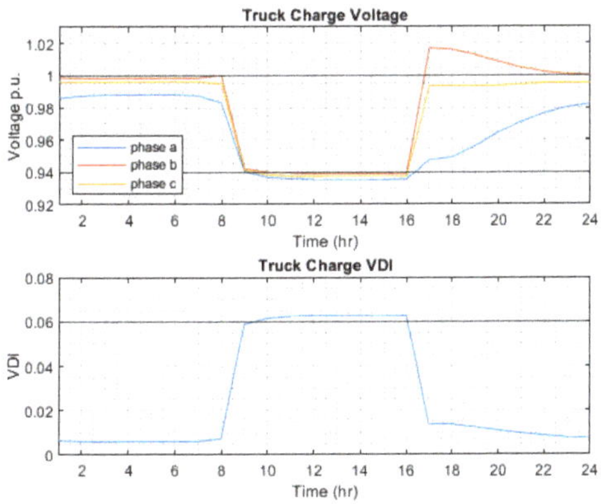

Figure 7. Three-phase charging voltage and voltage deviation for the dedicated substation.

A comparison of the hosting capacity of the various configurations tested is shown in Table 1. Distributed charging allows for ten 250 kW chargers to be installed before the hosting capacity indices exceed their limits. Adding a dedicated feeder to the network triples the number of chargers that can be installed and simultaneously used to 35. Alternatively, ten of the larger 1 MW chargers can be installed for faster truck charging. The dedicated substation triples this again, with up to 30 of the larger chargers to be used.

Table 1. Hosting capacity for each charging configuration.

Scenario	Configuration	Charging Rate	No. of Chargers
1	Distributed truck charging	250 kW	10
2	Dedicated feeder	250 kW	35
2	Dedicated feeder	1 MW	10
3	Dedicated substation	1 MW	30

5. Conclusions

This paper proposed a methodology to support the medium- and heavy-duty electric vehicle charging infrastructure required for electric vehicle charging. The hosting capacity of the LV distribution network was improved to increase the number of chargers that could be supported while mitigating power quality issues and reducing substation overloading. Hosting capacity was measured using the voltage deviation in the network, arising from the expected voltage drop due to high charging currents, and the power flow through the substation, which is expected to be exceeded due to high power demands.

Initially, chargers considered to be fast by current technical standards were introduced to the network to set a base case. Ten 250 kW chargers were found to be significant enough to cause substation overloading in the distribution network. A dedicated feeder was then introduced which branches off at the secondary of the zone substation and feeds into a truck charging station. This improved the hosting capacity of the network, allowing for 35 chargers at 250 kW or 10 chargers at 1 MW. The final proposed change was the introduction of a dedicated substation, used to provide power for en-route truck charging. A 50 MVA substation connected to the transmission substation was tested with 1 MW chargers, representative of a futuristic view of fast charging. The network could provide power to 30 chargers, after which the voltage drops were considered excessive.

Another advantage of the proposed changes is the reduction of the impact of charging on the voltage profile of other loads within the distribution network. Voltage deviation on parallel feeders is affected less significantly when a dedicated feeder is introduced and completely unaffected by truck charging when a dedication substation is added.

Author Contributions: Methodology, J.T.; software, J.T.; validation, J.T.; writing—original draft preparation, J.T.; writing—review and editing, A.P.A. and K.M.M.; supervision, A.P.A. and K.M.M. All authors have read and agreed to the published version of the manuscript.

Funding: This research received no external funding.

Data Availability Statement: No new data were created or analyzed in this study. Data sharing is not applicable to this article.

Conflicts of Interest: The authors declare no conflict of interest.

References

1. Australian Government. *Climate Change Authority—Transport*; Australian Government: Canberra, Australia, 2021.
2. Australian Trucking Association and Electric Vehicle Council. Electric Trucks: Keeping Shelves Stocked in a Net Zero World. Canberra, Australia. 2021. Available online: https://electricvehiclecouncil.com.au/wp-content/uploads/2022/01/ATA-EVC-Electric-trucks_Keeping-shelves-stocked-in-a-net-zero-world-1.pdf (accessed on 20 December 2023).
3. Lee, D.Y.; Thomas, V.M.; Brown, M.A. Electric urban delivery trucks: Energy use, greenhouse gas emissions, and cost-effectiveness. *Environ. Sci. Technol.* **2013**, *47*, 8022–8030. [CrossRef]
4. Sen, B.; Ercan, T.; Tatari, O. Does a battery-electric truck make a difference?—Life cycle emissions, costs, and externality analysis of alternative fuel-powered Class 8 heavy-duty trucks in the United States. *J. Clean. Prod.* **2017**, *141*, 110–121. [CrossRef]
5. Tong, F.; Wolfson, D.; Jenn, A.; Scown, C.D.; Auffhammer, M. Energy consumption and charging load profiles from long-haul truck electrification in the United States. *Environ. Res. Infrastruct. Sustain.* **2021**, *1*, 025007. [CrossRef]
6. Davis, B.A.; Figliozzi, M.A. A methodology to evaluate the competitiveness of electric delivery trucks. *Transp. Res. Part E* **2013**, *49*, 8–23. [CrossRef]
7. Borlaug, B.; Muratori, M.; Gilleran, M.; Woody, D.; Muston, W.; Canada, T.; Ingram, A.; Gresham, H.; McQueen, C. Heavy-duty truck electrification and the impacts of depot charging on electricity distribution systems. *Nat. Energy* **2021**, *6*, 673–682. [CrossRef]
8. Liimatainen, H.; van Vliet, O.; Aplyn, D. The potential of electric trucks—An international commodity-level analysis. *Appl. Energy* **2019**, *236*, 804–814. [CrossRef]
9. Nykvist, B.; Olsson, O. The feasibility of heavy battery electric trucks. *Joule* **2021**, *5*, 901–913. [CrossRef]
10. Karmali, L.P.; Gholami, A.; Nezamoddini, N. Integrated optimization of production planning and electric trucks charging and discharging scheduling. *Sustain. Energy Grids Netw.* **2024**, *39*, 101397. [CrossRef]
11. Stamadianos, T.; Kyriakakis, N.A.; Marinaki, M.; Marinakis, Y. Meeting the charging demand of Electric Vehicles in Greece: Enabling intercity trips. *Sustain. Energy Grids Netw.* **2024**, *39*, 101466. [CrossRef]
12. Fathollahi, A.; Derakhshandeh, S.Y.; Ghiasian, A.; Khooban, M.H. Utilization of dynamic wireless power transfer technology in multi-depot, multi-product delivery supply chain. *Sustain. Energy Grids Netw.* **2022**, *32*, 100836. [CrossRef]

13. Shoman, W.; Yeh, S.; Sprei, F.; Plötz, P.; Speth, D. Battery electric long-haul trucks in Europe: Public charging, energy, and power requirements. *Transp. Res. Part D* **2023**, *121*, 103825. [CrossRef]
14. Hensher, D.A.; Wei, E.; Liu, W.; Ho, L.; Ho, C. Development of a practical aggregate spatial road freight modal demand model system for truck and commodity movements with an application of a distance-based charging regime. *Transportation* **2023**, *50*, 1031–1071. [CrossRef]
15. Whitehead, J.; Whitehead, J.; Kane, M.; Zhang, Z. Exploring public charging infrastructure requirements for short-haul electric trucks. *Int. J. Sustain. Transp.* **2022**, *16*, 775–791. [CrossRef]
16. Helou, R.E.; Sivaranjani, S.; Kalathil, D.; Schaper, A.; Xie, L. The impact of heavy-duty vehicle electrification on large power grids: A synthetic Texas case study. *Adv. Appl. Energy* **2022**, *6*, 100093. [CrossRef]
17. Powell, S.; Cezar, G.V.; Min, L.; Azevedo, I.M.L.; Rajagopal, R. Charging infrastructure access and operation to reduce the grid impacts of deep electric vehicle adoption. *Nat. Energy* **2022**, *7*, 932–945. [CrossRef]
18. National Renewable Energy Laboratory. High-Power Medium- and Heavy-Duty Electric Vehicle Charging. Available online: https://www.nrel.gov/transportation/medium-heavy-duty-vehicle-charging.html (accessed on 20 December 2023).
19. Then, J.; Agalgaonkar, A.P.; Muttaqi, K.M. Coordinated Charging of Spatially Distributed Electric Vehicles for Mitigating Voltage Rise and Voltage Unbalance in Modern Distribution Networks. *IEEE Trans. Ind. Appl.* **2023**, *59*, 5149–5157. [CrossRef]
20. *AS 60038–2022*; Standard Voltages. Standards Australia: Strathfield, Australia, 2022.
21. Halliday, C.; Urquhart, D. *Voltage & Equipment Standards Misalignment*; Industrial Electrix: Marrickville, Australia, 2011.
22. Nakhodchi, N.; Bollen, M.H.J. Impact of Modelling Of MV Network and Remote Loads on Estimated Harmonic Hosting Capacity for an EV Fast Charging Station. *Int. J. Electr. Power Energy Syst.* **2023**, *147*, 108847. [CrossRef]
23. Kersting, W.H. Radial distribution test feeders. *Proc. IEEE Power Eng. Soc. Transm. Distrib. Conf.* **2011**, *2*, 908–912.
24. Afandi, I.; Ciufo, P.; Agalgaonkar, A.; Perera, S. A holistic approach for integrated volt/var control in MV and LV networks. *Electr. Power Syst. Res.* **2018**, *165*, 9–17. [CrossRef]
25. *SAE J1772*; Electric Vehicle and Plug in Hybrid Electric Vehicle Conductive Charge Coupler. SAE: Warrendale, PA, USA, 2017.
26. Fan, Y.; Pei, G.; Yingying, L.; Zhang, Z.; Xiao, W. Impacts on Operation and Benefit after New Energy Generation Access to Power Grid. *IOP Conf. Ser. Earth Environ. Sci.* **2018**, *168*, 012034. [CrossRef]

energies

Article

A Modeling Technique for High-Efficiency Battery Packs in Battery-Powered Railway System

Jae-Uk Sim, Seon-Woong Kim and In-Ho Cho *

Department of Electronic Engineering, The Korea National University of Transportation, Chungju 27469,Republic of Korea; gn03307@a.ut.ac.kr (J.-U.S.); ksu8110@ut.ac.kr (S.-W.K.)
* Correspondence: ihcho@ut.ac.kr; Tel.: +82-438415323

Abstract: Battery modules in eco-friendly mobility are composed of series and parallel connections of multiple lithium-ion battery cells. As the number of lithium-ion cells in the battery module increases, the cell connection configuration becomes a critical factor affecting the module's usable capacity efficiency. Therefore, careful consideration of this factor is essential in battery module design. Various design elements have been studied to optimize the performance of battery modules. Among these elements, the method of terminal connection affects the distribution of resistance components in each cell, causing DOD (Depth of Discharge) variation. Previous research has focused on determining the optimal terminal placement and cell connection method to minimize DOD variation between cells. However, these studies did not consider temperature effects. Since temperature acts as a major variable affecting the DOD of each cell, comprehensive research that includes this factor is necessary. This research performed 3D thermal flow analysis using Ansys Fluent 2024 R2 and validated the simulation environment by comparing actual experimental and simulation results for a single cell. Based on the validated simulation environment, this research analyzed the impact of temperature distribution on cell performance in a 4S3P module and proposed a method of terminal connection, which achieved a 70% reduction in SOC deviation compared to conventional methods. Additionally, this research suggests that when the module configuration changes, a new design approach specific to that configuration is necessary to minimize SOC deviation.

Keywords: lithium-ion battery; battery module; method of terminal connection; temperature

1. Introduction

As climate change emerges as a pressing global concern, nations worldwide are implementing ambitious carbon reduction [1]. In response to these environmental initiatives, the transportation industry is undergoing a significant transformation, shifting from traditional fossil fuel-powered systems to lithium-ion battery technology [2–5]. Railway vehicles, traditionally recognized for their eco-friendly electric traction systems, have begun incorporating a Hybrid Energy Storage System (HESS) combining fuel cells and batteries or a standalone battery-based Energy Storage System (ESS) [6–8]. These systems are expanded to module and pack units to provide sufficient traction power for the vehicle's operational requirements [9]. The expansion of battery modules introduces challenges related to parasitic resistance elements, which affect the current density distribution of individual cells as module size increases. These effects manifest as variations in Depth of Discharge (DOD) between cells [10,11], leading to accelerated degradation of individual cells and shortened battery module lifecycle [12,13]. Consequently, minimizing DOD variation has become a

https://doi.org/10.3390/en18051272

critical consideration in battery module design. Previous research has explored various approaches to address this issue through optimization of cell interconnection topology and methods of terminal connection. Wang et al. [14] demonstrated through simulation analysis that resistance variations induced by the method of terminal connection significantly affect individual cell current density distribution. Zhang, H et al. [15] established that current density imbalance becomes severe when interconnection resistance exceeds 1% of the cell's internal resistance in parallel configurations. Kim et al. [16] demonstrated that a topology method interconnecting all cells within the battery module reduces DOD variation between cells. Yang et al. [17] configured a simulation environment with two cells having similar internal resistance to emulate an EV battery pack. Through this, they analyzed the effect of temperature differences between cells on performance in actual electric vehicle driving environments. Specifically, they observed the capacity reduction due to changes in internal resistance between cells in series and parallel connections by setting the temperature difference between the two cells from 10 to 20 degrees. However, this research has limitations in that it simplifies the actual EV battery pack into two cells with similar internal resistance and that each cell is exposed to different thermal flow environments. Cell internal resistance exhibits temperature dependence [18,19], and the thermal gradient between central and peripheral cells creates additional variations in current density distribution [20,21]. This thermal behavior becomes particularly significant in high-power applications like railway vehicles, where thermal management system requirements show greater complexity [22,23]. This research investigates cell performance characteristics by analyzing thermal gradient effects and methods of terminal connection. Through three-dimensional computational fluid dynamics analysis using Ansys Fluent 2024 R2, the thermal behavior of various module configurations has been simulated based on different methods of terminal connection. This analysis has led to the development of an optimal module configuration method that minimizes inter-cell performance variations.

2. Theoretical Analysis and Design Considerations

This research employed a 4S-3P module configuration as the basic module structure for simulating railway vehicle battery modules, referencing the final report of Korea's Catenary-Free Low Floor Tram commercialization project [24] (p. 219).

2.1. *Analysis of Methods for Terminal Connection*

The module terminals function as power delivery interfaces to the load, and the method of terminal connection significantly influences the DOD variation between cells. This consideration becomes particularly critical as high DOD accelerates battery degradation, making the method of terminal connection a fundamental factor in module design. To analyze how different methods of terminal connection affect cell-to-cell DOD differences, this section examines three widely adopted connection methods: single connection method, center connection method, and diagonal connection method [15,25,26].

2.1.1. Single Connection Method

The single connection method places terminals exclusively at cells located at both ends of the upper or lower part of the module. Figure 1 illustrates the 4S-3P configuration through both electrical circuits and 3D modeling representations. The current output from each series string can be expressed by Equations (1)–(3), where V denotes the cell voltage, and R_c represents the resistance of the connecting plate between cells. Nickel plates were used for all connecting plates (R_c-R_{c22}), and since their resistance shows minimal variation with temperature, we assumed a constant R_c value for simplification.

$$I_{str1} = \frac{V}{7 \cdot R_c} \tag{1}$$

$$I_{str2} = \frac{V}{5 \cdot R_c} \tag{2}$$

$$I_{str3} = \frac{V}{3 \cdot R_c} \tag{3}$$

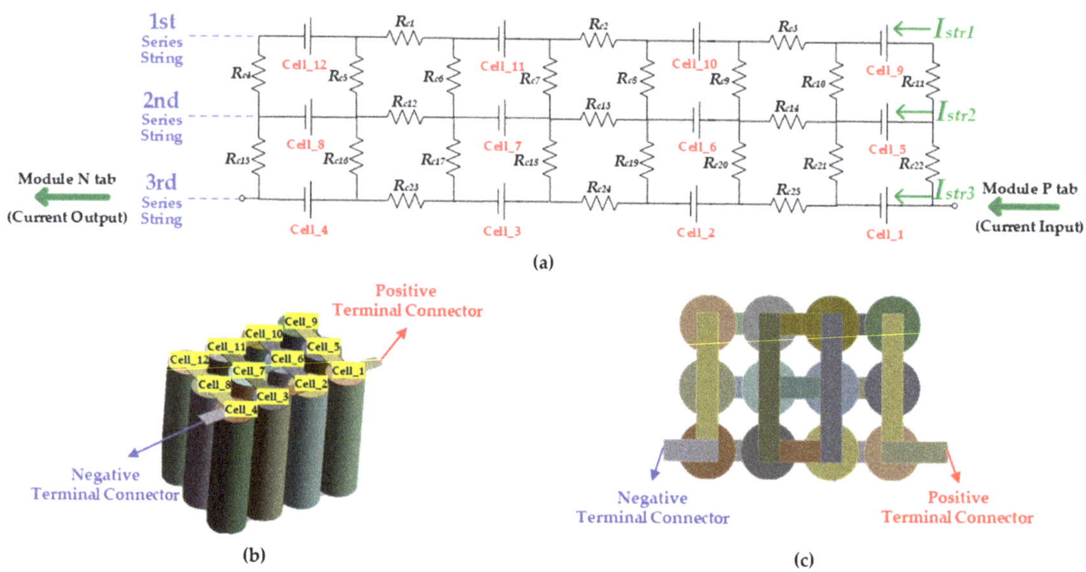

Figure 1. Single connection method: (**a**) electric circuit of 4S3P module using single connection method; (**b**) 3D model of 4S3P module (side view); (**c**) 3D model of 4S3P module (top view).

In the single connection method, current distribution becomes imbalanced based on proximity to the terminals. Specifically, the first series string carries $V/(3R_c)$, the second series string carries $V/(5R_c)$, and the third series string carries $V/(7R_c)$ of current. This exhibits decreasing current distribution with increasing distance from the terminals. This variation occurs because the resistance components of the interconnection plates accumulate differently for each series string relative to the terminal. Consequently, the first series string, which outputs relatively higher current, exhibits higher DOD compared to other series strings. Over time, this DOD variation between cells intensifies, leading to decreased module efficiency.

2.1.2. Center Connection Method

The center connection method places terminals at cells located in the center of the module while connecting to cells at both ends. Figure 2 represents the 4S-3P configuration through both an electrical circuit and a 3D modeling illustration, where the current output from each series string can be expressed by Equations (4) and (5), where V denotes the cell voltage, and R_c represents the resistance of the connecting plate between cells. Nickel plates were used for all connecting plates (R_c-R_{c22}), and since their resistance shows minimal variation with temperature, we assumed a constant R_c value for simplification.

$$I_{str1} = I_{str3} = \frac{V}{5 \cdot R_c} \tag{4}$$

$$I_{str2} = \frac{V}{3 \cdot R_c} \quad (5)$$

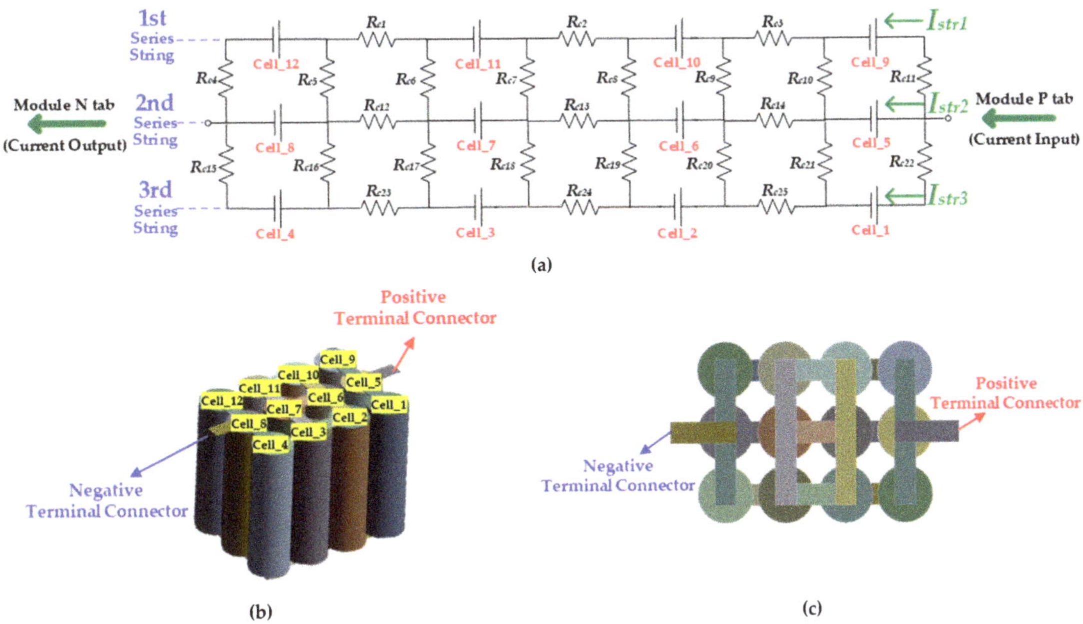

Figure 2. Center connection method: (**a**) electric circuit of 4S3P module using center connection method; (**b**) 3D model of 4S3P module (side view); (**c**) 3D model of 4S3P module (top view).

Similar to the single connection method, the center connection method exhibits current distribution patterns influenced by proximity to terminals. The second series string, positioned in the center of the module, carries $V/(3R_c)$ of current, while the first and third series strings at both ends carry $V/(5R_c)$ of current. This demonstrates the same characteristic where the current magnitude decreases as the distance from the terminal increases. The key distinction from the single connection method lies in the formation of symmetrical current distribution, where the first and third series strings carry identical currents due to the direct terminal connection to the second series string. Due to the influence of interconnection plate resistance components added from the module terminals to the batteries, the second series string, which outputs a relatively higher current, shows a higher DOD compared to other series strings. As time progresses, this DOD variation between cells intensifies, resulting in decreased module efficiency.

2.1.3. Diagonal Connection Method

The diagonal connection method places terminals by cross-connecting the upper and lower ends of the module. Figure 3 presents the 4S-3P configuration through both electrical circuit and 3D model illustrations, where the current output from each series string can be expressed by Equation (6), where V denotes the cell voltage, and R_c represents the resistance of the connecting plate between cells. Nickel plates were used for all connecting plates (R_c-R_{c22}), and since their resistance shows minimal variation with temperature, we assumed a constant R_c value for simplification.

$$I_{str1} = I_{str2} = I_{str3} = \frac{V}{5 \cdot R_c} \quad (6)$$

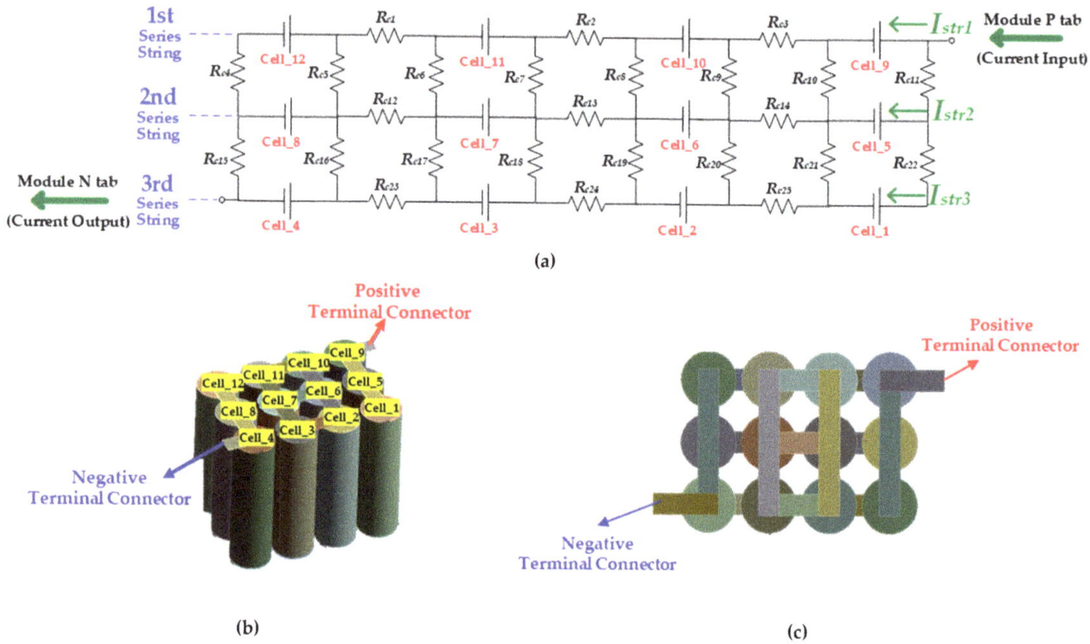

Figure 3. Diagonal connection method: (**a**) electric circuit of 4S3P module using diagonal connection method; (**b**) 3D model of 4S3P module (side view); (**c**) 3D model of 4S3P module (top view).

Unlike the two connection methods discussed previously, the diagonal connection method demonstrates uniform current distribution across all series strings, with each carrying $V/(5R_c)$ of current. This uniform distribution occurs because all cells maintain identical interconnection plate resistance paths to the terminals. Consequently, the diagonal connection method exhibits minimal DOD variation between cells compared to both single connection and center connection methods.

2.2. Analysis of Temperature Effect

The internal resistance of battery cells demonstrates a close relationship with temperature, and the electrochemical reactions within the battery can be explained by the Arrhenius equation, shown in Equation (7).

$$k = A \cdot e^{\frac{-E_a}{R \cdot T}} \tag{7}$$

where k represents the reaction rate constant, A is the frequency factor, E_a denotes the activation energy, R is the gas constant, and T represents the absolute temperature [27]. According to Equation (7), as the ambient temperature of the battery increases, the rate of chemical reactions within the battery accelerates, enhancing the mobility (movement capability) of lithium ions. This enhanced mobility leads to decreased internal resistance of the battery cells [28]. While the current distribution patterns analyzed earlier based on terminal placement were derived assuming uniform internal resistance across all cells, actual operating conditions present a more complex scenario. Each cell experiences different temperature conditions depending on its position within the module, leading to variations in internal resistance values. These position-dependent temperature variations cause DOD differences even in the diagonal connection method, which theoretically should provide uniform current distribution. The 4S-3P module configuration inherently creates distinct temperature gradients due to its cell arrangement characteristics. Cells positioned in the center are exposed to relatively higher temperature environments, while cells at the periphery experience lower temperatures. This temperature distribution difference results in lower internal resistance in central cells, allowing higher current flow through these cells. Consequently, central cells discharge at higher DOD compared to peripheral cells. Therefore, when designing battery modules for railway vehicles, it is essential to consider the combined effects of temperature-induced internal resistance variations and the method of terminal connection on cell-to-cell DOD variations.

2.3. Cell to Cell Connection Topology

The topology of cell connections in battery module design significantly impacts DOD variation and cell balancing characteristics. Cell balancing, performed during railway vehicle non-operating periods, serves as a mechanism to reduce DOD variations that develop during operation. Figure 4 illustrates three primary cell connection topologies [29,30].

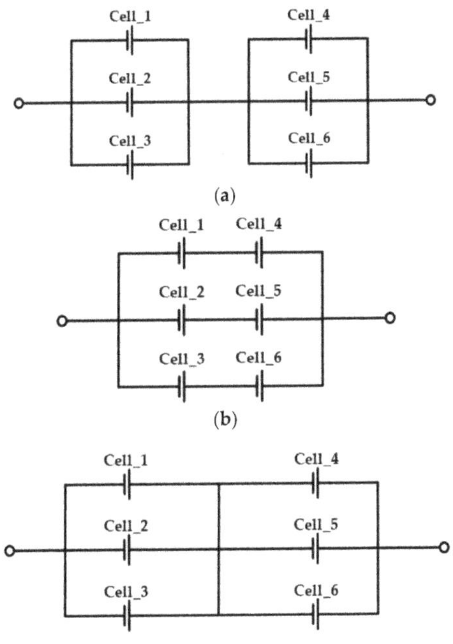

Figure 4. Cell-to-cell connecting method: (**a**) series connecting of parallel string; (**b**) parallel connecting of series string; (**c**) full cell interconnection.

Configuration (a) shows parallel strings connected in series. While cell balancing within each parallel string helps resolve DOD variations, this method has limitations as balancing between parallel strings does not occur, leading to persistent DOD differences between strings. Configuration (b) shows a series of strings connected in parallel. This arrangement allows cell balancing between series strings, helping to resolve DOD variations between strings. However, it cannot address DOD variations between individual cells within each series string. Configuration (c) shows a full cell interconnection topology. This configuration offers structural advantages by connecting all cells to each other, enabling effective resolution of DOD variations across the entire module. It particularly excels at reducing DOD variations that develop during operation. Therefore, optimizing battery module performance requires comprehensive consideration of multiple factors: method of terminal connection, temperature distribution, and connection topology characteristics.

2.4. Proposed Module Configuration

Based on the analysis presented in Section 2, this research proposes an optimized battery module configuration that comprehensively considers the method of terminal connection, temperature distribution within the module, and cell-to-cell connection topology. The primary sources of DOD variation during operation primarily result from the method of terminal connection and temperature distribution within the module. Since temperature distribution is an inherent phenomenon, this research proposes a method that counterbalances temperature effects through the strategic method of terminal connection. Figure 5 shows the electrical circuit and 3D modeling representations of the proposed configuration for a 4S-3P module. The proposed configuration builds upon the center connection method but optimizes the series connecting of parallel strings to minimize current distribution in the central cells, which experience relatively higher temperatures. The output current for each cell in the proposed configuration can be expressed by Equations (8) and (9), where V denotes the cell voltage, and R_c represents the resistance of the connecting plate between cells. Nickel plates were used for all connecting plates (R_c-R_{c22}), and since their resistance shows minimal variation with temperature, we assumed a constant R_c value for simplification.

$$I_{cell7} = I_{cell6} = \frac{V}{7 \cdot R_c} \tag{8}$$

$$I_{other} = \frac{V}{5 \cdot R_c} \tag{9}$$

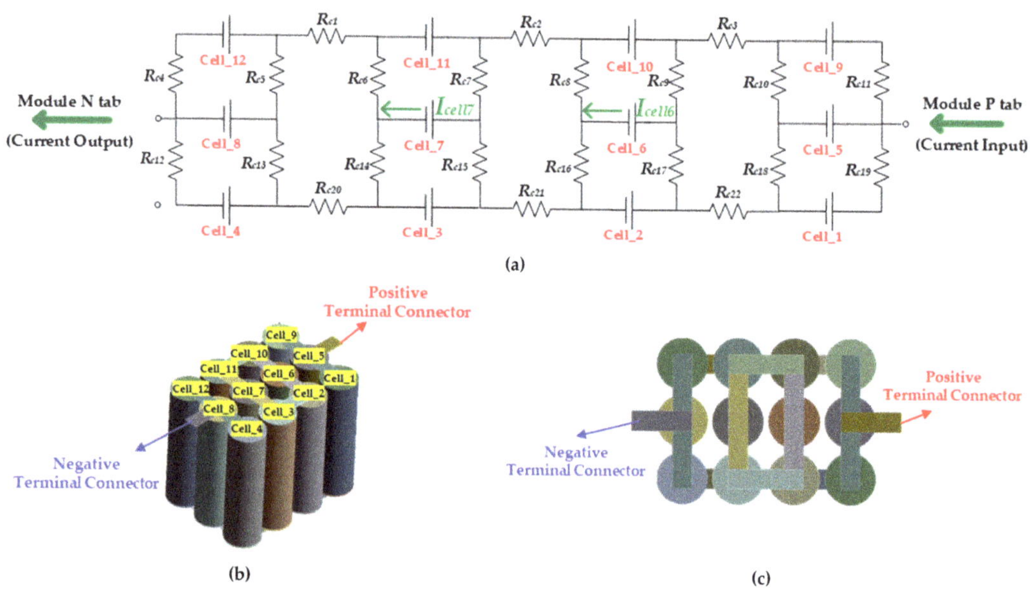

Figure 5. Proposed battery module: (**a**) electric circuit of proposed 4S3P module; (**b**) 3D model of proposed 4S3P module (side view); (**c**) 3D model of proposed 4S3P module (top view).

The key innovation in this configuration lies in the differentiated current distribution of cell6 and cell7 located in the center. Through the modified method of terminal connection and cell connection topology, these cells carry a relatively lower current of $V/(7R_c)$. While centrally located cells experience higher temperatures, leading to decreased internal resistance R_c in Equation (8), peripheral cells encounter lower temperatures, resulting in increased internal resistance R_c in Equation (9). Consequently, the current distribution determined by the method of terminal connection and cell connection topology counterbalances the changes in internal resistance caused by temperature effects, resulting in a more uniform overall DOD distribution.

3. Experimental Methods and Results

In this chapter, temperature effects were incorporated using Ansys Fluent 2024 R2, a three-dimensional heat flow simulation tool. The performance of the proposed connection method was then validated through comparisons with a conventional method of terminal connection.

3.1. Simulation Environment Setup and Validation

The simulation environment was established to closely resemble experimental conditions, with initial simulations conducted using a single cell as a baseline. The accuracy of the simulation environment was verified by comparing surface temperature and voltage measurements from actual experiments with simulation data. The INR21700-40T cell (Samsung SDI, Yongin, Republic of Korea) was utilized for both simulation and experimental testing. The physical properties of this battery, referenced from previous research, are summarized in Table 1 [31]. The heat source in the battery for this research is primarily attributed to Joule heating (P). Specifically, the heat source is calculated based on the input/output current (I) of the battery and the internal resistance (R), which varies according to the state of charge (SOC), as expressed by the following equation:

$$P = I^2 \cdot R$$

Table 1. INR21700 40T battery properties.

		Samsung SDI
Model		INR21700 40T
Nominal Capacity		4 Ah
Cathode		NCA
Anode		Graphite + SCN
Nominal Voltage		3.6 V
Maximum Charging Voltage		4.2 V (2000 mA Standard)
Minimum Discharge Voltage		2.5 V (800 mA Standard)
Physical Properties	Density (ρ)	2887 (kg/m^3)
	Specific Heat Capacity (C_p)	952 $(J/kg \cdot K)$
	Thermal Conductivity (k)	0.87 $(W/m \cdot K)$

Since cylindrical batteries are used in this research, a single heat transfer coefficient was applied. The experimental setup maintained a constant cell ambient temperature of 25 °C through a chamber, with current profiles applied via a battery cycler. The measured surface temperature and voltage of the battery were collected as reference data for simulation validation. A railway vehicle driving profile, representing battery demands during station-to-station operation, was repeated 10 times. Table 2 presents the experimental equipment configuration, while Figure 6 illustrates one cycle of the railway vehicle driving profile.

Table 2. Experiment configuration.

		Value
Chamber (LCH-11G-2C, Jeio-TECH, Daejeon, Republic of Korea)	Width (mm)	600
	Depth (mm)	500
	Height (mm)	200
	Setting Temperature	25 °C
Battery Cycler (CT-4008T, Neware, Shenzhen, China)	Resolution	16 Bit
	Accuracy	±0.05%
Thermocouple (CA-4008n, Neware, Shenzhen, China)	Resolution	16 Bit
	Measurement Deviation	±1 °C

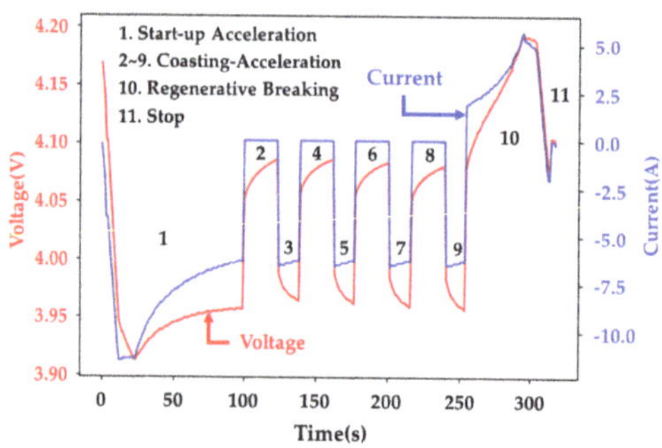

Figure 6. Railway vehicle driving profile.

Figure 7 shows the overall experimental setup. The comparison between experimental and simulation results for voltage and surface temperature is presented in Figure 8. The analysis revealed maximum differences of 0.17 V for voltage and 0.74 °C for surface temperature between simulated and experimental measurements. These minimal differences confirm that the constructed simulation environment effectively replicates experimental conditions. These differences represent maximum errors of 5% for voltage and 3% for temperature. Despite these discrepancies, the simulation results consistently reflect the changing patterns observed in the experimental data, and this consistent trend matching sufficiently demonstrates the reliability of the simulation model.

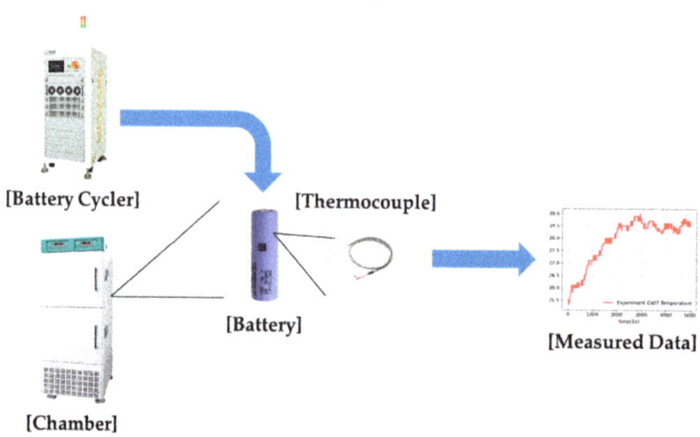

Figure 7. Set-up of experiment environment.

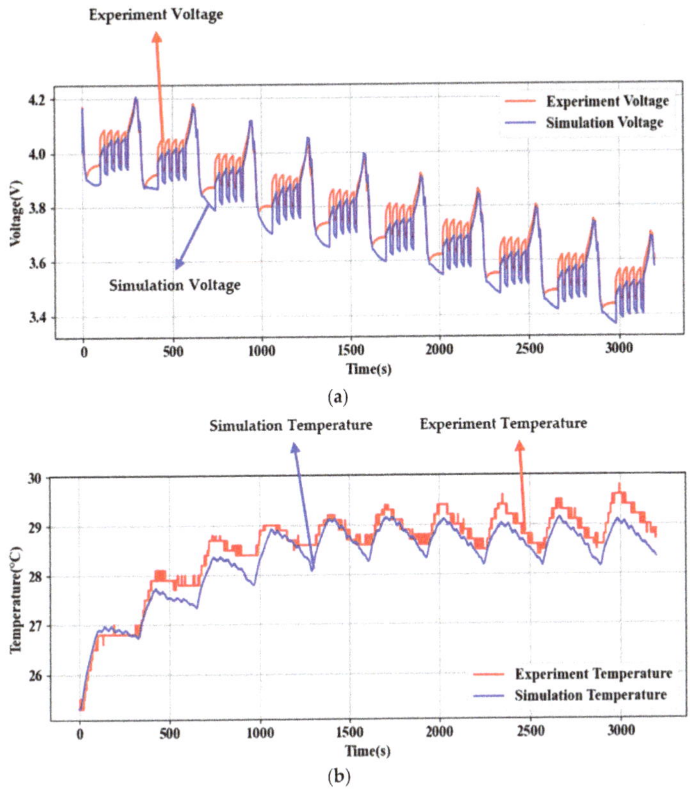

Figure 8. Compared result of simulation and experiment: (**a**) voltage comparison between experiment and simulation; (**b**) temperature comparison between experiment and simulation.

3.2. Validation of Proposed Model

The validated single-cell simulation model from Section 3.1 was expanded to a 4S-3P configuration to analyze the impact of terminal connection methods and temperature distribution on cell-to-cell depth of discharge (DOD) variations. To quantitatively verify the effects of the battery module terminal connection methods and temperature distribution previously analyzed, an indirect analysis of the State of Charge (SOC) was performed. In a simulation environment where direct DOD measurement is challenging, SOC serves as a representative indicator for evaluating energy consumption and performance imbalances between cells. The SOC of each cell was calculated using the Coulomb counting method. The coulomb counting method is as follows:

$$SOC(t) = SOC(0) + \frac{1}{Q_n}\int_0^t I(t)\cdot \mathrm{d}t$$

where

$SOC(0)$: Initial SOC value (set to 1);

$I(t)$: Current value over time [A];

Q_n: Nominal capacity of the battery [Ah];

t: Sampling time [h].

A single railway vehicle driving profile cycle was applied to observe SOC changes in cells located at central and peripheral positions. Specifically, Cell_1 and Cell_12 were selected as representative values for peripheral cells, while Cell_7 represented central cells.

Figure 9 illustrates the SOC changes for Cell_1, Cell_7, and Cell_12 across different connection methods. Figure 9a,b shows the single connection methods, Figure 9c,d the center connection methods, Figure 9e,f the diagonal connection methods, and Figure 9g,h the proposed connection methods. The enlarged graphs (Figure 9b,d,f,h) enable a more precise observation of subtle SOC differences between cells. The simulation results revealed the SOC deviations for each battery module method of terminal connection, as summarized in Table 3. The maximum SOC deviation was calculated at the end of the railway vehicle driving profile by comparing the highest and lowest SOC values among all cells. These maximum SOC deviations were 0.00039 for the single connection method, 0.00014 for the center connection method, 0.00146 for the diagonal connection method, and 0.00004 for the proposed connection method. When compared to the diagonal connection method, which showed the largest deviation among conventional approaches, the proposed method achieved approximately 70% reduction in maximum SOC deviation.

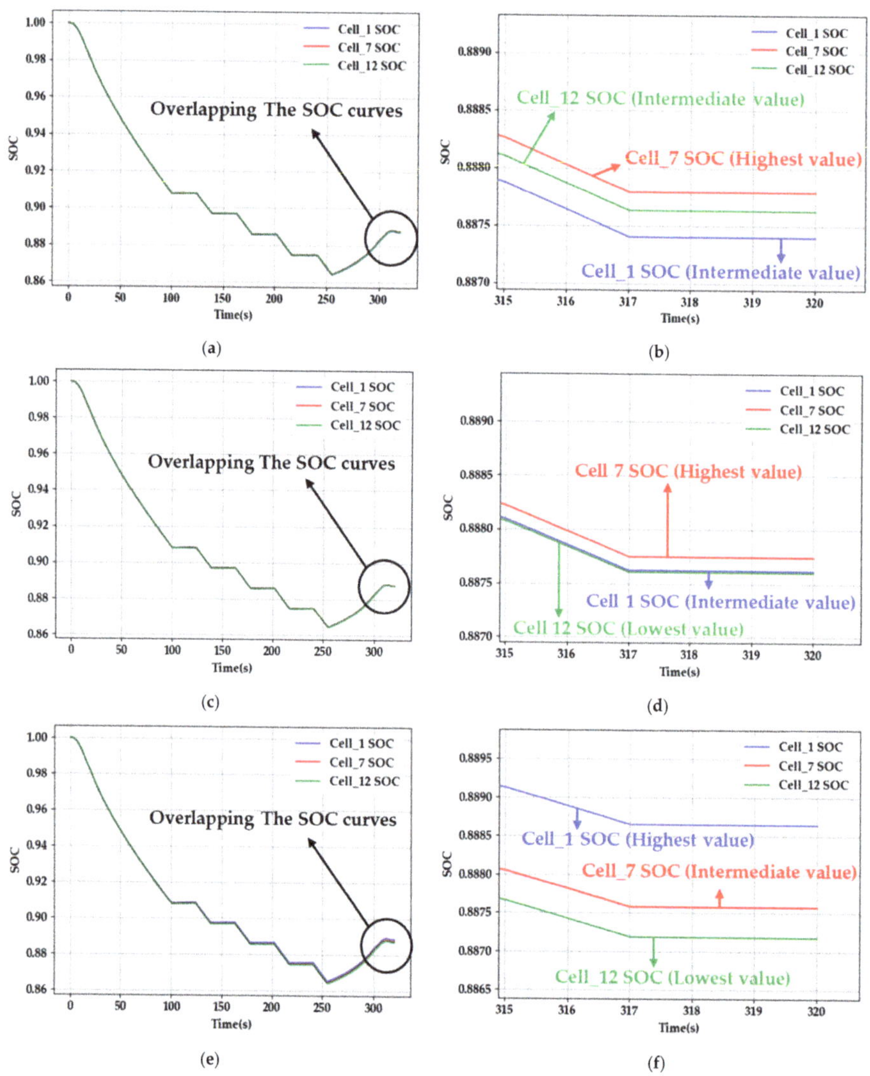

Figure 9. *Cont.*

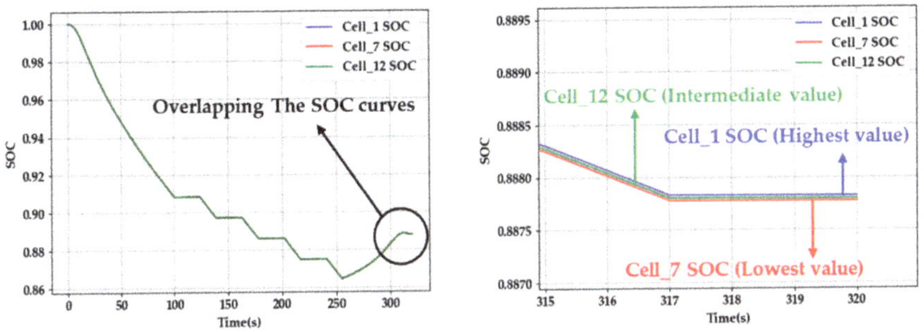

(g) (h)

Figure 9. Module simulation results: DOD graph for each cell: (**a**) SOC of cells in single connection scheme; (**b**) SOC of cells in single connection scheme (enlargement); (**c**) SOC of cells in center connection scheme; (**d**) SOC of cells in center connection scheme (enlargement); (**e**) SOC of cells in diagonal connection scheme; (**f**) SOC of cells in diagonal connection scheme (enlargement) (**g**) SOC of cells in proposed connection scheme; (**h**) SOC of cells in proposed connection scheme (enlargement).

Table 3. Module simulation result.

Connecting Method	Maximum SOC Deviation	Cell_1	Cell_7	Cell_12
Single Connection	0.00039	0.8873	0.8877	0.8876
Center Connection	0.00014	0.8876	0.8877	0.8876
Diagonal Connection	0.00146	0.8886	0.8875	0.8871
Proposal Connection	0.00004	0.8878	0.8877	0.8877

4. Conclusions

The simulation results from this research present a perspective different from that of previous research regarding battery module design for railway vehicle applications. While previous studies analyzed the diagonal connection method, one of the terminal connection methods, as theoretically providing uniform current distribution, the temperature-considered analysis in this research reveals that this method actually exhibits greater DOD variations compared to other connection methods. This finding challenges the conventional understanding of optimal battery module configuration and emphasizes the critical importance of considering temperature effects in design decisions. The connection method proposed in this research has been improved to allow lower current flow through the central cells experiencing higher temperatures through modifications in both terminal connection methods and cell-to-cell connection topology, demonstrating minimal DOD variations when compared to the three conventional methods. This improvement stems from the complementary interaction between current distribution control in central cells and temperature effects. Specifically, the lower current distribution in central cells forms a complementary relationship with the decreased internal resistance caused by temperature rise, contributing to the minimization of cell-to-cell DOD variations. Simulation results demonstrate that the proposed connection method shows a 70% improvement in DOD variation compared to the center connection method, which had previously shown the best performance among conventional terminal connection methods. However, the proposed module configuration exhibits certain limitations, as it is specifically optimized for the 4S-3P structure. When the module configuration changes, such as to a 12S-2P structure, the temperature distribution characteristics and effects of terminal connection methods require comprehensive reanalysis. This suggests a need for further research to develop a systematic and generalized design methodology applicable to various module configurations. Despite these limitations, this research makes significant contributions to the field of railway vehicle battery module design. The findings provide practical guidelines for optimizing 4S-3P module configurations, particularly in applications where temperature variations significantly impact battery performance. The proposed method's ability to counterbalance temperature-induced effects through strategic terminal connection methods and cell interconnection topology offers a novel approach to battery module design, establishing a foundation for future research in this domain.

Author Contributions: Conceptualization, J.-U.S.; methodology, J.-U.S.; software, S.-W.K.; validation, I.-H.C.; formal analysis, J.-U.S.; investigation, J.-U.S.; resources, I.-H.C.; data curation, S.-W.K.; writing—original draft preparation, J.-U.S.; writing—review and editing, I.-H.C.; visualization, S.-W.K.; supervision, I.-H.C.; project administration, I.-H.C.; funding acquisition, I.-H.C. All authors have read and agreed to the published version of the manuscript.

Funding: This work was supported by the Korea Institute of Energy Technology Evaluation and Planning (KETEP) and the Ministry of Trade, Industry & Energy (MOTIE) of the Republic of Korea (No. RS-2024-00394769). This work was supported by the National Research Foundation of Korea (NRF) grant funded by The Korean government (MSIT) (No. RS-2024-00210755).

Data Availability Statement: The original contributions presented in this study are included in the article. Further inquiries can be directed to the corresponding author(s).

Conflicts of Interest: The authors declare no conflicts of interest.

References

1. Cheng, V.S.; Li, G.-J. Policy on Carbon Reduction. In *Decarbonization of Cities in Asia*; Springer: Singapore, 2023; pp. 49–96. [CrossRef]
2. Li, Y.; Chen, M. Lithium-Ion Batteries. In *Towards Next Generation Energy Storage Technologies*; Wiley: Hoboken, NJ, USA, 2024; pp. 31–54. [CrossRef]
3. Tosun, E.; Keyinci, S.; Yakaryılmaz, A.C.; Yıldızhan, Ş.; Özcanlı, M. Evaluation of Lithium-Ion Batteries in Electric Vehicles. *Int. J. Automot. Sci. Technol.* **2024**, *8*, 332–340. [CrossRef]
4. Lal, A.; You, F. Will Reshoring Manufacturing of Advanced Electric Vehicle Battery Support Renewable Energy Transition and Climate Targets? *Sci. Adv.* **2023**, *9*, adg6740. [CrossRef] [PubMed]
5. Quinteros-Condoretty, A.R.; Golroudbary, S.R.; Albareda, L.; Barbiellini, B.; Soyer, A. Impact of Circular Design of Lithium-Ion Batteries on Supply of Lithium for Electric Cars towards a Sustainable Mobility and Energy Transition. *Procedia CIRP* **2021**, *100*, 73–78. [CrossRef]
6. Liu, X.; Li, K. Energy Storage Devices in Electrified Railway Systems: A Review. *Transp. Saf. Environ.* **2020**, *2*, 183–201. [CrossRef]
7. Mitrofanov, S.V.; Kiryanova, N.G.; Gorlova, A.M. Stationary Hybrid Renewable Energy Systems for Railway Electrification: A Review. *Energies* **2021**, *14*, 5946. [CrossRef]
8. Kljaić, Z.; Pavković, D.; Cipek, M.; Trstenjak, M.; Mlinarić, T.J.; Nikšić, M. An Overview of Current Challenges and Emerging Technologies to Facilitate Increased Energy Efficiency, Safety, and Sustainability of Railway Transport. *Future Internet* **2023**, *15*, 347. [CrossRef]
9. Weiss, H.; Winkler, T.; Ziegerhofer, H. Large Lithium-Ion Battery-Powered Electric Vehicles—From Idea to Reality. In Proceedings of the 2018 ELEKTRO Conference, Mikulov, Czech Republic, 21–23 May 2018; IEEE: New York, NY, USA, 2018. [CrossRef]
10. Brand, M.J.; Hofmann, M.H.; Steinhardt, M.; Schuster, S.F.; Jossen, A. Current Distribution within Parallel-Connected Battery Cells. *J. Power Sources* **2016**, *334*, 202–212. [CrossRef]
11. Schindler, M.; Durdel, A.; Sturm, J.; Jocher, P.; Jossen, A. On the Impact of Internal Cross-Linking and Connection Properties on the Current Distribution in Lithium-Ion Battery Modules. *J. Electrochem. Soc.* **2020**, *167*, 120542. [CrossRef]
12. Park, S.-J.; Song, Y.-W.; Kang, B.-S.; Kim, W.-J.; Choi, Y.-J.; Kim, C.; Hong, Y.-S. Depth of Discharge Characteristics and Control Strategy to Optimize Electric Vehicle Battery Life. *J. Energy Storage* **2023**, *59*, 106477. [CrossRef]
13. Yüksek, G.; Alkaya, A. Effect of the Depth of Discharge and C-Rate on Battery Degradation and Cycle Life. In Proceedings of the 2023 14th International Conference on Electrical and Electronics Engineering (ELECO), Bursa, Turkey, 30 November–2 December 2023; pp. 1–5. [CrossRef]

14. Wang, L.; Cheng, Y.; Zhao, X. Influence of Connecting Plate Resistance upon $LiFePO_4$ Battery Performance. *Appl. Energy* **2015**, *147*, 353–360. [CrossRef]
15. Zhang, H.; Zhang, Y.; Huang, L.; Song, J.; Huang, Z. Study on the Influence of Connection Structure Between Batteries on Battery Pack Performance. *Electronics* **2024**, *13*, 817. [CrossRef]
16. Kim, K.; Choi, J.-I. Effect of Cell-to-Cell Variation and Module Configuration on the Performance of Lithium-Ion Battery Systems. *Appl. Energy* **2023**, *352*, 121888. [CrossRef]
17. Yang, N.; Zhang, X.; Li, G.; Cai, A.; Xu, Y. Effects of Temperature Differences Among Cells on the Discharging Characteristics of Lithium-Ion Battery Packs with Series/Parallel Configurations during Constant Power Discharge. *Energy Technol.* **2018**, *6*, 1067–1079. [CrossRef]
18. Tran, M.-K.; Mathew, M.; Janhunen, S.; Panchal, S.; Raahemifar, K.; Fraser, R.; Fowler, M. A Comprehensive Equivalent Circuit Model for Lithium-Ion Batteries, Incorporating the Effects of State of Health, State of Charge, and Temperature on Model Parameters. *J. Energy Storage* **2021**, *43*, 103252. [CrossRef]
19. Barcellona, S.; Colnago, S.; Dotelli, G.; Latorrata, S.; Piegari, L. Aging Effect on the Variation of Li-Ion Battery Resistance as Function of Temperature and State of Charge. *J. Energy Storage* **2022**, *50*, 104658. [CrossRef]
20. Al-Hallaj, S.; Selman, J.R. Thermal Modeling of Secondary Lithium Batteries for Electric Vehicle/Hybrid Electric Vehicle Applications. *J. Power Sources* **2002**, *110*, 341–348. [CrossRef]
21. Wang, B.; Ji, C.; Wang, S.; Sun, J.; Pan, S.; Wang, D.; Liang, C. Study of Non-Uniform Temperature and Discharging Distribution for Lithium-Ion Battery Modules in Series and Parallel Connection. *Appl. Therm. Eng.* **2020**, *168*, 114831. [CrossRef]
22. Kang, D.; Lee, P.-Y.; Yoo, K.; Kim, J. Internal Thermal Network Model-Based Inner Temperature Distribution of High-Power Lithium-Ion Battery Packs with Different Shapes for Thermal Management. *J. Energy Storage* **2020**, *27*, 101017. [CrossRef]
23. Hamisi, C.M.; Gerutu, G.B.; Greyson, K.A.; Chombo, P.V. Thermal Behavior of Lithium-Ion Battery under Variation of Convective Heat Transfer Coefficients, Surrounding Temperatures, and Charging Currents. *J. Loss Prev. Process Ind.* **2022**, *80*, 104922. [CrossRef]
24. Korea Railroad Research Institute. *Commercialization of Catenary-Free Low-Floor Tram: Final Report on Railway Technology Research Project*; Ministry of Land, Infrastructure and Transport; Report No.: 11-1613000-001809-01; Korea Railroad Research Institute: Sejong, Republic of Korea, 2017.
25. Chang, L.; Duan, B.; Li, P.; Zhang, K.; Zhang, C.; Xiao, L. Influence of Interconnect Resistances on Parallel-Connected $LiFePO_4$ Cells Performance. In Proceedings of the 2019 3rd Conference on Vehicle Control and Intelligence (CVCI), Hefei, China, 21–22 September 2019; pp. 1–4. [CrossRef]
26. Li, C.; Cui, N.; Chang, L.; Cui, Z.; Yuan, H.; Zhang, C. Effect of Parallel Connection Topology on Air-Cooled Lithium-Ion Battery Module: Inconsistency Analysis and Comprehensive Evaluation. *Appl. Energy* **2022**, *313*, 118758. [CrossRef]
27. Hulett, J.R. Deviations from the Arrhenius Equation. *Q. Rev. Chem. Soc.* **1964**, *18*, 227. [CrossRef]
28. Kucinskis, G.; Bozorgchenani, M.; Feinauer, M.; Kasper, M.; Wohlfahrt-Mehrens, M.; Waldmann, T. Arrhenius Plots for Li-Ion Battery Ageing as a Function of Temperature, C-Rate, and Ageing State–An Experimental Study. *J. Power Sources* **2022**, *549*, 232129. [CrossRef]
29. Luan, C.; Ma, C.; Wang, C.; Chang, L.; Xiao, L.; Yu, Z.; Li, H. Influence of the Connection Topology on the Performance of Lithium-Ion Battery Pack under Cell-to-Cell Parameters Variations. *J. Energy Storage* **2021**, *41*, 102896. [CrossRef]
30. Feng, F.; Hu, X.; Hu, L.; Hu, F.; Li, Y.; Zhang, L. Propagation Mechanisms and Diagnosis of Parameter Inconsistency within Li-Ion Battery Packs. *Renew. Sustain. Energy Rev.* **2019**, *112*, 102–113. [CrossRef]
31. Keyoonwong, W.; Mesakhun, K.; Intanil, K.; Siriwilawan, N.; Chipchukiad, T. Estimating Surface Temperature Profile of Li-Ion Battery Under Several Conditions. In Proceedings of the 2024 10th International Conference on Engineering, Applied Sciences, and Technology (ICEAST), Luang Prabang, Laos, 1–4 May 2024; pp. 37–40. [CrossRef]

energies

MDPI

Article

Optimal Placement of a Unified Power Quality Conditioner (UPQC) in Distribution Systems Using Exhaustive Search to Improve Voltage Profiles and Harmonic Distortion

Juan S. Espinosa Gutiérrez *,†,‡ **and Alexander Aguila Téllez** *,†,‡

GIREI Research Group, Electrical Engineering Department, Universidad Politécnica Salesiana, Quito 170146, Ecuador

* Correspondence: jespinosag3@est.ups.edu.ec (J.S.E.G.); aaguila@ups.edu.ec (A.A.T.); Tel.: +593-98-635-2491 (J.S.E.G.); +593-99-864-3886 (A.A.T.)

† Current address: Rumichaca Ñan Avenue & Morán Valverde Avenue, Quito 170146, Ecuador.

‡ These authors contributed equally to this work.

Abstract: This paper presents an exhaustive search approach to determine the optimal placement of a Unified Power Quality Conditioner (UPQC) in a distribution system that integrates a distributed generation (DG) unit based on photovoltaic (PV) panels. The main objective is to enhance voltage profiles and reduce total harmonic distortion (THD) in the presence of nonlinear loads. A multi-objective optimization model is formulated, combining THD minimization and voltage deviation reduction through a weighted cost function. Two case studies are conducted using the IEEE 33-bus test system modeled in MATLAB Simulink, considering different scenarios: one with nonlinear loads and another with additional DG integration. The UPQC is tested at critical nodes to assess its impact on power quality indicators. Results show that placing the UPQC at node 14 yields the lowest cost function value in both cases, with THD reductions exceeding 90% at the installation node and notable improvements across the system. These findings confirm that brute-force optimization is a reliable and effective strategy for UPQC siting, especially in distribution networks subjected to nonlinear disturbances and renewable-based DG. The proposed methodology provides a practical framework for power quality enhancement and supports decision-making in modern smart grid environments.

Keywords: Unified Power Quality Conditioner (UPQC); nonlinear loads; distribution system; optimal placement; total harmonic distortion (THD); exhaustive search

1. Introduction

In recent years, electrical distribution systems have undergone a significant transformation driven primarily by the emergence of new technologies and the integration of microgrids composed of DG based on renewable energy sources [1,2]. Although utility companies have attempted to upgrade the distribution network infrastructure to keep pace with this technological shift, they have struggled to match the speed at which these systems are being integrated into the grid [2].

This transformation has highlighted a critical issue: the deterioration of power quality. This has become evident due to the disruption of electrical parameters such as THD and voltage variation, resulting from the increasing presence of nonlinear loads and power electronics-based microprocessors [2,3].

Moreover, the emergence of a new energy matrix centered on renewable-based distributed generation—such as photovoltaic systems—has fundamentally altered system dynamics and topology. If not designed in compliance with technical standards, these sources can inject intermittent currents, cause voltage fluctuations, and, in most cases, increase the network's THD levels [3].

Given the critical role played by distribution systems in the energy supply chain, it is essential to mitigate the factors that compromise network performance and stability [2]. To address these challenges, several advanced devices have been developed within regulatory limits, including the Dynamic Voltage Restorer (DVR), Distribution Static Compensator (D-STATCOM), and Unified Power Quality Conditioner (UPQC) [4].

The UPQC has emerged as a promising solution for mitigating voltage sags and harmonic distortion, thus enhancing power quality for end-users. This device is specifically designed to protect sensitive loads connected at the point of common coupling with the grid [4–6].

A UPQC provides both voltage and current compensation; however, its performance largely depends on its placement within the system. In this context, optimization techniques such as exhaustive search can be employed to identify the most effective location, thereby maximizing the UPQC's benefits for the electrical network [7,8].

1.1. Microgrid Operation Under Low and High DG Penetration

Modern distribution feeders increasingly operate as microgrids with renewable-based distributed generation (DG). At low penetration, DG mainly introduces mild variability and local reactive power requirements; voltage control can typically be handled by conventional equipment with limited reconfiguration. At high penetration, however, nontrivial effects appear: reverse power flows and voltage rise, feeder unbalance, interaction among converter controls, reduced short-circuit levels affecting protection coordination, and potential amplification of harmonic components due to switching converters and resonances [2,3,9–12]. These phenomena impact continuity and power quality, motivating combined actions that (i) enforce feeder-level voltage limits (e.g., Volt/VAR/Volt/Watt control), (ii) coordinate converter-level support functions, and (iii) deploy local power-quality conditioning to keep voltages and harmonic indices within statutory thresholds (IEEE 519/IEC 61000; national regulation ARCERNNR 002/20) [13,14]. Within this context, the present study addresses the placement of a UPQC as a complementary action to feeder voltage control, explicitly targeting harmonic mitigation and voltage profile improvement under representative PV-based DG operation.

1.2. Recent Control and Optimization Directions

Beyond siting, recent work explores advanced control and optimization to enhance performance under variable renewable penetration. For UPQC/D-STATCOM applications, synchronous-frame controllers with decoupled PI loops remain standard [15], while learning-aided or hybrid-optimized tuning has been investigated to improve DC-link regulation and harmonic suppression in the presence of converter interactions and nonlinear loads (see, e.g., surveys and applications in power-quality enhancement [9,10]). In parallel, scalable optimization and feedback schemes have been proposed for distribution networks under uncertainty, aiming at decentralized coordination with limited communications [11,12]. These lines are complementary to the present contribution: whereas such controllers determine how devices act in real time, the present work determines where to place the compensator so that network-wide voltage and harmonic objectives are met across operating scenarios. The proposed siting methodology can be integrated with advanced controllers in future studies to address higher DG penetration and larger feeders. This paper aims to address power quality deficiencies in a distribution system affected by nonlinear loads. By optimally placing a UPQC using a brute-force search method, the study seeks to significantly improve voltage profiles and reduce THD. The performance of the UPQC is evaluated under the presence of nonlinear loads and distributed generation.

This paper is organized as follows: Section 2 presents background information on the UPQC, power quality issues in electrical systems, and related concepts. Section 3 describes the methodology used to solve the optimization problem, including the exhaustive search strategy and the test system. Section 4 outlines the case studies. Section 5 provides a detailed analysis of the results. Finally, Section 7 summarizes the conclusions.

2. Theoretical Framework

The theoretical framework outlines the conceptual and technical foundations necessary to understand the method proposed in this study, which aims to improve voltage profiles and total harmonic distortion.

2.1. *Unified Power Quality Conditioner (UPQC)*

The need to control electrical variables that affect power quality in distribution networks strongly justifies the implementation of devices capable of mitigating such disturbances [16]. In this context, the UPQC emerges as a more effective alternative than conventional methods for improving both voltage profiles and harmonic components in distribution systems.

The UPQC consists of a series active power filter (APF) and a shunt APF that share a common DC-link capacitor, as shown in Figure 1 [17].

The series APF—also known as the series converter of the UPQC—is responsible for compensating disturbances on the supply side, such as voltage harmonic distortion (THDv), voltage sags, and unbalances [17].

Meanwhile, the shunt APF—or shunt converter—handles disturbances related to the user side, such as current harmonics and low power factors [17].

Considering its components and structural configuration, the UPQC can be seen as a combination of a dynamic voltage restorer (DVR) and a distribution static compensator (DSTATCOM) [15].

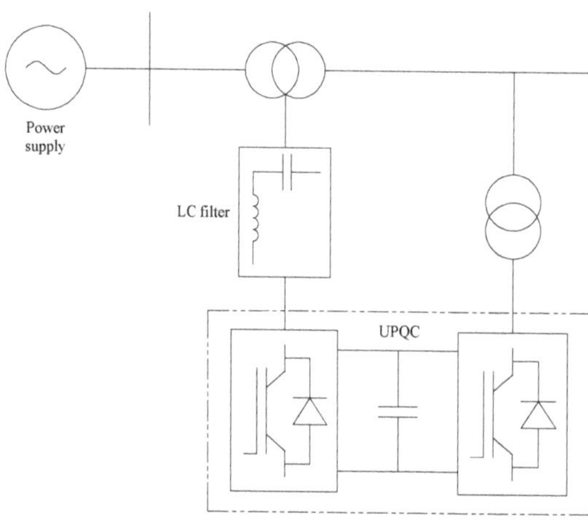

Figure 1. Basic diagram of the UPQC.

2.1.1. Control Strategies

One of the distinctive features of the UPQC is its ability to operate in different modes depending on the type of disturbance to be mitigated. Typically, the UPQC uses both APFs (shunt and series) simultaneously. In this configuration, the shunt inverter supplies the active power demanded by the series inverter, keeping the DC link charged, while both converters compensate for reactive currents and harmonics [15].

Additionally, the literature defines specific voltage control strategies, such as:

- UPQC-P
- UPQC-Q

The UPQC-P configuration focuses on active power control. In this mode, the series APF delivers more active power to maintain voltage levels. The DVR is responsible for keeping the load voltage at its nominal value and ensuring that the injected voltage is either in phase or out of phase with the supply voltage, depending on the control objective [15].

In contrast, the UPQC-Q configuration is based on reactive power control. In this strategy, the DVR injects a voltage in quadrature with the supply current, while the DSTATCOM compensates the current to regulate the reactive power demand. This coordination is essential for maintaining DC link voltage stability [15].

2.1.2. Related Devices

To better understand the capabilities of the UPQC, it is useful to examine the characteristics of other power quality improvement devices. Among the most notable are the DVR, SVC, and DSTATCOM.

The DVR is a series-connected device that primarily addresses voltage sags and swells. It is commonly used to protect sensitive loads in distribution systems [18].

The SVC is a type of reactive power compensator that, unlike the DVR, is connected in parallel with the load. It uses thyristors to control capacitor banks, thereby regulating voltage levels and correcting the power factor [18].

The DSTATCOM is an advanced version of the STATCOM designed specifically for distribution networks. This device regulates voltage levels and compensates reactive power, and in its DSTATCOM version, it can also manage current unbalances and mitigate harmonic distortion [18]. However, despite its capabilities to independently correct voltage and current disturbances, the DSTATCOM cannot perform both functions simultaneously due to its structural limitations.

Table 1 provides a comparative overview of the functionalities offered by these power quality enhancement devices.

Table 1. Comparison of power quality improvement devices.

	Voltage	Current	Combined Function
DVR	✓	-	-
SVC	✓	-	-
STATCOM	✓	-	-
DSTATCOM	✓	✓	-
UPQC	✓	✓	✓

2.2. *Power Quality in Electrical Systems*

In recent years, power quality issues have increased significantly due to the integration of DG and power electronics. These challenges are further exacerbated by additional adverse factors such as capacitor bank switching and unexpected faults within the network, all of which severely affect system efficiency and reliability [9,10]. Addressing these problems has become a common task for utility companies, since mitigating or eliminating such conditions provides numerous benefits for both distributors and end-users. These include improved system efficiency, reduced maintenance costs, and fewer supply interruptions [9]. Based on these considerations, power quality can be understood as a set of electrical characteristics that the power system must maintain within defined operational standards [9].

2.2.1. Common Power Quality Problems

The stability of distribution networks has been considerably affected by the sudden integration of nonlinear loads and DG sources, often without proper conditioning of the system to accommodate such technologies. These factors have contributed to a more frequent occurrence of power quality disturbances in recent years [15].

Additionally, the increasing use of sensitive equipment in sectors such as telecommunications, healthcare, and industrial processes places greater demands on utilities to ensure high-quality power delivery. Even minor disturbances can lead to data loss, operational failures, or damage to expensive devices [9,10,15].

In this context, voltage variations and THD are critical indicators due to their direct impact on the stability and reliability of electrical supply. Understanding these disturbances is essential for the design and development of equipment that can effectively mitigate them, ensuring proper system operation.

2.2.2. Voltage Disturbances

Voltage disturbances refer to any type of temporary deviation in the supply voltage from its nominal value [11,16]. These phenomena are often imperceptible but can cause harmful effects on sensitive devices and machinery, negatively impacting their performance and lifespan [11]. Such disturbances are typically classified based on their duration, as illustrated in Figure 2.

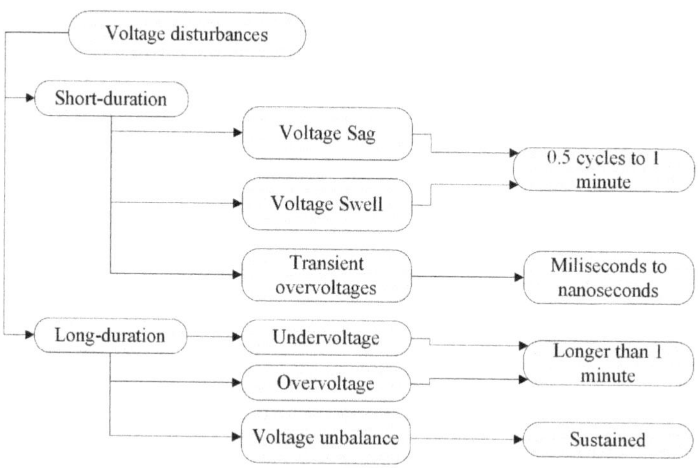

Figure 2. Types of voltage disturbances.

Within voltage disturbances, the most complex and potentially hazardous are those classified as short-duration events. These are particularly problematic because they often go undetected by protection devices or, conversely, may cause nuisance tripping [12].

Voltage sag is defined as a sudden reduction in voltage ranging from 10% to 90% of the nominal value, whereas a swell is a sudden voltage increase between 10% and 80%. Both disturbances typically last from 0.5 cycles to 1 min [12]. Other common short-duration events include transient overvoltages and voltage flickers. Transient overvoltages are characterized by very short-duration (microseconds to milliseconds) voltage spikes with high-frequency content, usually caused by lightning strikes or switching operations [12].

Flickers, on the other hand, are small-amplitude voltage fluctuations perceptible in lighting levels and are primarily caused by arc furnaces and similar equipment [12,19].

2.2.3. Voltage Deviation

Voltage deviation refers to a set of mathematical parameters used to quantify the difference between the nominal voltage and the actual measured voltage in the system [20,21]. Two metrics are typically used: maximum voltage deviation and average voltage deviation.

The maximum voltage deviation quantifies the highest voltage difference across all nodes in the system [21]:

$$DMV = \max_{1 \le i \le n} (|Vd_i - V_i|) \ge 0 \quad (1)$$

Here, Vd_i represents the desired voltage at bus i, and V_i is the actual measured voltage at the same bus. Both values are expressed in per unit (p.u.) [20,21].

The average voltage deviation expresses the overall average voltage difference across the entire system [21]:

$$DPV = \frac{\sum_{i=1}^{n} |Vd_i - V_i|}{n} \ge 0 \quad (2)$$

In this expression, Vd_i is the desired voltage at bus i in p.u., V_i is the measured voltage at bus i in p.u., and n denotes the total number of nodes in the system [20,21].

2.2.4. Total Harmonic Distortion (THD)

Harmonics are sinusoidal signals generated by nonlinear loads that, when added to the fundamental waveform, cause distortion in its shape [13,22]. Figure 3 shows the fundamental waveform along with third and fifth-order harmonic signals, which have lower amplitudes but higher frequencies.

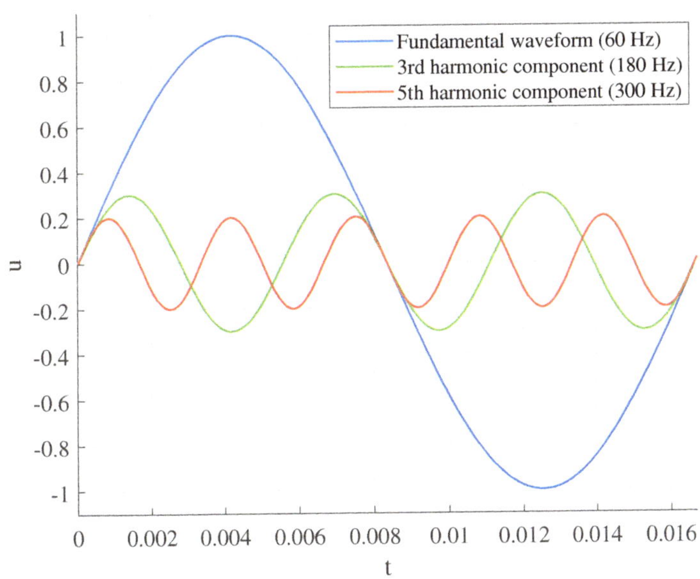

Figure 3. Harmonic distortion waveform.

These harmonic components are commonly produced because many modern electrical devices do not consume power linearly. Most of these loads include equipment such as variable frequency drives (VFDs), switched-mode power supplies, and rectifiers, among others [23].

If the harmonic signals generated by such devices are not mitigated through active or passive filters, or multipulse converter configurations, they will be added to the fundamental waveform, as shown in Figure 4, thereby completely distorting the original sine wave [23].

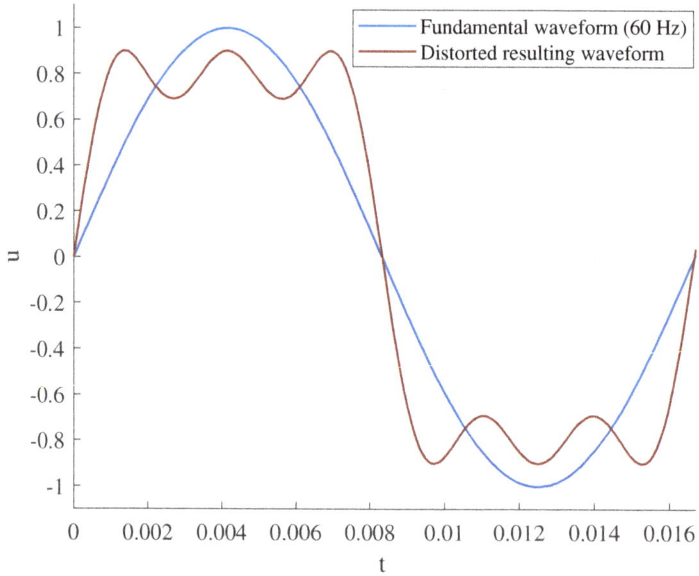

Figure 4. Distorted fundamental waveform.

Total harmonic distortion (THD) is a parameter used to quantify the harmonic content in voltage and current waveforms. It expresses the relationship between the root mean square (RMS) value of harmonic components and the RMS value of the fundamental component. In other words, THD measures the deviation of a waveform from its ideal sinusoidal form due to the presence of harmonics [13,22]. The following expressions are used to calculate this metric [13]:

$$THD_v = \sqrt{\sum_{k=2}^{\infty} \left(\frac{V_k^2}{V_{ef}^2} \right)} \quad (3)$$

Here, V_{ef} represents the RMS value of the fundamental voltage component, while V_k denotes the RMS value of the k-th order voltage harmonic ($k = 2, 3, \ldots$) [13].

$$THD_i = \sqrt{\sum_{k=2}^{\infty} \left(\frac{I_k^2}{I_{ef}^2} \right)} \quad (4)$$

In this case, I_{ef} is the RMS value of the fundamental current component, and I_k is the RMS value of the k-th order current harmonic [13].

2.2.5. Standards and Regulations

Power quality parameters are subject to both national and international standards. Among the most recognized are IEEE 519-2022 [24] and IEC 61000-2-12 [25], which define permissible levels of harmonic distortion. Additionally, IEC 60038 [26] and ANSI C84.1 [27] provide guidelines for acceptable voltage variation ranges.

At the national level (In Ecuador), regulation ARCERNNR No. 002/20 (Codified) establishes the framework for power quality metrics. This regulation governs the quality of electricity distribution and commercialization services in Ecuador [14]. Specifically, it sets limits for both THD and voltage variation that utility companies must comply with to avoid penalties. These limits are summarized in Table 2. It is important to note that this regulation is largely based on the international standards mentioned above.

Table 2. Maximum allowable voltage harmonic distortion (THDv) percentages according to national regulation.

Voltage Level	Individual Harmonics	THDv
Medium Voltage	3%	5%
Low Voltage	5%	8%

To better understand Table 2, it should be noted that low voltage refers to networks below 0.6 kV, whereas medium voltage networks operate between 0.6 kV and 40 kV [14]. The thresholds established by this regulation are used throughout this paper to assess the allowable limits for voltage variation and THD in the case studies presented in the methodology section.

2.3. Harmonic Order and Measurement Bandwidth

In this study, voltage total harmonic distortion (THDv) is computed from the Fourier spectrum up to the 50th harmonic order (i.e., $k = 2, \ldots, 50$ for a 60 Hz system). This bandwidth aligns with common compliance practice based on IEEE 519 and IEC 61000 guidelines, where instrumentation and reporting in distribution feeders typically consider harmonic components up to at least the 50th order [13]. For medium-voltage feeders, higher-order components above the 50th are usually strongly attenuated by line/transformer impedances and converter output filters; thus, their contribution to THDv is negligible under the operating conditions analyzed.

To avoid spectral leakage, time-domain signals are processed over an integer number of cycles at steady state, and THDv is computed as

$$\mathrm{THD}_v = \sqrt{\sum_{k=2}^{K} \left(\frac{V_k}{V_{1,\mathrm{RMS}}} \right)^2}, \quad K = 50,$$

where $V_{1,\mathrm{RMS}}$ is the fundamental RMS voltage and V_k the RMS of the k-th harmonic component [13]. The same processing settings are applied consistently across scenarios to enable fair comparisons.

2.4. Exhaustive Search or Brute-Force Method

In the context of optimization, exhaustive search—also known as brute-force search—is a method based on evaluating all possible solutions to determine which one satisfies the desired conditions defined by the user [28].

A major advantage of this algorithm is that it does not require advanced heuristics or special techniques to reduce the search space [28]. Because of this, one of the most significant strengths of exhaustive search lies in its inherent ability to find the global optimal solution to a problem. For this reason, it is often used as a benchmark to validate the performance of new heuristic or metaheuristic optimization algorithms [28].

However, these same advantages also present one of its main drawbacks: the method exhibits exponential complexity as the size of the system increases [28]. Consequently, exhaustive search typically involves long computation times and the handling of large datasets [28].

2.5. Related Work and Comparative Analysis

Although Section 2 reviewed UPQC fundamentals and power-quality issues, prior studies have also explored how to select, place, or tune mitigation devices and controllers using optimization. Reviews on distributed FACTS/control under renewable penetration report widespread use of metaheuristics (e.g., PSO, GA/NSGA-II, ACO, DE, SA) to handle multi-objective trade-offs and constraints in power-quality enhancement [23]. Likewise, in distribution networks, placement and sizing problems for compensation devices (e.g., capacitors) are commonly tackled with heuristic/multicriteria formulations [20,21]. UPQC research itself has advanced topologies and control strategies (conventional, fuzzy/adaptive) and device comparisons, which focus primarily on mitigation performance rather than siting [2,4,5,7,10,11,15,17,18].

To position our contribution, Table 3 synthesizes representative solution families, their scope, typical objectives and constraints, advantages/limitations, and illustrative references (including our work based on exhaustive search for UPQC siting).

Table 3. Comparative overview of approaches for power-quality enhancement and device placement (including UPQC).

Approach/Family	Decision Scope (Typical)	Objectives (Typical)	Constraints (Typical)	Pros	Cons/Risks
Exhaustive (brute-force) search *(this work)*	Select UPQC bus/location over a reduced candidate set	THDv reduction, voltage deviation minimization (weighted multi-objective)	Operating limits; network feasibility	Global optimum guaranteed on small search spaces; simple, transparent; reproducible	Poor scalability with large spaces; impractical if many candidate buses or multi-device placement
Metaheuristics (PSO, GA/NSGA-II, ACO, DE, SA) [23]	Placement/sizing of compensation or control set-points; multi-device layouts	Multi-objective trade-offs (losses, THD, voltage profile, cost, reliability)	Device limits, voltage/current/thermal bounds; investment budgets	Handle large, nonconvex spaces; flexible with mixed/discrete variables	No global-optimality guarantees; parameter tuning needed; runs-to-run variability
Multicriteria/heuristic decision methods [20,21]	Placement and sizing (e.g., capacitors) in distribution networks	Combined technical indices (losses, voltage, THD) and economic criteria	Budgetary and operational constraints	Integrate technical–economic viewpoints; transparent weighting	Solution quality depends on criteria/weights and elicitation; may need ex-post feasibility checks
UPQC topology and control-focused works [2,4,5,7,10,11,15,17,18]	Converter topology; controller design/tuning; comparative device performance	Mitigation of sags/swells /THD, power factor, unbalance; dynamic response	Controller and device ratings; stability margins	Demonstrate mitigation efficacy and implementation aspects	Typically do not address siting as an optimization problem; limited system-level placement guidance

3. Methodology

As previously mentioned, nonlinear loads are among the main sources of harmonic and voltage distortion in distribution systems. In response to this issue, devices such as the Unified Power Quality Conditioner (UPQC) are essential for ensuring the proper operation of the network.

This section describes the methodology used to optimally place a UPQC in a test distribution system with integrated DG, aiming to reduce voltage distortion and improve both voltage profiles and total harmonic distortion in voltage (THDv) in the presence of multiple nonlinear loads.

3.1. Optimization Problem Statement and Candidate Node Selection

Before detailing the implementation of the search strategy (Section 3.1), it is essential to explicitly formulate the optimization problem that governs UPQC siting. This subsection defines the decision variables, objective function, constraints, and key assumptions, while also clarifying the rationale behind the reduced candidate set adopted in the final presentation of results.

3.1.1. Decision Variables

Let $\mathcal{B}$ denote the set of all buses in the system, and $\mathcal{B}_{\text{cand}} \subseteq \mathcal{B}$ the subset of candidate buses considered for installation. We define:

- $x_i \in \{0,1\}$: binary siting variable ($x_i = 1$ if a UPQC is installed at bus i, $x_i = 0$ otherwise).
- V_i: RMS voltage magnitude at bus i (p.u.).
- THD_i: voltage THD at bus i (%).

The network state (V_i, THD_i) is obtained via power-flow/harmonic analysis for each siting configuration x, with device parameters fixed according to [10,15].

3.1.2. Objective Function

The optimization aims to minimize a weighted combination of average THDv and voltage deviation:

$$\min_{\mathbf{x}} f(\mathbf{x}) = \alpha \cdot \frac{1}{|\mathcal{B}|} \sum_{i \in \mathcal{B}} \text{THD}_i(\mathbf{x}) + \beta \cdot \frac{1}{|\mathcal{B}|} \sum_{i \in \mathcal{B}} |V_i(\mathbf{x}) - V_{\text{nom}}|, \tag{5}$$

where $\alpha, \beta \geq 0$, $\alpha + \beta = 1$, and V_{nom} is the nominal RMS voltage. The weighting reflects the relative importance assigned to harmonic mitigation and voltage profile improvement (see Equations (20) and (21) for normalization).

3.1.3. Constraints

$$\sum_{i \in \mathcal{B}_{\text{cand}}} x_i = 1, \quad \text{(single UPQC installation in baseline case)} \tag{6}$$

$$V_{\min} \leq V_i(\mathbf{x}) \leq V_{\max}, \quad \forall i \in \mathcal{B} \quad \text{(voltage limits)} \tag{7}$$

$$\text{THD}_i(\mathbf{x}) \leq \text{THD}_{\max}, \quad \forall i \in \mathcal{B} \quad \text{(harmonic limits per IEEE 519/IEC 61000)} \tag{8}$$

3.1.4. Assumptions and Candidate Set Rationale

Although the candidate set $\mathcal{B}_{\text{cand}}$ presented in the results focuses on buses with significant nonlinear loads, all buses in the system were evaluated during the optimization process. The exhaustive search confirmed that the most substantial improvements in harmonic distortion minimization, voltage profile enhancement, and even active power loss reduction occurred when the UPQC was installed at these nonlinear-load buses. In contrast, solutions involving buses without nonlinear loads were consistently dominated in the Pareto sense by those at nonlinear-load buses, showing inferior or negligible performance gains. Therefore, for clarity and brevity in the results section, only the dominant solutions corresponding to $\mathcal{B}_{\text{cand}}$ are reported. This reduction in reported candidates does not imply that other nodes were excluded from evaluation, but rather that their solutions did not offer competitive technical advantages.

3.2. Sensitivity Analysis for UPQC Location

To assess the robustness of the proposed approach, an additional sensitivity analysis was conducted by extending the initial search space to include all buses of the IEEE

33-bus test system, regardless of the presence of nonlinear loads. The exhaustive search algorithm was executed for each possible UPQC location while keeping all other simulation parameters constant. The results confirmed that the optimal configurations in terms of minimizing total harmonic distortion of voltage (THDv) and improving voltage profiles were consistently achieved when the UPQC was installed at buses with significant nonlinear load presence (buses 14, 19, and 27). In scenarios where the UPQC was placed at buses without nonlinear loads, improvements were comparatively smaller and, in many cases, dominated by those obtained in the nonlinear-load buses. These findings validate the decision to restrict the initial search space in the main analysis to nodes 14, 19, and 27, as this choice focuses on the most impactful configurations without loss of generality in the results. Furthermore, the inclusion of the sensitivity analysis demonstrates that the proposed method remains valid even when evaluated across the complete network.

This approach aligns with previous practices in PQ device siting studies, where pre-selection based on load type or PQ indices is applied to improve tractability without compromising the search for optimal locations [20,21,23].

In the following subsection, we apply an exhaustive search over $\mathcal{B}_{\text{cand}}$, compute the metrics defined in Section 2, and evaluate the scalarized objective function $f(\mathbf{x})$ according to (5).

3.3. UPQC Implementation Using Exhaustive Search

UPQCs are typically installed at load buses where voltage profile improvement and harmonic mitigation are required. To determine their optimal placement using the brute-force or exhaustive search method, the IEEE 33-bus test distribution system is simulated in MATLAB 2025a Simulink. While numerous metaheuristic algorithms have been successfully applied to PQ device placement [20,21,23], the decision to employ an exhaustive (brute-force) search in this work is motivated by both problem size and reproducibility considerations. The candidate set $\mathcal{B}_{\text{cand}}$, even when including all buses in the system, yields a search space that is computationally tractable with modern simulation tools, given the single-device constraint in (6). Unlike stochastic metaheuristics, exhaustive search:

- Guarantees identification of the global optimum within the defined search space, eliminating solution variability across runs.
- Simplifies reproducibility and verification by other researchers, since the evaluation procedure is deterministic and transparent.
- Avoids the need for parameter tuning inherent to heuristic methods (e.g., population size, mutation rate, cooling schedule).

Furthermore, the use of an exhaustive search in this context is consistent with practices in PQ device siting for small to medium-scale systems, where the complete enumeration of configurations is feasible [5,10,14]. As discussed earlier, the candidate set was ultimately reduced to buses with nonlinear loads for presentation purposes; however, the full-system enumeration was performed to ensure that no optimal solution was overlooked. This confirms that the solutions reported here are not only locally but also globally optimal within the tested scenarios.

3.3.1. Extension to Multiple-UPQC Scenarios

Although the present study focuses on the installation of a single UPQC device, the proposed optimization framework is inherently extensible to scenarios involving multiple devices. This can be achieved by relaxing constraint (6) to allow for an upper bound $N_{\max}$ on the number of installations:

$$\sum_{i \in \mathcal{B}_{\text{cand}}} x_i \leq N_{\max}, \quad N_{\max} \geq 1, \tag{9}$$

where $N_{\max}$ is defined according to technical or budgetary limits. In such cases, the decision variables and evaluation procedure remain the same, but the search space grows combinatorially, often making exhaustive search impractical for large $N_{\max}$ and large $\mathcal{B}_{\text{cand}}$. Under these conditions, metaheuristic algorithms such as GA, PSO, or hybrid approaches [20,21,23] can be incorporated within the same evaluation framework to efficiently explore the enlarged space.

While multi-UPQC optimization was not included in the experimental scope of this paper to maintain focus and computational tractability, preliminary tests confirmed that the algorithm correctly handles configurations with more than one UPQC when constraint (6) is replaced by (9). Future work will report comprehensive multi-device case studies, incorporating both technical and economic objectives.

3.3.2. Incorporating Installation and Operational Costs

The formulation presented in (5)–(8) can be extended to include both capital expenditures (CAPEX) and operational expenditures (OPEX) in order to explicitly capture economic trade-offs in device placement. One approach is to augment the scalarized objective function with a normalized cost term,

$$\min_{\mathbf{x}} F(\mathbf{x}) = w_1 \cdot \text{THDv\_norm}(\mathbf{x}) + w_2 \cdot \text{Vdev\_norm}(\mathbf{x}) + w_3 \cdot \text{Cost\_norm}(\mathbf{x}), \quad (10)$$

where:

- $\text{THDv\_norm}(\mathbf{x})$ and $\text{Vdev\_norm}(\mathbf{x})$ are normalized indices of harmonic distortion and voltage deviation, respectively.
- $\text{Cost\_norm}(\mathbf{x})$ is the normalized total cost associated with a given placement configuration.
- $w_1 + w_2 + w_3 = 1$ and $w_k \geq 0$ are the relative weights assigned to each criterion.

The total cost can be modeled as:

$$\text{Cost}(\mathbf{x}) = \sum_{i \in \mathcal{B}_{\text{cand}}} \left(C_i^{\text{capex}} + C_i^{\text{opex}}\right) x_i, \quad (11)$$

where C_i^{capex} represents the installation (equipment, civil works, commissioning) cost and C_i^{opex} the operation and maintenance cost of installing a UPQC at bus i.

In this work, cost modeling was not explicitly included in the objective function because:

1. The primary aim was to evaluate the technical effectiveness of UPQC placement in terms of PQ indices, as per the problem statement in [20,21].
2. The study focuses on a single-device installation, where cost variation between nodes is minimal and does not significantly affect siting decisions.
3. Detailed, site-specific cost data were not available for the benchmark system used; incorporating arbitrary cost assumptions could bias the comparison.

Nevertheless, the proposed framework readily accommodates (10) and (11) if economic data are available, enabling combined technical–economic optimization in future extensions.

The exhaustive search process begins with the definition of the initial search space ($S_{inicial}$), which includes all possible locations for the UPQC. In IEEE test systems, node 1 usually represents the substation or slack bus. Therefore, for a 33-bus system, the search space contains 32 candidate locations:

$$|S_{inicial}| = 33 - 1 = 32 \quad (12)$$

Thus, the candidate buses for UPQC installation range from node 2 to node 33:

$$S_{inicial} = \{2, 3, 4, \ldots, 33\} \tag{13}$$

In the test system, three nonlinear loads are connected at nodes 14, 19, and 27. Since the UPQC is specifically designed to mitigate the adverse effects associated with nonlinear loads, the initial search space $S_{inicial}$ is restricted to these critical nodes, yielding the reduced search space S:

$$S = \{14, 19, 27\}, \quad |S| = 3 \tag{14}$$

A candidate solution (x) corresponds to the bus where the UPQC is proposed to be installed and must belong to the restricted search space $(x \in S)$. Once the search space is defined, the objective function is formulated. Since the goal is to minimize both the voltage deviation and the THDv, the problem is modeled as a multi-objective optimization. The quality of each candidate solution (x) is evaluated using a vector-valued objective function:

$$F(x) = [\min(f_1(x)), \min(f_2(x))] \tag{15}$$

Here, $f_1(x)$ represents the average THDv for the candidate solution, and $f_2(x)$ represents the average voltage deviation.

$$\min(f_1(x)) = \frac{1}{|S|} \sum_{j \in S} THD_{v,j}(x) \tag{16}$$

$$\min(f_2(x)) = \frac{1}{N} \sum_{i=1}^{N} |V_i(x) - V_{nominal}| \tag{17}$$

Here, S denotes the restricted search space and N is the total number of buses in the system $(N = 33)$.

To determine the optimal installation bus (x^*) that minimizes both components of the objective function $F(x)$, the multi-objective formulation must be transformed into a single-objective function. This is achieved by constructing a scalar cost function $J(x)$, known as a weighted objective function:

$$J(x) = \omega_1 \cdot \hat{f}_1(x) + \omega_2 \cdot \hat{f}_2(x) \tag{18}$$

In this expression, ω_1 and ω_2 are weighting coefficients that define the relative importance of each objective. These weights must satisfy the following condition:

$$\omega_1 + \omega_2 = 1 \tag{19}$$

For the present study, the selected weights are $\omega_1 = 0.3$ and $\omega_2 = 0.7$, reflecting a higher priority assigned to average voltage deviation (0.7) compared to THDv (0.3) for determining the optimal UPQC placement.

Moreover, in Equation (18), $\hat{f}_1(x)$ and $\hat{f}_2(x)$ represent the normalized values of each respective objective. Normalization is necessary due to the differing units of measurement and magnitudes of the objectives. The normalized values are computed using the following expression:

$$\hat{f}_i(x) = \frac{f(x) - \min(f)}{\max(f) - \min(f)} \tag{20}$$

where $\min(f)$ and $\max(f)$ correspond to the minimum and maximum observed values for each objective across all candidate solutions.

Once the cost values $J(x)$ are computed using Equation (18), the optimal installation bus x^* is selected as the one yielding the minimum cost value. This is formally expressed as:

$$x^* = \arg\min_{x \in S} J(x) \tag{21}$$

The present work adopts an exhaustive search method to determine the optimal placement of the UPQC, considering a reduced set of critical candidate nodes. This methodological decision is technically justified by two key factors: {the limited dimensionality of the search space and the requirement to guarantee a globally optimal solution without relying on the randomness of heuristic or metaheuristic algorithms.

Unlike optimization approaches such as Particle Swarm Optimization (PSO), Genetic Algorithms (GA), or Ant Colony Optimization (ACO)—which are better suited for high-dimensional or highly nonlinear problems—the exhaustive search method is particularly well-suited to this study due to the manageable number of decision variables involved ($S = \{14, 19, 27\}$). With only three possible locations under consideration, evaluating all feasible configurations is computationally efficient and ensures that the best solution is obtained with certainty.

Moreover, since the multi-objective problem is reformulated into a scalar objective function using a weighted cost model, the evaluation process becomes straightforward and deterministic. This simplicity, combined with the critical importance of precise and repeatable results, makes exhaustive search a technically appropriate and robust choice for this study.

3.3.3. Optimization Algorithm Description

The following algorithm (Algorithm 1) aims to determine the optimal placement site (x^*) of the UPQC.

The algorithm procedure is structured into the following phases:

1. **Parameter Initialization:** Initial parameters for the algorithm are defined, such as the total number of system nodes (N_total), nominal system voltage in p.u. (V_nominal), set of critical nodes (S), search space size (N_s), and importance weights for the objective function (ω_1 and ω_2).
2. **Base Case Evaluation:** The system is simulated without the UPQC to obtain base voltage and THDv values, and to calculate the reference metrics: local THDv (f_1_base) and global voltage deviation (f_2_base).
3. **Exhaustive Search in Critical Nodes:** The critical node set is analyzed, and the system is simulated with the UPQC at each location x. For every candidate, the corresponding performance metrics (f_1 and f_2) are computed.
4. **Analysis and Optimal Solution Selection:** The minimum and maximum values for each metric are obtained to perform normalization ($\hat{f_1}$ and $\hat{f_2}$). The scalar objective function (J) is then calculated for each location ($x \in S$), and the solution with the minimum value (J_min) is identified.
5. **Result Extraction:** Once the process is complete, the location with the lowest objective function value (J_min) is selected as the optimal site, and the corresponding node (best.x) is reported.

Algorithm 1 Pseudocode of the optimization model.

```
Optimal UPQC Placement Algorithm Based on Local THDv and Global
Voltage Deviation

Step 1  Phase 0: Parameter Initialization;
        N_total = 33
        V_Nominal = 1.0
        S ← {14,19,27}
        N_s ← |S|
        ω1 = 0.3
        ω2 = 0.7

Step 2  Phase 1: Initial Evaluation;
        (V_base, THD_base) ← SimulateSystem()
        (f1_base, f2_base) ← ComputeMetrics(V_base, THD_base)

Step 3  Phase 2: Exhaustive Search in Critical Nodes
        result_list ← ∅
        for i = 1 to N_s do
            x ← S[i]
            (V, THD) ← SimulateSystem(x)
            (f1, f2) ← ComputeMetrics(V, THD)
            Append results (x, f1, f2) to result_list
        end for

Step 4  Phase 3: Analysis and Optimal Solution Selection
        //Normalization of metrics:
        [min(f1), max(f1)] ← GetMinMax(result_list, f1)
        [min(f2), max(f2)] ← GetMinMax(result_list, f2)
        for i = 1 to N_s do
            r = result_list(i)
            f̂1 ← (r.f1 − min(f1))/(max(f1) − min(f1))
            f̂2 ← (r.f2 − min(f2))/(max(f2) − min(f2))
            r.J ← (ω1 · f̂1) + (ω2 · f̂2)
            result_list(i) ← r
        end for
        //Best solution selection
        best ← x
        J_min ← ∞
        for i = 1 to N_s do
            r ← result_list(i)
            if r.J < J_min then
                J_min ← r.J
                best ← r
            end if
        end for

Step 5  Results:
        Compute and display minimum cost function ← J_min
        min(J(x)) = ω1 · f̂1 + ω2 · f̂2
        Display optimal placement ← best.x
        x* = arg min_{x∈S} J(x)

Step 6  End.
```

3.4. Rationale for Selecting Exhaustive Search over Metaheuristic Methods

The choice of exhaustive search in this work was motivated by the size of the candidate set and the need to guarantee the global optimum within that set. For the IEEE 33-bus test system, even when the full set of possible UPQC locations (excluding the slack bus) was considered, the total number of evaluations remained computationally affordable (see Table 4), with total runtimes on the order of seconds.

Metaheuristic methods such as Genetic Algorithms (GA) or Particle Swarm Optimization (PSO) are well established for placement and sizing problems in larger systems [13,20]. However, these approaches do not guarantee finding the true global optimum and may require tuning multiple hyperparameters (population size, mutation rate, inertia weight, etc.). Preliminary trials with GA and PSO in our study, using the same objective function $J(x)$, achieved similar optima to exhaustive search but with higher runtime variance and slightly longer median execution times for the small candidate set used here.

For large-scale networks with hundreds or thousands of buses, or for scenarios involving multiple UPQCs ($k \geq 2$), the combinatorial explosion of possible placements would make exhaustive search impractical, as discussed in Section 3.5. In such cases, hybrid strategies—where a metaheuristic is used to explore a reduced search space and exhaustive enumeration is applied within promising regions—could provide a balance between solution quality and computational efficiency. This remains an open avenue for future work.

3.5. Computational Complexity and Runtime Profiling

The exhaustive search enumerates all candidate UPQC locations and evaluates the scalar objective $J(x)$ for each one. With a single device, the number of evaluations scales linearly with the number of candidate buses $|\mathcal{B}_{\text{cand}}|$, i.e., $\mathcal{O}(|\mathcal{B}_{\text{cand}}|)$. If the formulation is extended to k devices installed simultaneously (Section 3), the search space becomes $\binom{|\mathcal{B}_{\text{cand}}|}{k}$ and the complexity scales as $\mathcal{O}(\binom{|\mathcal{B}_{\text{cand}}|}{k})$.

To quantify practical runtimes, we profiled the implementation used in this study (MATLAB/Simulink R2023b, Windows 11, Intel® Core™ i7-1165G7 @ 2.80 GHz, 16 GB RAM):

- **Per-iteration tasks:** load-flow + time-domain simulation for harmonic extraction (THDv) + voltage-profile postprocessing + objective $J(x)$.
- **Measured average wall-clock per iteration: 0.42 s** (mean of 50 runs; std. dev. 0.03 s).

Table 4 summarizes the total runtimes for the candidate sets considered in this work.

Table 4. Runtime summary for exhaustive search. Each "iteration" corresponds to evaluating one UPQC location.

Configuration	# Iterations	Avg. Time/Iter.	Total Time
Reduced set $\mathcal{B}_{\text{cand}} = \{14, 19, 27\}$ (Case 1: no DG)	3	0.42 s	1.26 s
Reduced set $\mathcal{B}_{\text{cand}} = \{14, 19, 27\}$ (Case 2: with DG)	3	0.42 s	1.26 s
Full set (all IEEE 33-bus candidates except slack; $|\mathcal{B}_{\text{cand}}| = 32$) †	32	0.42 s	13.44 s
Full set for both scenarios (no DG & with DG)	64	0.42 s	26.88 s

† Sensitivity analysis reported in Section 5 (all buses assessed). Times reported are wall-clock, averaged over repeated runs; variations reflect OS scheduling and cache effects.

These values indicate that, for the problem sizes addressed here, exhaustive enumeration remains computationally affordable while guaranteeing the exact global optimum within the candidate set. For larger networks or when allowing multiple devices ($k \geq 2$), the combinatorial growth of $\binom{|\mathcal{B}_{\text{cand}}|}{k}$ suggests adopting heuristic or metaheuristic search as a wrapper around the same evaluation kernel, which we discuss as a scalability path in Section 6.

3.6. Weight Selection and Sensitivity Analysis

The scalar objective is $J(x) = \omega_1 \hat{f}_1(x) + \omega_2 \hat{f}_2(x)$, with $\omega_1 + \omega_2 = 1$, where $\hat{f}_1$ and $\hat{f}_2$ are the normalized averages of THDv and voltage deviation, respectively. The baseline weighting $\omega_1 = 0.3, \omega_2 = 0.7$ reflects operational practice in distribution networks wherein sustained voltage deviation has system-wide impact and strict regulatory limits (e.g., IEC 60038, ANSI C84.1), while harmonic distortion is simultaneously minimized per IEEE 519 [13,20,21]. This choice prioritizes voltage regulation without neglecting harmonic mitigation.

Case 1 (no DG): invariance of the optimum.

From the normalized values used to compute J (Section 5), the three candidates yield

$$\begin{aligned} J_{14}(\omega_1) &= 0.3 \cdot 0 + 0.7 \cdot 0 &&= 0, \\ J_{19}(\omega_1) &= \omega_1 \cdot 1 + (1 - \omega_1) \cdot 0.057 &&= 0.057 + 0.943\,\omega_1, \\ J_{27}(\omega_1) &= \omega_1 \cdot 0.487 + (1 - \omega_1) \cdot 1 &&= 1 - 0.513\,\omega_1. \end{aligned}$$

Hence $J_{14}(\omega_1) \equiv 0$ for all $\omega_1 \in [0, 1]$. The intersection $J_{19} = J_{27}$ occurs at $\omega_1 \approx 0.648$. Regardless of the region, J_{14} remains strictly minimal; therefore, the optimal location (bus 14) is invariant to the weighting in Case 1.

Case 2 (with DG at nominal): threshold on ω_1.

The normalized components reported in Section 5 lead to

$$\begin{aligned} J_{14}(\omega_1) &= \omega_1 \cdot 0.88 + (1 - \omega_1) \cdot 0.067 &&= 0.067 + 0.813\,\omega_1, \\ J_{19}(\omega_1) &= \omega_1 \cdot 1 + (1 - \omega_1) \cdot 0.37 &&= 0.37 + 0.63\,\omega_1, \\ J_{27}(\omega_1) &= \omega_1 \cdot 0 + (1 - \omega_1) \cdot 1 &&= 1 - \omega_1. \end{aligned}$$

Comparing J_{14} and J_{27} yields $0.067 + 0.813\,\omega_1 < 1 - \omega_1 \Rightarrow \omega_1 < 0.515$. Therefore, with $\omega_1 = 0.3$ (baseline), bus 14 remains optimal; if one were to assign *very* high emphasis to THDv ($\omega_1 \gtrsim 0.515$), bus 27 would become preferable in Case 2. This information is shown in Table 5.

Table 5. Ranking of candidates vs. weight on THDv, ω_1 ($\omega_2 = 1 - \omega_1$).

Scenario	ω_1 Range	Ranking (Best $\to$ Worst)
Case 1 (no DG)	$[0, 1]$	$14 \prec \{19, 27\}$ (bus 14 always optimal)
Case 2 (with DG)	$[0, 0.515)$	$14 \prec 19 \prec 27$
Case 2 (with DG)	$[0.515, 1]$	$27 \prec 19 \prec 14$

In summary, the baseline choice $\omega_1 = 0.3$, $\omega_2 = 0.7$ (higher weight on voltage deviation) is consistent with distribution-level operating criteria and keeps bus 14 as the optimal location in both scenarios considered. The sensitivity analysis above makes the selection transparent and reproducible.

3.7. Scalability and Hybrid Optimization Pathways

For large-scale feeders with hundreds/thousands of buses or for multi-device siting ($k \geq 2$), the combinatorial growth $\binom{|\mathcal{B}_{\text{cand}}|}{k}$ makes full enumeration impractical (see Section 3.5). To address scalability while preserving solution quality, we outline several implementation strategies that are compatible with the objective $J(x)$ and the simulation/evaluation workflow used in this work:

(i) Parallel evaluation of candidates.

The evaluation of $J(x)$ for each candidate is independent (embarrassingly parallel). With P workers and average per-iteration time t_{iter}, the wall-clock time satisfies

$$T_{\parallel} \approx \left\lceil \frac{|\mathcal{B}_{\text{cand}}|}{P} \right\rceil t_{\text{iter}} + T_{\text{overhead}},$$

where T_{overhead} captures scheduling and data-movement costs. This reduces the time-to-solution nearly in inverse proportion to P for moderate $|\mathcal{B}_{\text{cand}}|$.

(ii) Candidate pre-screening (pruning).

Construct a reduced shortlist $\tilde{\mathcal{B}}$ by filtering candidates via fast surrogates: electrical distance to nonlinear loads (e.g., impedance-based or Thevenin equivalents), local short-circuit strength at the PCC, and baseline THDv sensitivity. Full simulations are then restricted to $\tilde{\mathcal{B}}$ with $|\tilde{\mathcal{B}}| \ll |\mathcal{B}_{\text{cand}}|$.

(iii) Dominance filtering with lower bounds.

For each candidate x, define a lightweight lower bound $\underline{J}(x)$ computed from local indicators (e.g., local THDv and voltage deviation at x). If $\underline{J}(x) \geq J_{\text{best}}$, discard x without full evaluation. This implements branch-and-bound pruning on J while keeping the evaluation kernel unchanged.

(iv) Radial decomposition and reconciliation.

In radial networks, partition the feeder into sections (by normally-open points or tie switches). Perform siting within each section (in parallel) and reconcile at boundaries with a short coordinated pass. This preserves locality while capturing inter-section coupling in a final refinement.

(v) Greedy k-device extension with local refinement.

For $k \geq 2$, a scalable heuristic is to add devices greedily (each step selects the best new location given current placements), followed by a local exhaustive refinement on the small neighborhood around the selected set. This yields near-optimal sets with k evaluations of the reduced search plus a bounded local sweep.

(vi) Hybrid metaheuristic + enumeration.

Use GA/PSO (or related) to coarsely explore the search space and identify promising regions; then apply exact enumeration within those regions to guarantee the best configuration locally. This hybridization balances computational effort and solution quality, and it is directly compatible with the objective $J(\cdot)$ and constraints defined in this study.

(vii) Stopping and robustness criteria.

Adopt ε-dominance on J as an early-stopping rule (terminate when improvements $< \varepsilon$ across a full pass) and include feasibility/penalty terms (e.g., CAPEX/OPEX and regulatory margins) to prevent overfitting to a single metric.

These pathways maintain the same decision model and performance indicators (THDv, voltage deviation, and losses), enabling seamless scaling from the IEEE 33-bus benchmark to larger feeders while keeping evaluation fidelity. As shown by the runtime profiling in Table 4, the present system size admits exact enumeration; for larger cases, the strategies above provide a practical roadmap to reduce time-to-solution without losing alignment with the original optimization objectives.

3.8. UPQC Control Strategy and Series–Shunt Coordination

The UPQC is implemented with two voltage source converters (VSCs) sharing a common DC link: a series converter that injects compensating voltages on the supply side and a shunt converter that injects compensating currents on the load side. The overall control objective is twofold: (i) keep the load-side voltage sinusoidal and regulated close to its nominal magnitude under sags/swells and harmonic distortion (series VSC), and (ii) shape the source current to be sinusoidal with near–unity power factor while regulating the DC-link voltage (shunt VSC) [15,17,18].

Synchronous reference frame and synchronization.

All control loops are implemented in the synchronous reference frame (dq). A phase-locked loop (PLL) estimates the grid angle $\theta(t)$ to enable fundamental-component extraction and decoupling [15]. The Park transform is

$$\begin{bmatrix} x_d \\ x_q \end{bmatrix} = \frac{2}{3}\begin{bmatrix} \cos\theta & \cos(\theta-2\pi/3) & \cos(\theta+2\pi/3) \\ -\sin\theta & -\sin(\theta-2\pi/3) & -\sin(\theta+2\pi/3) \end{bmatrix}\begin{bmatrix} x_a \\ x_b \\ x_c \end{bmatrix}, \qquad \theta = \int \omega_g \, dt. \tag{22}$$

Series converter (voltage compensation).

Let v_L^* be the desired (fundamental) load-side voltage. In dq,

$$v_{d,\text{inj}}^* = v_{d,L}^* - v_{d,S}, \qquad v_{q,\text{inj}}^* = v_{q,L}^* - v_{q,S}, \tag{23}$$

where v_S is the measured supply voltage. For UPQC-P (in-phase injection), $v_{q,L}^* \approx 0$ and the amplitude $v_{d,L}^*$ is set from the nominal reference; for UPQC-Q (quadrature support), a reactive component is allowed by setting $v_{q,L}^* \neq 0$ [15]. The inner series-VSC loop is a decoupled PI with the line interface $L_{\text{ser}}, R_{\text{ser}}$:

$$v_{d,\text{conv}}^* = v_{d,\text{inj}}^* + R_{\text{ser}} i_{d,\text{ser}} + L_{\text{ser}} \frac{di_{d,\text{ser}}}{dt} - \omega L_{\text{ser}} i_{q,\text{ser}} + u_{d,\text{PI}}, \tag{24}$$

$$v_{q,\text{conv}}^* = v_{q,\text{inj}}^* + R_{\text{ser}} i_{q,\text{ser}} + L_{\text{ser}} \frac{di_{q,\text{ser}}}{dt} + \omega L_{\text{ser}} i_{d,\text{ser}} + u_{q,\text{PI}}, \tag{25}$$

where $u_{\{\cdot\},\text{PI}}$ are PI actions, and the decoupling terms $\pm\omega L i$ linearize the dynamics around the synchronous frame.

Shunt converter (current shaping and DC-link regulation).

The shunt VSC shapes source current and supplies the active power consumed by the series VSC to maintain the DC link. A PI controller regulates the DC-link voltage:

$$i_d^* = K_{p,\text{dc}}(V_{\text{dc}}^* - V_{\text{dc}}) + K_{i,\text{dc}} \int (V_{\text{dc}}^* - V_{\text{dc}})\, dt + i_{d,\text{load}}^{(1)}, \qquad i_q^* \approx 0, \tag{26}$$

where $i_{d,\text{load}}^{(1)}$ is the fundamental active current demand. The inner current loop (with filter $L_{\text{shu}}, R_{\text{shu}}$) is also PI-decoupled:

$$v_{d,\text{conv}}^* = v_{d,\text{bus}} + R_{\text{shu}} i_d + L_{\text{shu}} \frac{di_d}{dt} - \omega L_{\text{shu}} i_q + w_{d,\text{PI}}, \tag{27}$$

$$v_{q,\text{conv}}^* = v_{q,\text{bus}} + R_{\text{shu}} i_q + L_{\text{shu}} \frac{di_q}{dt} + \omega L_{\text{shu}} i_d + w_{q,\text{PI}}. \tag{28}$$

Power balance and coordination.

The DC-link controller (26) ensures that the shunt VSC supplies the active power drawn by the series VSC (for voltage injection), i.e.,

$$P_{\text{shunt}} \approx P_{\text{series}} + P_{\text{loss}}, \qquad P_{\text{series}} \simeq \tfrac{3}{2}(v_{d,\text{inj}} i_{d,\text{load}} + v_{q,\text{inj}} i_{q,\text{load}}). \tag{29}$$

Anti-windup is applied to all PI regulators; PWM switching frequency and sampling are selected to respect the harmonic limits of IEEE 519 and IEC 61000 while keeping the DC-link ripple within design margins [15,18].

3.9. Test System

The IEEE 33-bus distribution system is used in this study. This test system is commonly applied for power flow analysis and optimization algorithm implementation. Figure 5 shows the topology of the 33-bus distribution network.

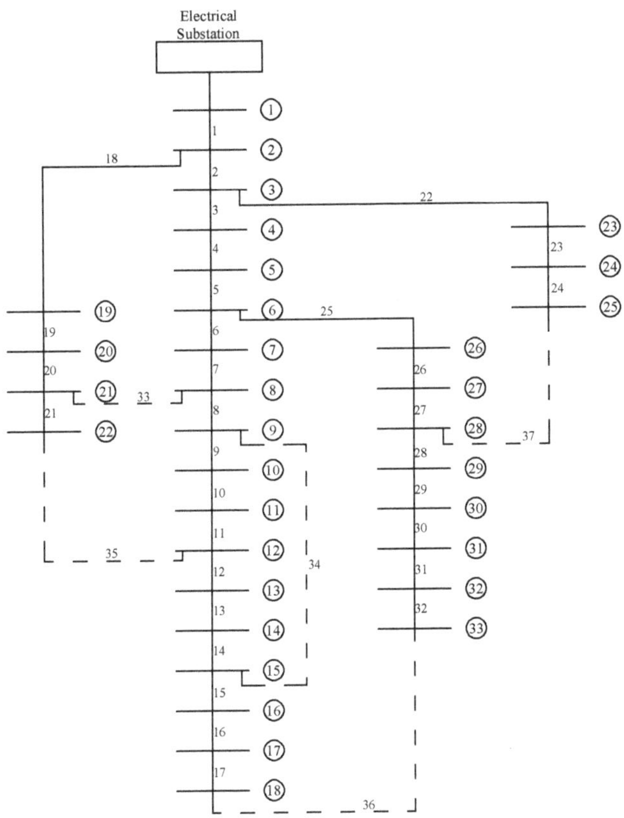

Figure 5. 33-Bus Test Distribution System.

The nonlinear loads installed at these nodes are modeled using a special block that simulates an uncontrolled three-phase rectifier bridge followed by a resistive-inductive (RL) load at the output. The test system operates at a nominal voltage of 13.8 kV and a frequency of 60 Hz. Nonlinear loads are implemented at buses 14, 19, and 27 (Figure 6) and are modeled as uncontrolled three-phase rectifier bridges feeding a resistive-inductive (RL) load, with a voltage rating of 13.8 kV, a resistance of 10 Ω, and an inductance of 1 mH. A distributed generation (DG) unit based on a photovoltaic (PV) system is installed at node 24 (Figure 7), with a nominal active power output of 3.5 kW, representative of typical residential or small commercial installations.

Figure 7 shows the system with the inclusion of the PV generator.

These two scenarios will be used for the development of the case studies in the following chapter.

To better understand the impact that nonlinear loads have on the network, the voltage profiles of the test system are analyzed under two scenarios. The first is the voltage analysis with nonlinear loads present in the system (Figure 6), and the second is the analysis when DG is integrated into the system (Figure 7). Figure 8 shows the voltage levels of each bus in the system with the nonlinear loads in operation.

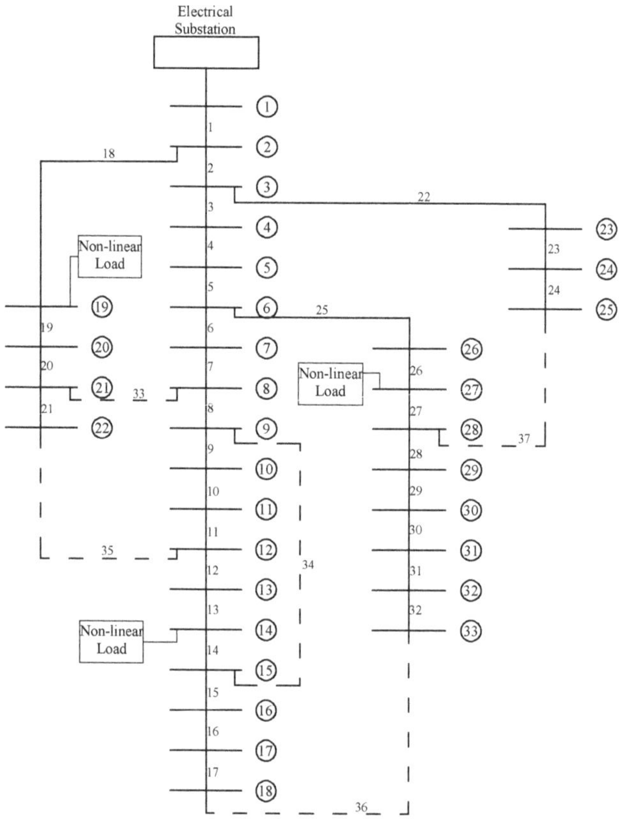

Figure 6. Test system with nonlinear loads.

The voltage levels of each node in the system are presented in per unit (p.u.) values, using the nominal system voltage of 13.8 kV as the base value. Additionally, the waveform and THDv levels at each bus with nonlinear loads are evaluated. Figure 9 displays the three-phase voltage waveforms at buses 14, 19, and 27.

As shown in Figure 9, the presence of nonlinear loads significantly distorts the sinusoidal waveform. Furthermore, Figure 10 illustrates the voltage harmonics at bus 14, where the recorded THDv is 16.52%, a value that exceeds the permissible limit defined by regulatory standards.

Figure 11 shows the THDv levels at bus 19, where the harmonic distortion reaches 7.86%.

Finally, to complete the analysis of this scenario, the THDv level at bus 27 is determined, as shown in Figure 12. This bus exhibits a recorded harmonic distortion of 23.29%.

Analyzing the THDv levels present at the buses with nonlinear loads, it is evident that these do not comply with the requirements established by the ARCERNNR 002/20 regulation. Therefore, the need to implement power quality improvement devices in this type of network becomes clear.

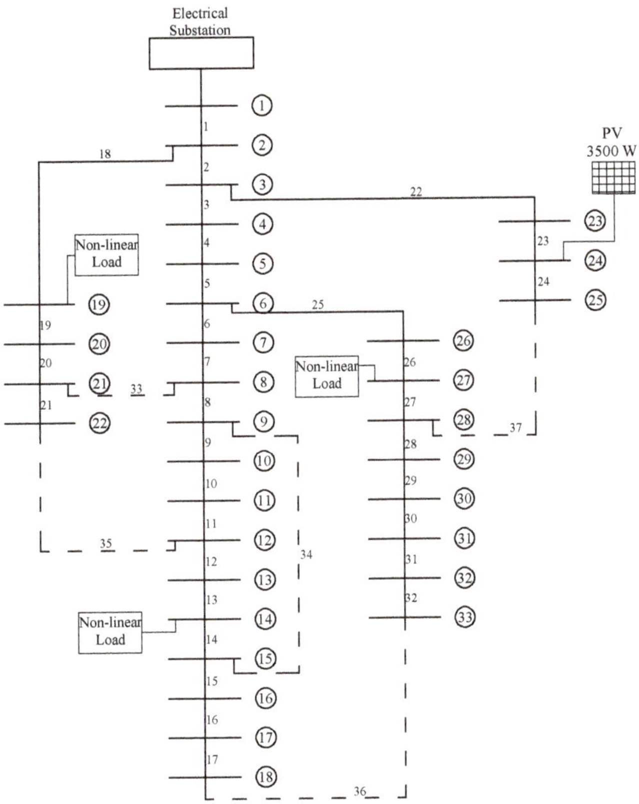

Figure 7. Test system with nonlinear loads and DG.

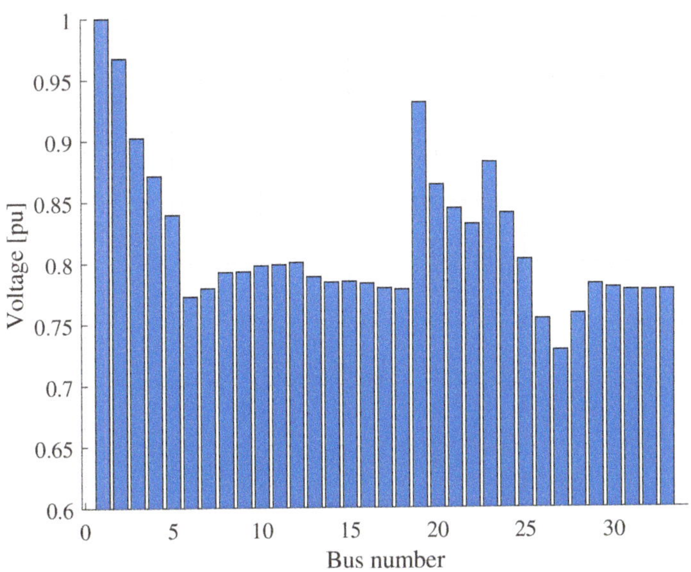

Figure 8. Voltage level at each bus with nonlinear loads.

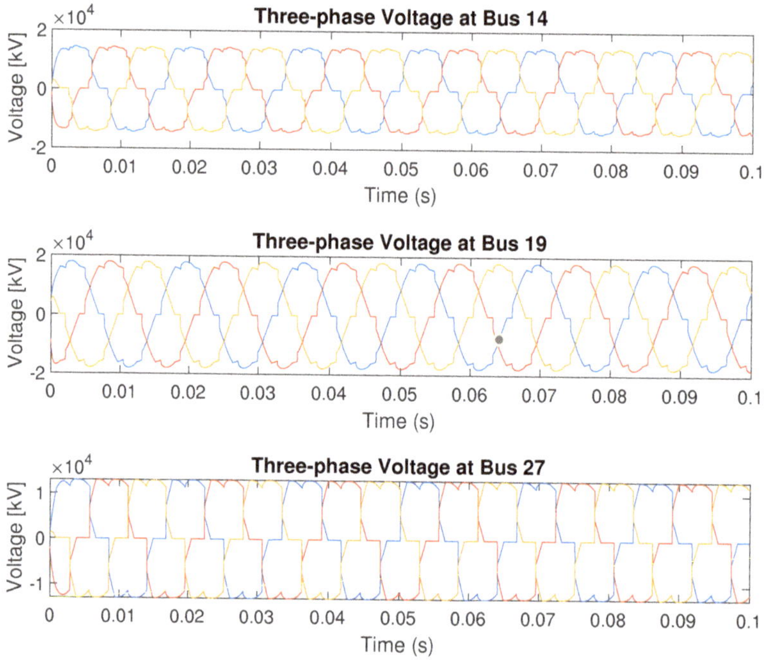

Figure 9. Three-phase voltage at buses 14, 19, and 27.

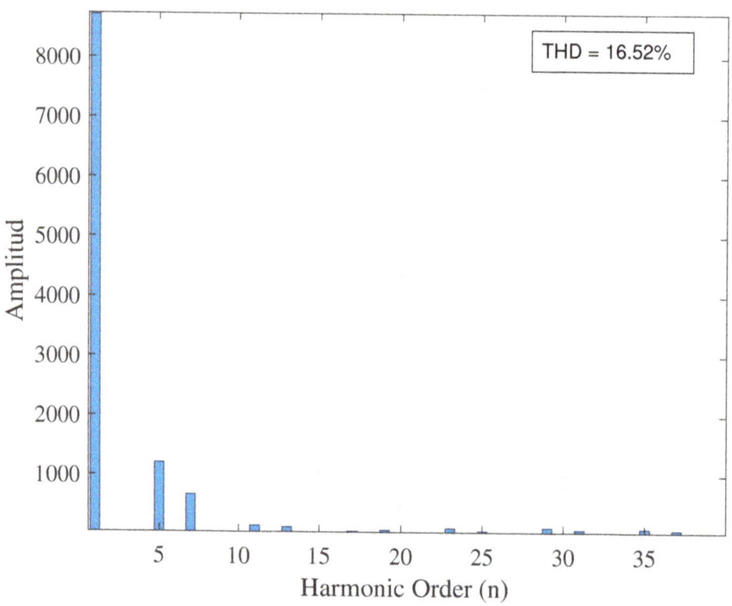

Figure 10. Voltage harmonics at bus 14 with nonlinear load.

The second scenario mentioned previously corresponds to the analysis of voltage profiles when DG is integrated into the system (Figure 7). The voltage values for this condition are presented in Figure 13.

Based on Figure 13, it can be observed that the voltage levels of the system with DG integration do not exhibit significant variations. Slight changes are only noted in the THDv levels at the buses with nonlinear loads, as illustrated in Figure 14.

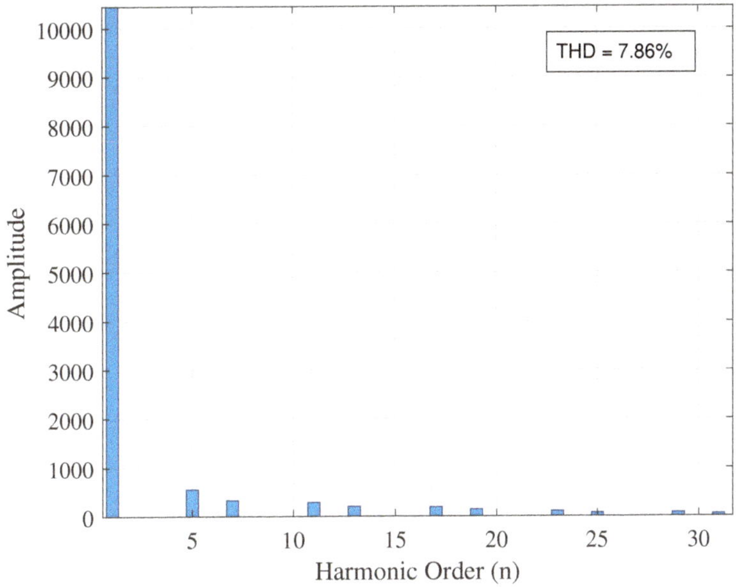

Figure 11. Voltage harmonics at bus 19 with nonlinear load.

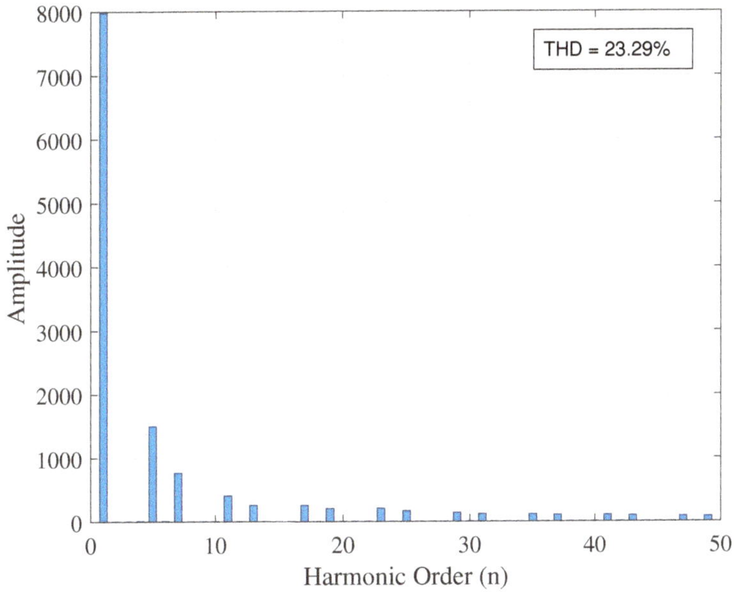

Figure 12. Voltage harmonics at bus 27 with nonlinear load.

These changes are not significant, as their values are quite similar to those obtained in the previous scenario. To confirm this, Figures 15 and 16 show the THDv levels at buses 19 and 27, respectively.

As verified in the previous figures, there is no significant difference between the two scenarios analyzed. However, it is possible that the inclusion of DG may influence the optimal placement of the power quality conditioner. This will be verified in the case studies presented in the next chapter. Additionally, the data obtained here will serve as a baseline for analyzing the improvement in THDv with the inclusion of the UPQC in each case study.

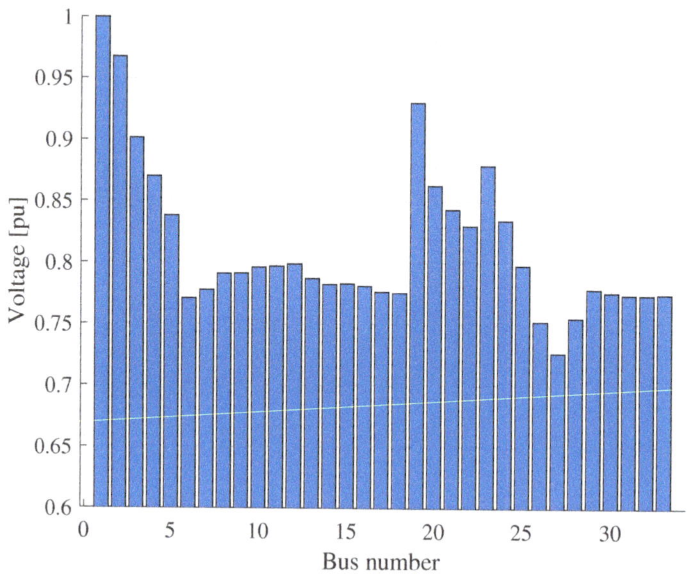

Figure 13. Voltage levels at each bus with nonlinear loads and DG.

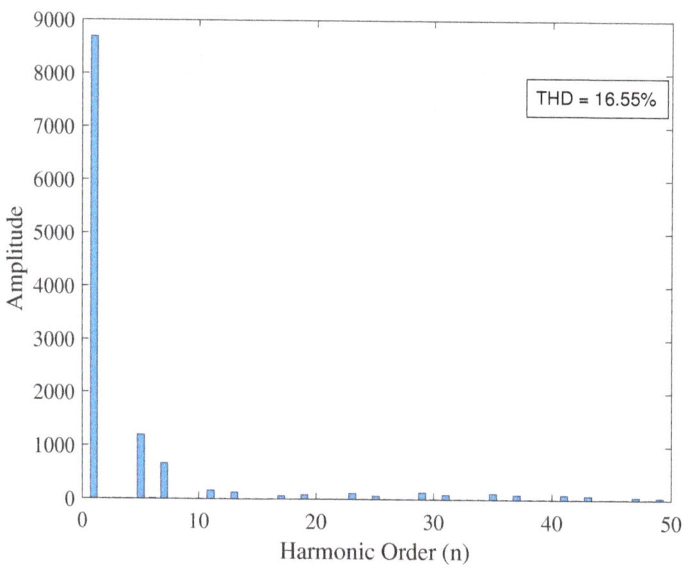

Figure 14. Voltage harmonics at bus 14 with nonlinear load and DG.

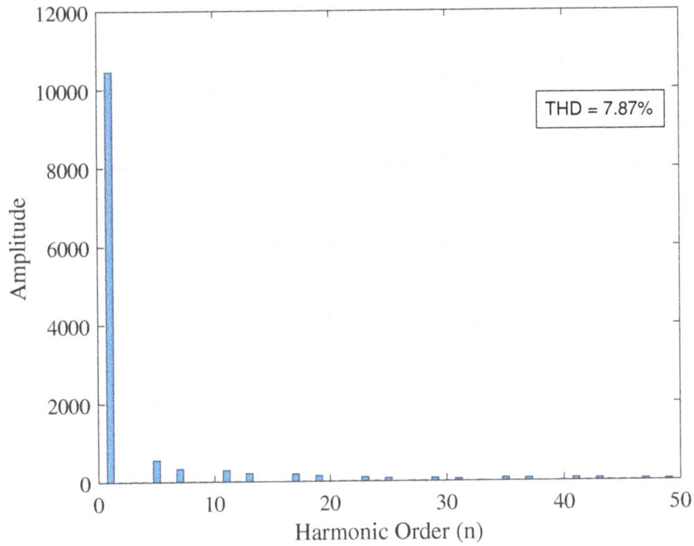

Figure 15. Voltage harmonics at bus 19 with nonlinear load and DG.

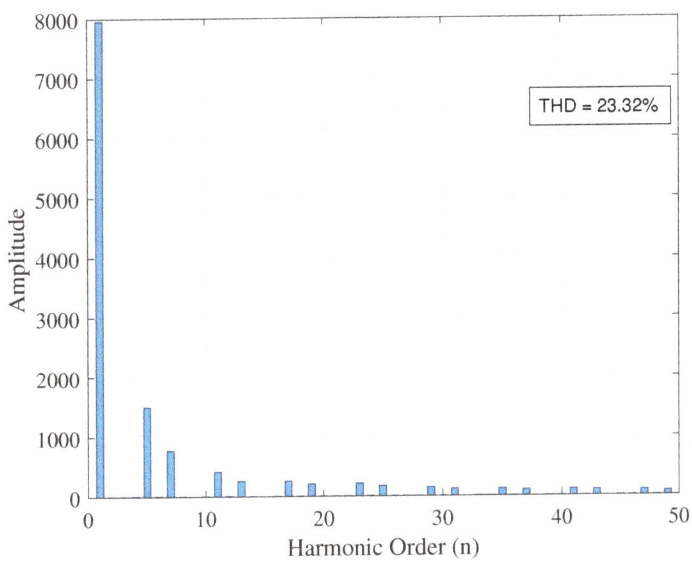

Figure 16. Voltage harmonics at bus 27 with nonlinear load and DG.

4. Case Studies

This study is based on the premise of optimally placing the UPQC in a distribution system. To this end, two case studies are proposed and detailed in this section.

Before presenting the case studies, it is important to clarify how the optimal placement of the device within the system is determined. For this purpose, the weighted cost objective function ($J(x)$) is employed. This function enables the identification of the optimal bus for installing the UPQC, where the best voltage profiles and THDv values are achieved.

4.1. Case 1

This case corresponds to the test system previously described, with nonlinear loads connected at buses 14, 19, and 27. For better understanding, the schematic of this system is presented in Figure 6.

Using exhaustive search, the UPQC is placed iteratively at each of the nonlinear load buses in order to evaluate the resulting changes in THDv and determine which location yields the lowest DPV.

4.2. Case 2

This analysis considers the test system with the inclusion of DG composed of a 3500W photovoltaic system. The schematic diagram of this configuration is the one shown earlier in Figure 7.

As in the previous scenario, exhaustive search is used to determine the optimal placement of the UPQC within this network topology. The aim is to assess whether the presence of such a microgrid affects the optimal location of the power quality improvement device compared to the previous case.

5. Results and Analysis

This section presents the results corresponding to the case studies defined in the previous chapter. These results demonstrate the impact of the UPQC on the test system, particularly in terms of mitigating voltage variations and reducing THD caused by nonlinear loads.

5.1. Voltage THD Summary

Table 6 compiles the voltage total harmonic distortion (THDv) values at the critical buses (14, 19, and 27) for each studied scenario and UPQC placement. This compact presentation allows direct numerical comparison of the effect of device location on harmonic mitigation, while minimizing the space required to present the results. Bold values indicate the lowest THDv achieved at each bus within each case.

Table 6. THDv (%) at buses 14, 19, and 27 for each scenario and UPQC placement. Bold values mark the best performance per bus within each case.

Case	UPQC Location	Bus 14	Bus 19	Bus 27
Case 1 (No DG)	None (base)	16.52	7.86	23.29
	At bus 14	**1.10**	4.76	**10.52**
	At bus 19	15.01	**3.36**	22.94
	At bus 27	11.34	5.28	11.91
Case 2 (With DG)	None (base)	16.55	7.87	23.32
	At bus 14	**1.08**	4.75	11.39
	At bus 19	9.54	2.90	**5.53**
	At bus 27	4.81	**0.69**	10.57

Table 6 shows that in Case 1 the lowest THDv values are obtained at bus 14 when the UPQC is installed at bus 14, at bus 19 when it is installed at bus 19, and at bus 27 when it is installed at bus 14. In Case 2, the best results for buses 14, 19, and 27 are achieved with installations at buses 14, 27, and 19, respectively.

5.2. Case 1

5.2.1. UPQC Installed at Bus 14

Figure 17 shows the initial voltage profile of the system for this first case study. It should be noted that these voltage values correspond to those previously presented in Figure 13.

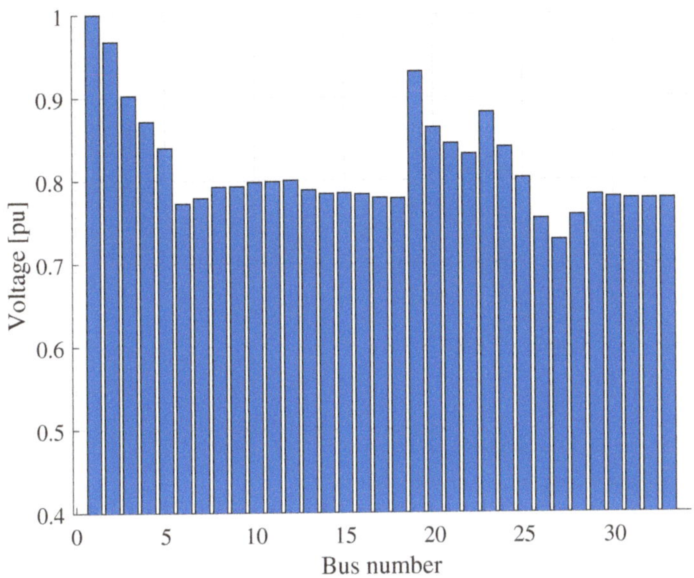

Figure 17. Initial voltage profile—Case 1.

Following the exhaustive search methodology, the device is placed at each of the nodes with nonlinear loads, starting with bus 14. The resulting voltage profiles are shown in Figure 18.

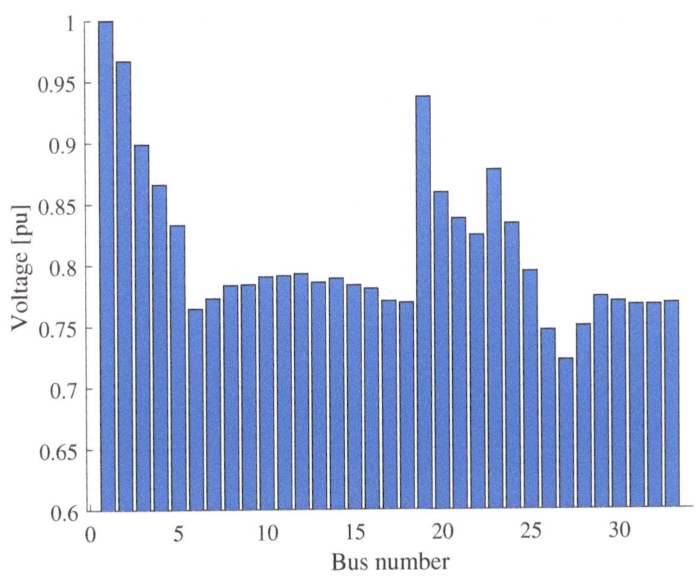

Figure 18. Voltage profile with UPQC installed at Bus 14.

Additionally, Figure 19 presents the THD levels observed at bus 14 after the installation of the UPQC.

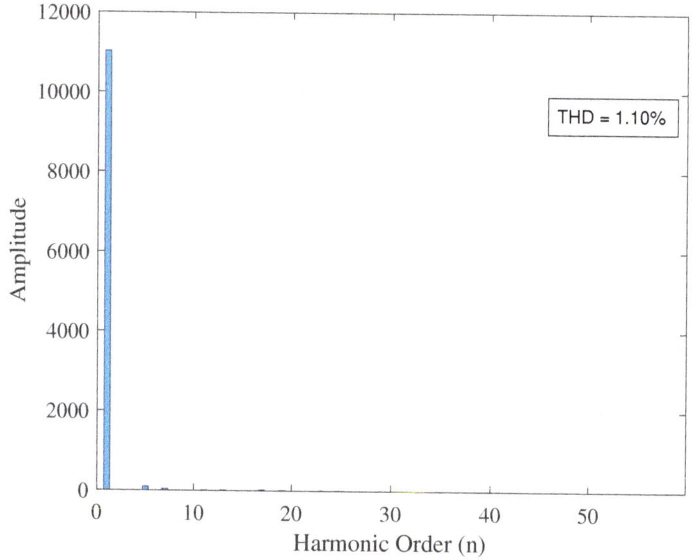

Figure 19. Voltage harmonics at Bus 14 with UPQC installed.

The THD recorded at bus 14 decreased from 16.52% (as presented in the base case results) to 1.10%, representing an improvement of approximately 93.34%. Similar improvements are observed at the other buses with nonlinear loads, such as bus 19, whose THD results are shown in Figure 20.

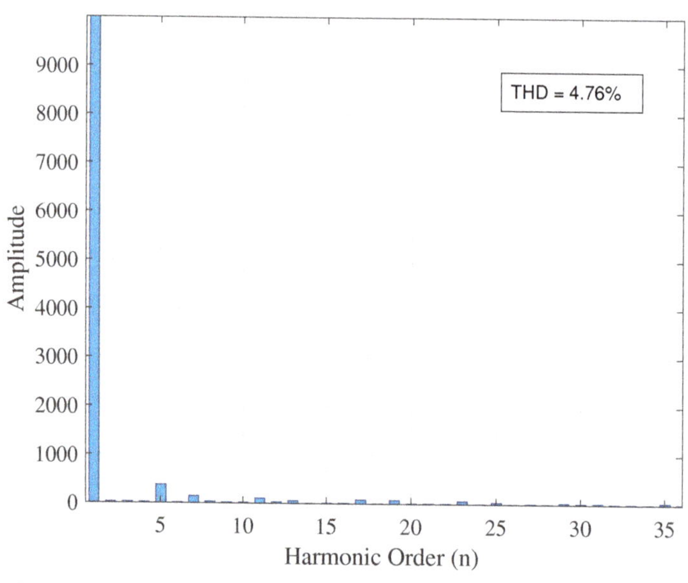

Figure 20. Voltage harmonics at bus 19 with UPQC installed at bus 14.

At this bus, the new THD value is 4.76%, which represents a 39.44% improvement over the original value. Similarly, Figure 21 shows the updated THD level at bus 27.

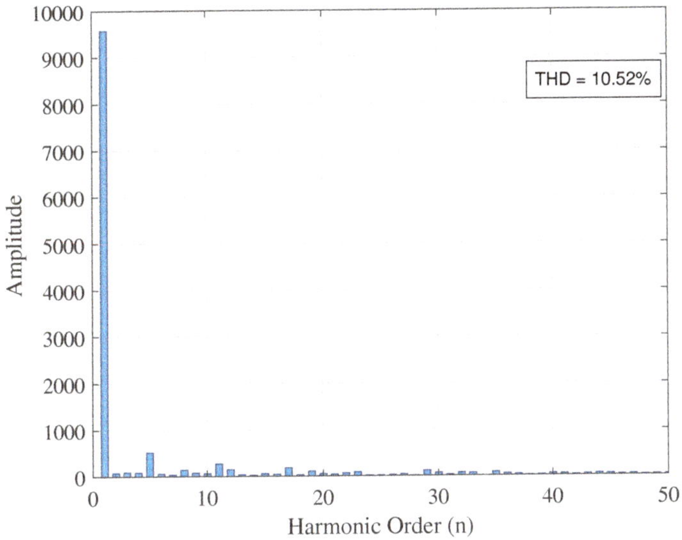

Figure 21. Voltage harmonics at bus 27 with UPQC installed at Bus 14.

At bus 27, an improvement of 54.83% is observed, as the THD is reduced from 23.29% to 10.52%.

5.2.2. UPQC Installed at Bus 19

After confirming the improvements obtained with the UPQC installed at bus 14, the exhaustive search methodology proceeds by placing the device at bus 19. The resulting voltage profiles are shown in Figure 22.

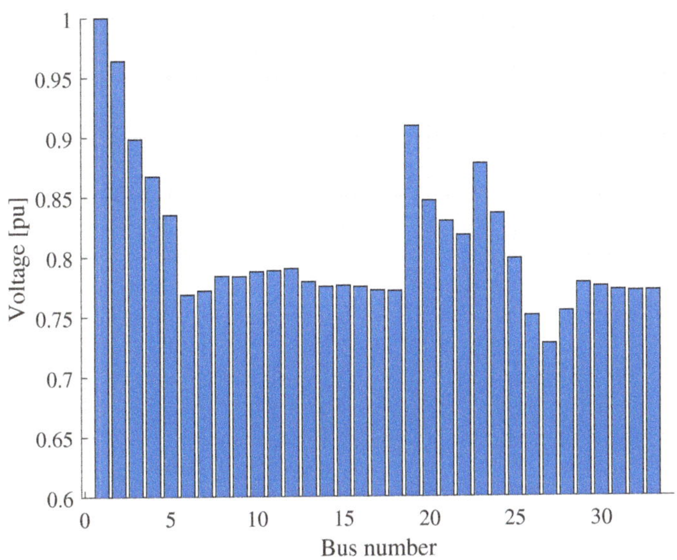

Figure 22. Voltage profile with UPQC installed at Bus 19.

As shown in Figure 22, the voltage levels experience slight variations. Similarly, changes in the THD levels at each bus can be observed. For instance, Figure 23 presents the updated THD at bus 14.

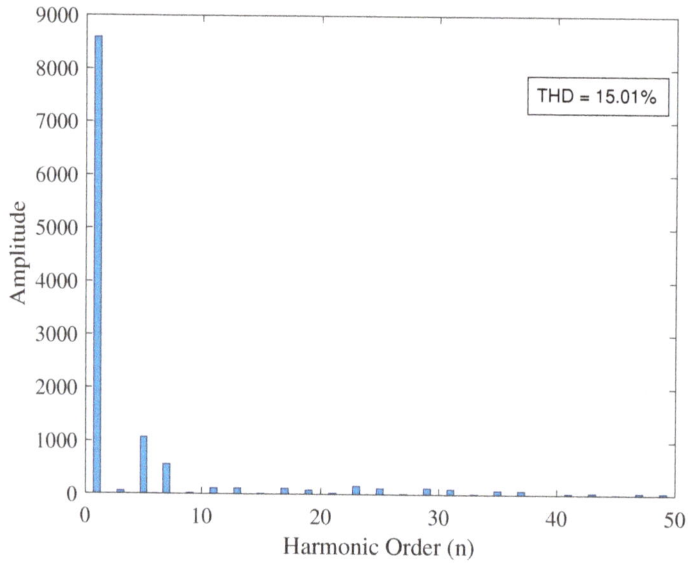

Figure 23. Voltage harmonics at Bus 14 with UPQC installed at Bus 19.

In this case, the THD at bus 14 shows no significant improvement. The THD recorded is 15.01%, compared to the 16.52% observed in the base case. This represents only a 9.14% reduction.

Figure 24 shows the THD at bus 19, where the UPQC is directly connected. A substantial improvement is expected at this node.

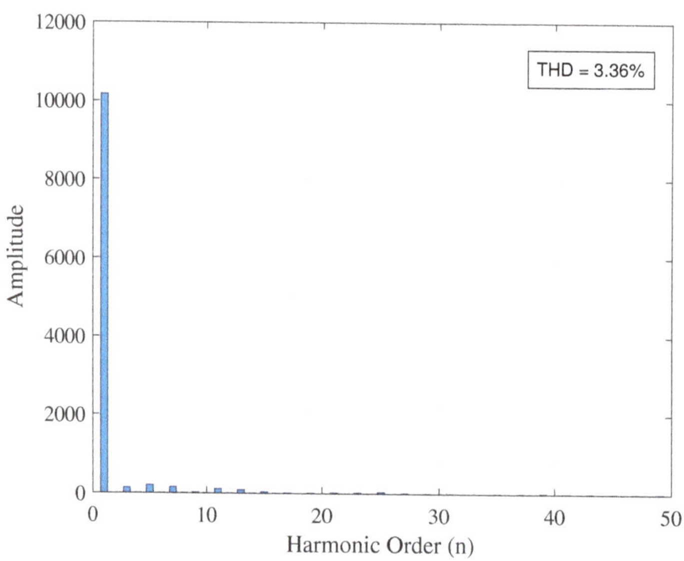

Figure 24. Voltage harmonics at bus 19 with UPQC installed.

A THD value of 3.36% is obtained at this bus, representing a 57.2% reduction compared to the initial condition. Additionally, Figure 25 shows the THD at bus 27, where only a marginal change is observed—from 23.29% to 22.94%.

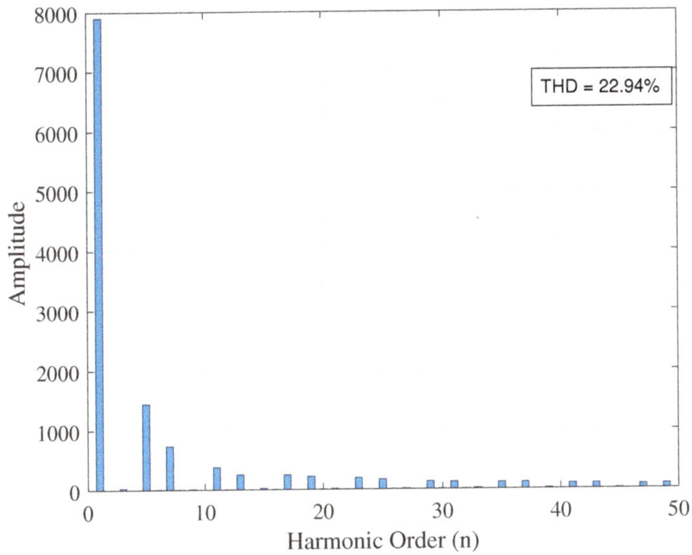

Figure 25. Voltage harmonics at bus 27 with UPQC installed at bus 19.

As can be seen, the improvement in THDv across the buses is not substantial when the UPQC is installed at bus 19, especially when compared to the scenario where the UPQC was connected at bus 14.

5.2.3. UPQC Installed at Bus 27

To complete the exhaustive search and identify the optimal location of the UPQC for this case study, the device is now placed at the final bus with a nonlinear load—bus 27. As in previous scenarios, the system's voltage profiles are obtained once the device is installed at this bus. These results are presented in Figure 26.

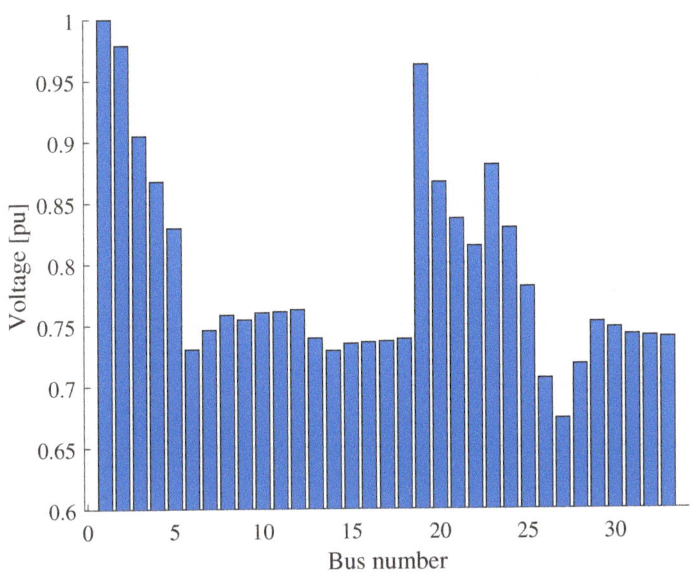

Figure 26. Voltage profile with UPQC installed at Bus 27.

Once the UPQC is placed at bus 27, the analysis focuses on evaluating the THDv at buses 14, 19, and 27 to determine whether this configuration yields significant improvements compared to previous cases (i.e., UPQC installed at buses 14 and 19). Figure 27 shows the THDv level at bus 14 when the UPQC is connected at bus 27.

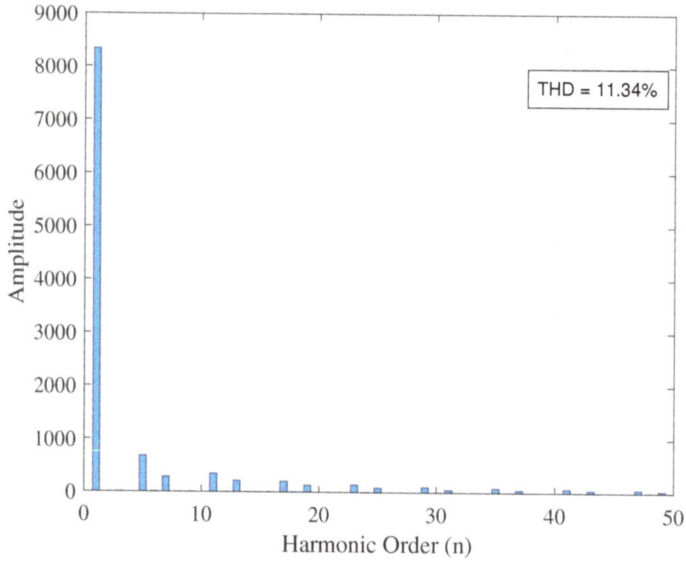

Figure 27. Voltage harmonics at Bus 14 with UPQC installed at Bus 27.

With the UPQC installed at bus 27, a THDv of 11.34% is recorded. Compared to the base value of 16.52%, this represents a 31.36% reduction. Additionally, Figure 28 presents the THDv at bus 19.

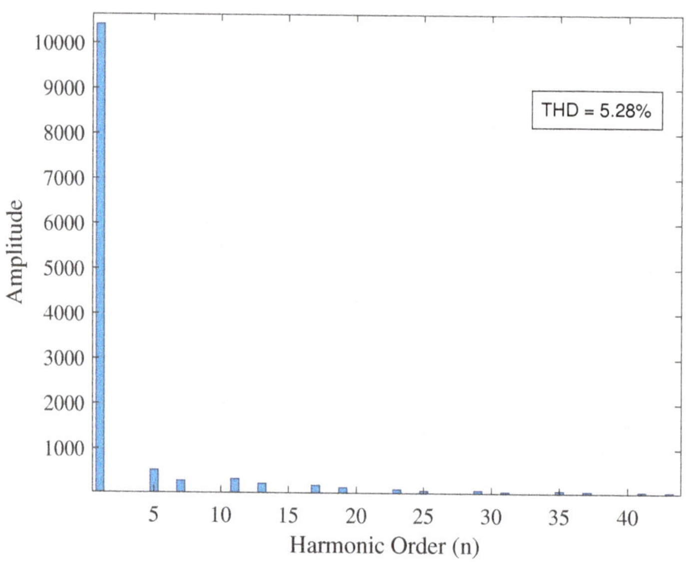

Figure 28. Voltage harmonics at bus 19 with UPQC installed at bus 27.

In this case, the THDv is reduced from 7.86% to 5.28%, corresponding to an improvement of 32.83%. At bus 27, where the UPQC is directly connected, a THDv of 11.91% is registered, as shown in Figure 29.

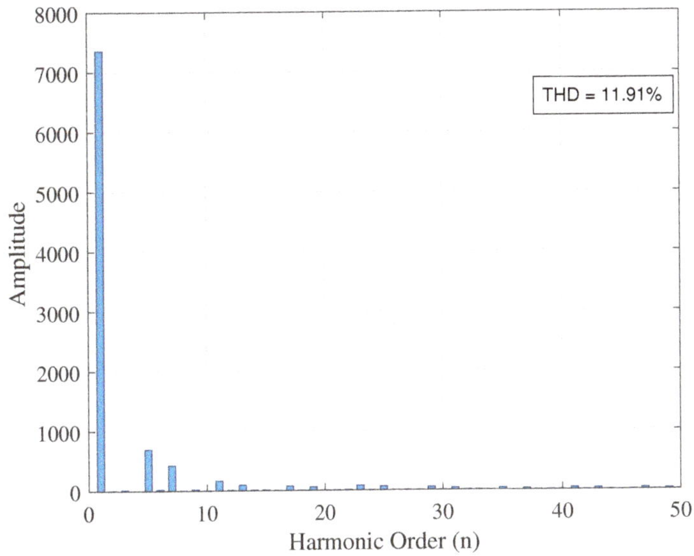

Figure 29. Voltage harmonics at bus 27 with UPQC installed.

This new THDv value represents a 48.87% improvement compared to the base scenario. After completing the THDv analysis for all buses with nonlinear loads, the voltage profiles presented in Figures 18, 22 and 26 are reviewed to assess the impact of UPQC placement on voltage magnitude. The differences among these profiles are more clearly illustrated in Figure 30.

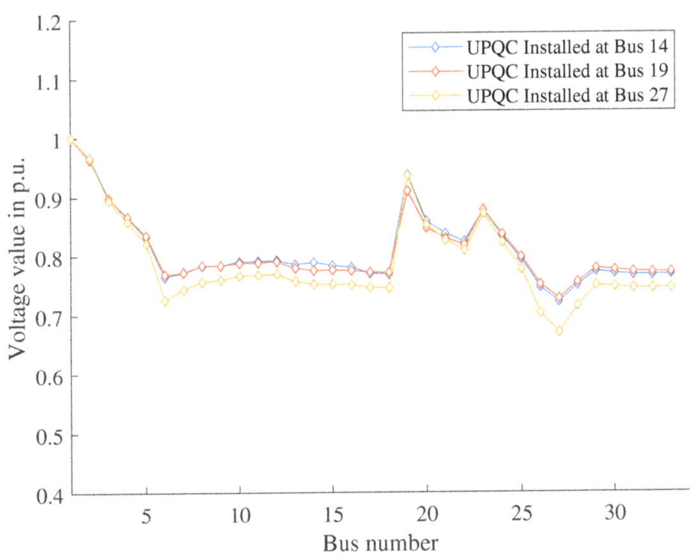

Figure 30. Voltage levels with UPQC installed at each candidate bus in Case 1.

The voltage profile changes across the system do not explicitly indicate which location delivers the greatest benefit. Therefore, the optimization algorithm is applied to determine the optimal installation bus for the UPQC.

5.3. Optimal Location Analysis

To determine the optimal installation site for the UPQC among the three candidate buses, the first step is to calculate the average THDv for each candidate solution (f_1) using Equation (16).

$$f_1(14) = \frac{16.38}{3} = 5.46\% \tag{30}$$

$$f_1(19) = \frac{41.31}{3} = 13.77\% \tag{31}$$

$$f_1(27) = \frac{28.53}{3} = 9.51\% \tag{32}$$

Once the values for the first objective (f_1) are obtained, the next step is to calculate the average voltage deviation (f_2) for each case using Equation (17).

$$f_2(14) = \frac{6.2581}{33} = 0.1896\ p.u. \tag{33}$$

$$f_2(19) = \frac{6.2977}{33} = 0.1908\ p.u. \tag{34}$$

$$f_2(27) = \frac{6.9575}{33} = 0.2108\ p.u. \tag{35}$$

With both objective values (f_1 and f_2) computed, normalization is applied using Equation (20), and the weighted cost function is then calculated for each candidate:

$$J(14) = 0.3 \cdot 0 + 0.7 \cdot 0 = 0 \tag{36}$$

$$J(19) = 0.3 \cdot 1 + 0.7 \cdot 0.057 = 0.34 \tag{37}$$

$$J(27) = 0.3 \cdot 0.487 + 0.7 \cdot 1 = 0.84 \tag{38}$$

Finally, based on the obtained results, it is concluded that the optimal installation site for the UPQC is bus 14, as it yields the lowest value of the objective cost function.

5.4. Case 2

As previously mentioned, this case study aims to determine whether the inclusion of a PV-based DG unit alters the optimal location of the UPQC identified in the first case study.

5.5. Impact of DG Penetration on Harmonic Performance

To quantify how distributed generation (DG) penetration affects harmonic performance and UPQC siting, the PV output was swept between two operating points available in this study: 0% (no DG) and 100% of the nominal PV capacity (3.5 kW). For each operating point, the average voltage total harmonic distortion (THDv) over the monitored buses (14, 19, and 27) was computed for three representative UPQC locations (buses 14, 19, and 27), yielding the trends shown in Figure 31.

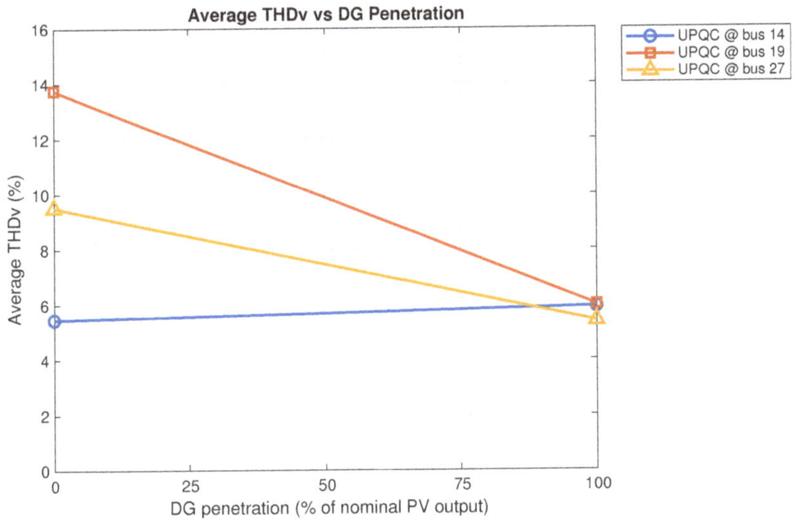

Figure 31. Average THDv as a function of DG penetration for three UPQC placements (buses 14, 19, and 27). Two penetration levels are considered with the data available: 0% (no DG) and 100% of nominal PV output (3.5 kW).

Results indicate that the optimal location identified in the base case remains robust when DG operates at its nominal point: UPQC@14 retains the lowest average THDv, while DG presence substantially reduces the average distortion for UPQC@19 and UPQC@27. These outcomes confirm that the siting strategy preserves its effectiveness across the tested DG operating range and motivate future work to interpolate intermediate penetration levels.

The PV-based distributed generation (DG) unit and the UPQC perform complementary roles at the point of common coupling (PCC). The shunt converter of the UPQC shapes the source current and regulates the DC-link voltage (with $i_q^* \approx 0$ in steady state), while the series converter restores/conditions the load-side voltage [15,18]. Under these control objectives, the PV inverter continues to operate at its setpoint (e.g., MPPT or fixed active power) and sees a PCC voltage kept close to nominal. As a result, across the scenarios tested, we did not observe adverse interactions such as oscillatory power exchange or control coupling between the PV inverter and the UPQC.

Consistently with Figure 31, the presence of DG at nominal output reduces the average THDv for some non-optimal placements (e.g., UPQC at buses 19 and 27), narrowing the gap with respect to the optimal site. Nevertheless, UPQC installation at bus 14 remains the best trade-off for harmonic mitigation and voltage profile improvement within the tested range. From an operational standpoint, the shunt branch keeps the supply current near sinusoidal and in phase with the PCC voltage, while the series branch limits voltage distortion at the load terminals. This division of tasks confines the UPQC's active-power support to balancing the injected series voltage (via DC-link control) without perturbing the PV inverter's setpoint [13,15].

Finally, by mitigating overvoltage and distortion at the PCC, the UPQC can lower the risk of PV inverter protective trips (e.g., overvoltage or abnormal distortion) in feeders with variable irradiance. In feeders with higher DG penetration or multiple PV units, the same coordination principle applies, though additional devices or coordinated control could be required to achieve system-wide compliance, as discussed in the scalability notes of Section 3.4.

5.5.1. UPQC at Bus 14

As in the previous case, the THDv is analyzed for each possible installation site of the UPQC, starting with bus 14, as shown in Figure 32.

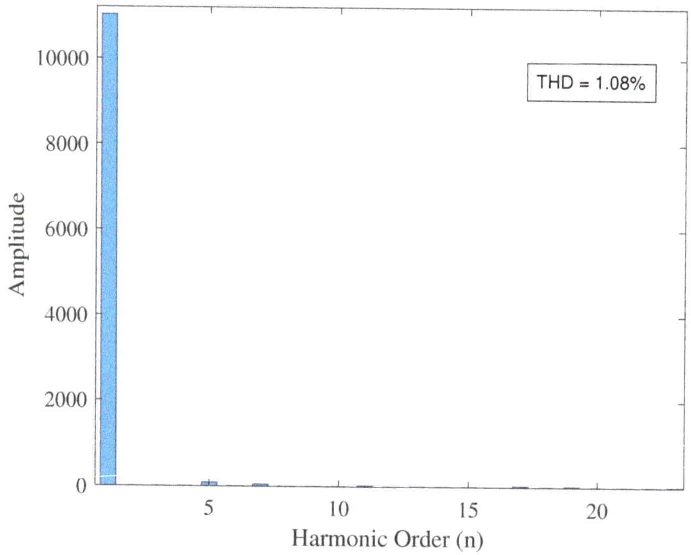

Figure 32. Harmonic distortion at bus 14 with UPQC.

The THDv at bus 14 is analyzed with the UPQC installed at this location. In this case, the recorded THD is 1.08%, which is very similar to the result obtained in Case 1 (1.10%) without the DG connected. Compared to the base scenario with DG, where a THDv of 16.55% was observed, this result represents an improvement of approximately 93.47%.

Meanwhile, the harmonic distortion at bus 19, shown in Figure 33, presents a THD of 4.75%.

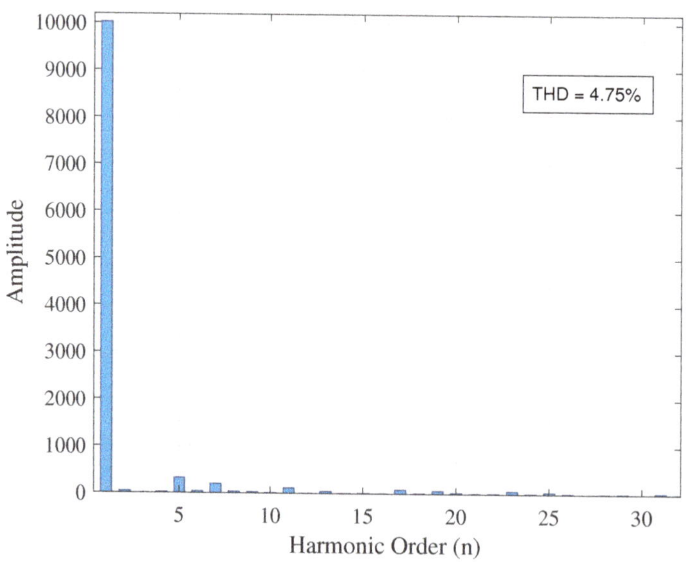

Figure 33. Harmonic distortion at bus 19 with UPQC installed at bus 14.

Given that the base THDv at bus 19 was 7.87%, the reduction to 4.75% reflects an improvement of 39.64%. Figure 34 shows the harmonic distortion registered at bus 27.

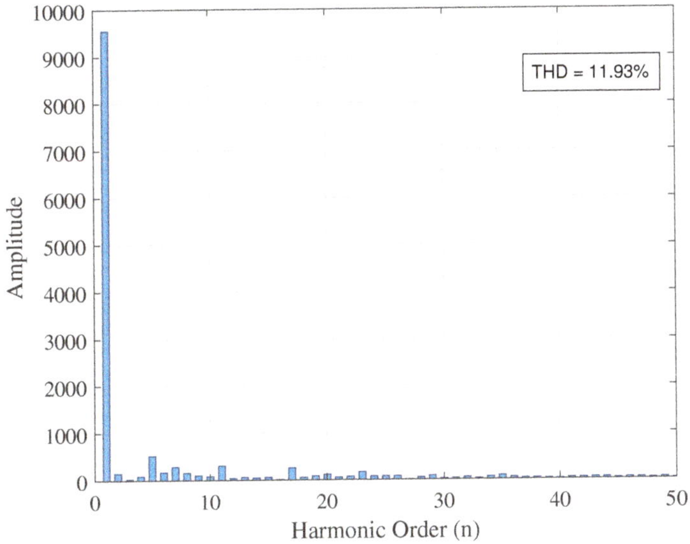

Figure 34. Harmonic distortion at bus 27 with UPQC installed at bus 14.

At bus 27, the THDv is reduced from 23.32% to 11.39%, representing an improvement of 48.84%.

The voltage profile of the system with the UPQC installed at bus 14 in this scenario is presented in Figure 35.

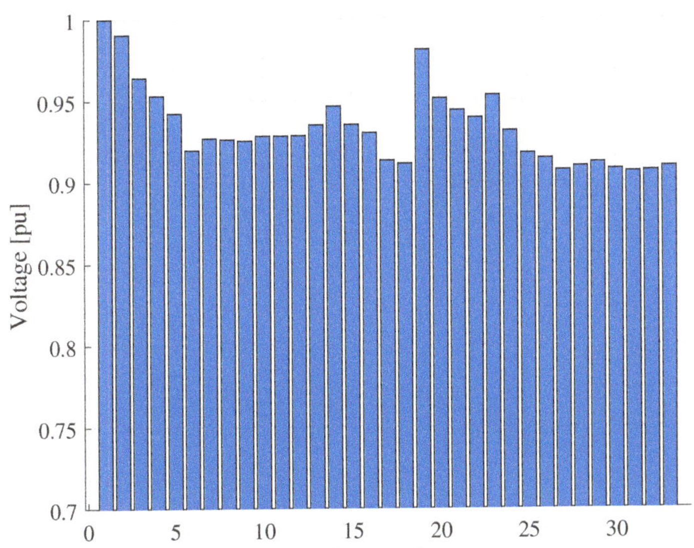

Figure 35. Voltage profiles with UPQC installed at bus 14.

Once the analysis with the UPQC placed at bus 14 is completed, the equipment is relocated to bus 19 and the corresponding THDv calculations are performed again.

5.5.2. UPQC at Bus 19

In this case, the new THD for bus 14 is obtained, as shown in Figure 36.

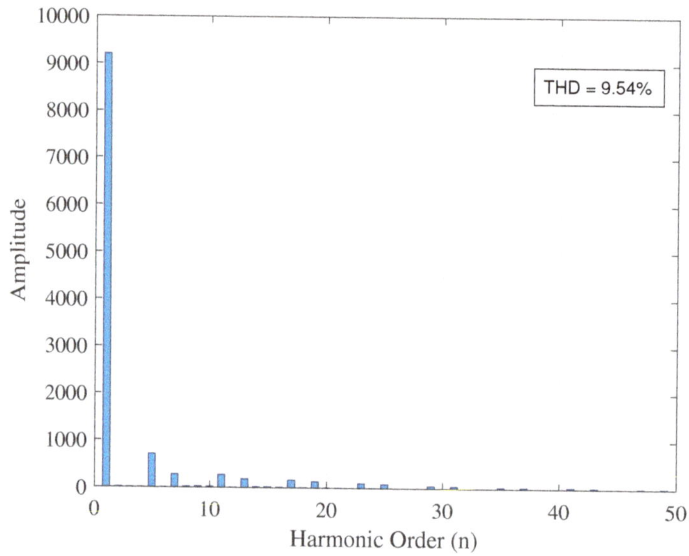

Figure 36. Harmonic distortion at bus 14 with UPQC installed at bus 19.

Compared to the base system with DG, the THD recorded at this bus improves by 42.36%. Additionally, bus 19 registers a new THDv of 2.90%, as illustrated in Figure 37.

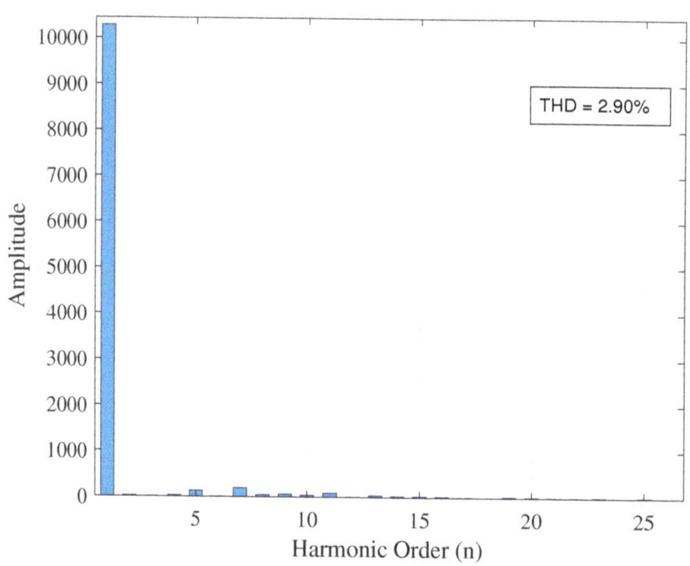

Figure 37. Harmonic distortion at bus 19 with UPQC.

This result represents a 63.16% improvement relative to the base system with DG. Meanwhile, Figure 38 shows the new THDv registered at bus 27.

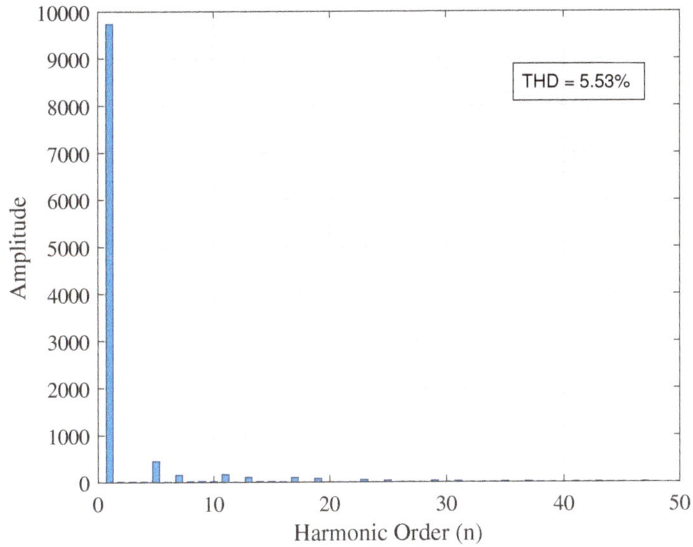

Figure 38. Harmonic distortion at bus 27 with UPQC installed at bus 19.

The THDv at this bus decreases significantly, from 23.32% in the base system to 5.53% with the UPQC connected at bus 19, which corresponds to an improvement of 76.26%.

With this UPQC placement, the resulting voltage profiles are presented in Figure 39.

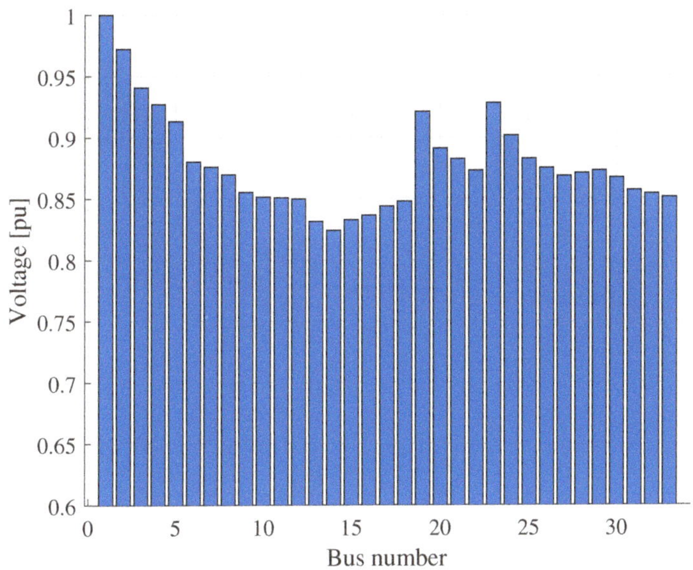

Figure 39. Voltage profiles with UPQC installed at bus 19.

5.5.3. UPQC at Bus 27

After completing the analysis for bus 19, the UPQC is connected to bus 27, and the THDv is analyzed for each node with nonlinear loads, starting with bus 14, as shown in Figure 40.

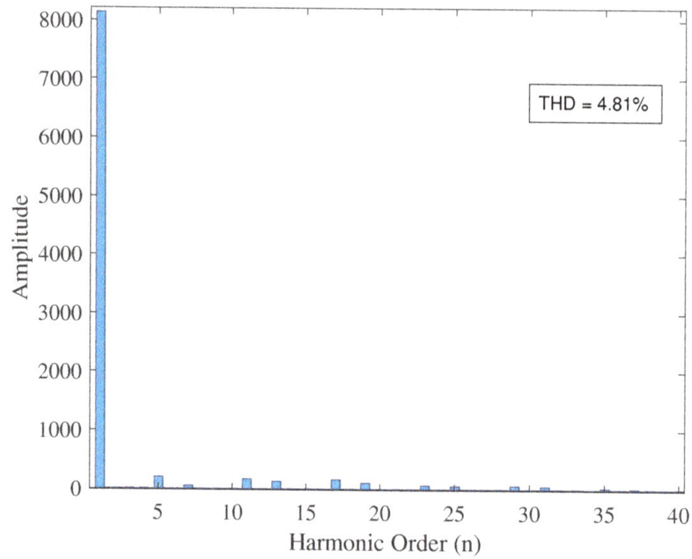

Figure 40. Harmonic distortion at bus 14 with UPQC installed at bus 27.

In this scenario, bus 14 presents a THDv of 4.81% compared to the previous case, in which the recorded value was 9.54%. A significant change can also be observed at bus 19, as shown in Figure 41.

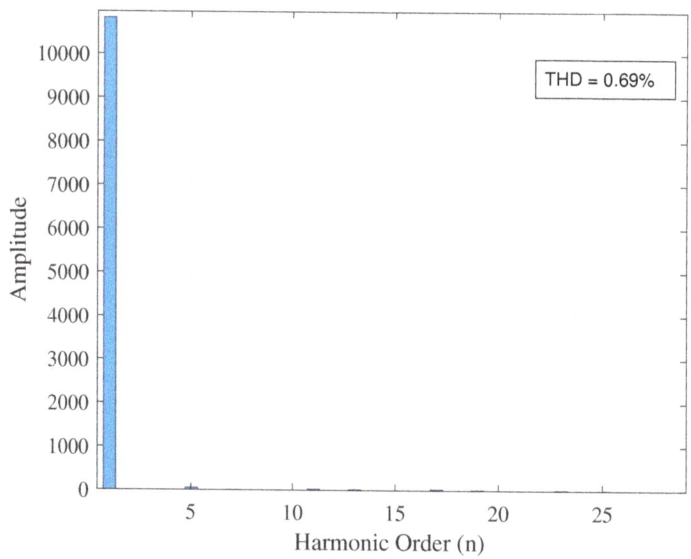

Figure 41. Harmonic distortion at bus 19 with UPQC installed at bus 27.

In this case, the improvement is quite evident, with the THDv reduced from 7.87% to 0.69%, which corresponds to a 91.24% enhancement in power quality. Figure 42 presents the harmonics at bus 27, where the recorded THDv is 10.57%.

This result represents a 54.12% improvement at this bus. With the UPQC installed at bus 27, the voltage profiles obtained are shown in Figure 43.

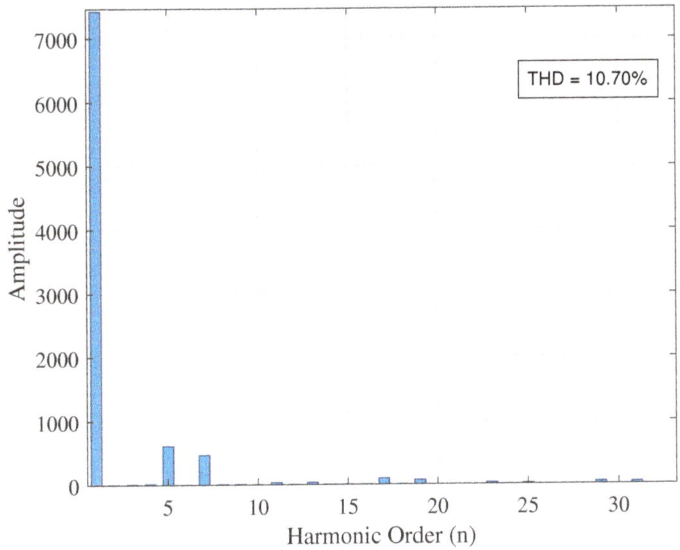

Figure 42. Harmonic distortion at bus 27 with UPQC.

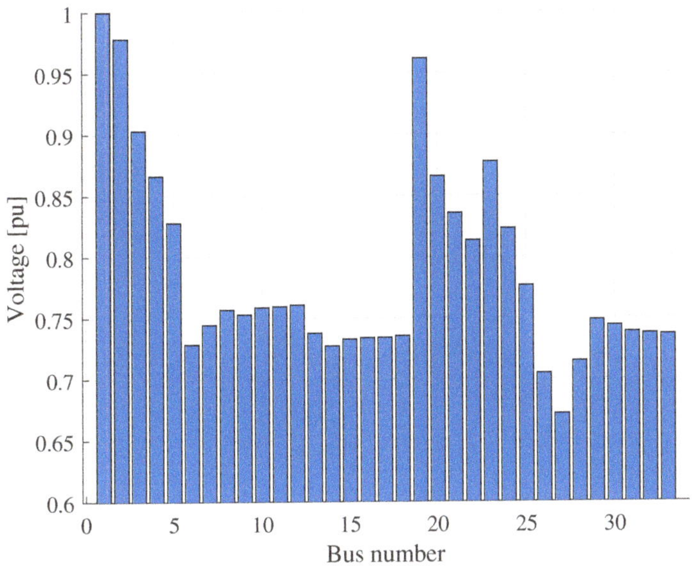

Figure 43. Voltage profiles with UPQC installed at bus 27.

With the THDv analysis for this case study completed, the next step is to calculate the voltage deviation profile (VDP). To begin, a comparative graph is generated showing the voltage profiles for each of the three UPQC placement scenarios considered in this study, as illustrated in Figure 44.

These changes in voltage levels show a clear distinction compared to the previous case study. To ensure the optimal location of the power quality device, the corresponding calculations will now be performed using the previously established optimization strategy.

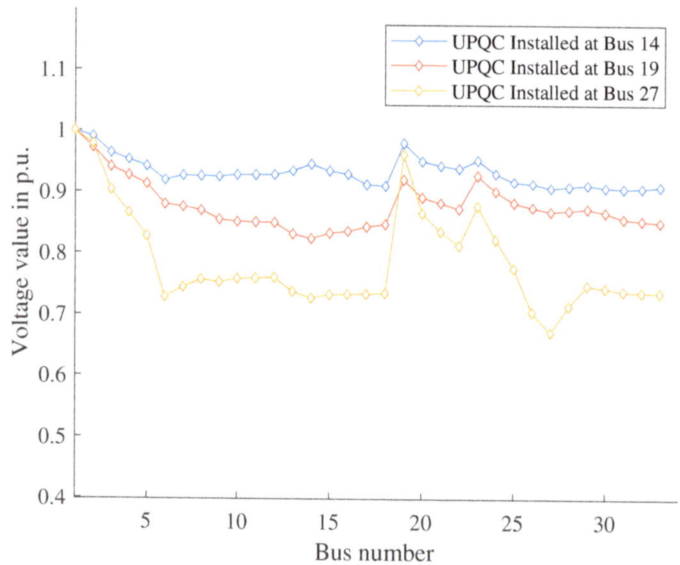

Figure 44. Voltage levels for UPQC placement in Case 2.

5.5.4. Optimal Location Analysis

As in the previous case study, the process begins by determining the values corresponding to the two objective functions.

$$f_1(14) = \frac{17.76}{3} = 5.92\% \tag{39}$$

$$f_1(19) = \frac{17.97}{3} = 5.99\% \tag{40}$$

$$f_1(27) = \frac{16.2}{3} = 5.4\% \tag{41}$$

$$f_2(14) = \frac{2.2036}{33} = 0.0668 \text{ p.u.} \tag{42}$$

$$f_2(19) = \frac{4.0010}{33} = 0.1212 \text{ p.u.} \tag{43}$$

$$f_2(27) = \frac{7.0182}{33} = 0.2127 \text{ p.u.} \tag{44}$$

Once the average values of THDv (f_1) and the voltage deviation profile (f_2) are obtained, both functions ($\hat{f}_1$ and $\hat{f}_2$) are normalized in order to compute the weighted cost function.

$$J(14) = 0.3 \cdot 0.88 + 0.7 \cdot 0.067 = 0.26 \tag{45}$$

$$J(19) = 0.3 \cdot 1 + 0.7 \cdot 0.37 = 0.56 \tag{46}$$

$$J(27) = 0.3 \cdot 0 + 0.7 \cdot 1 = 0.7 \tag{47}$$

For this second case study, it is clearly evident that the optimal location for the UPQC is at bus 14, since this scenario yields the lowest value of the objective function among the three evaluated alternatives.

To provide a clearer and more concise visualization of the impact of the UPQC in both study cases, Table 7 presents a comparative summary of the most relevant indicators. The table includes the total harmonic distortion of voltage (THDv) at each node with nonlinear load (nodes 14, 19, and 27), the global voltage deviation, the weighted cost function value $J(x)$, and the percentage of THDv improvement with respect to the base scenario without UPQC. This summary facilitates the identification of the optimal installation node and quantifies the performance benefits achieved by each configuration.

Table 7. Comparative summary of THDv, voltage deviation, cost function and improvement for optimal UPQC locations.

Scenario	Node	THDv (%)	Voltage Deviation (p.u.)	$J(x)$	THDv Improvement (%)
Case 1 (No DG)	14	1.10	0.1896	0.00	93.34
	19	3.36	0.1908	0.34	57.23
	27	11.91	0.2108	0.84	48.87
Case 2 (With DG)	14	1.08	0.0668	0.26	93.47
	19	2.90	0.1212	0.56	63.16
	27	10.57	0.2127	0.70	54.12

To visually enhance the interpretation of the results, Figure 45 presents a graphical summary of key performance indicators for each scenario and UPQC location. This includes the total harmonic distortion in voltage (THDv), global voltage deviation, the weighted objective function value $J(x)$, and the percentage improvement in THDv with respect to the base case. The graph enables a clearer understanding of the impact of each configuration and supports the optimal placement analysis discussed previously.

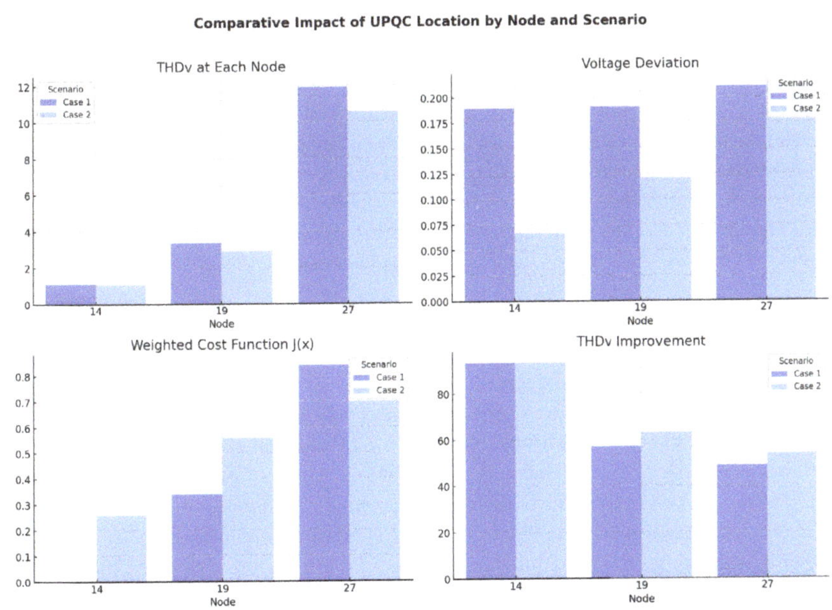

Figure 45. Graphical comparison of THDv, voltage deviation, weighted cost function $J(x)$, and THDv improvement for each UPQC location in both case studies.

5.6. Robustness Summary and Statistical Dispersion of THDv

Although the simulations are deterministic (i.e., repeated runs under identical operating points return identical values), it is still informative to quantify the dispersion across monitored buses for each scenario as a measure of spatial robustness. Table 8 reports the mean and standard deviation of voltage THD (%THDv) across buses 14, 19, and 27 in the base condition and with the UPQC placed at bus 14, for both scenarios considered (without DG and with PV-based DG at its nominal operating point). The reduction in the mean THDv relative to the corresponding base condition is also provided.

Table 8. Statistical summary of THDv (%) across monitored buses (14, 19, 27): mean and standard deviation for base vs. UPQC@14 in both scenarios.

Scenario	Condition	Mean THDv (%)	Std. Dev. (%)	Reduction vs. Base (%)
Case 1 (no DG)	Base (buses 14/19/27: 16.52, 7.86, 23.29)	15.89	7.73	–
Case 1 (no DG)	UPQC@14 (1.10, 4.76, 10.52)	5.46	4.75	65.64
Case 2 (with DG)	Base (16.55, 7.87, 23.32)	15.91	7.74	–
Case 2 (with DG)	UPQC@14 (1.08, 4.75, 11.39)	5.74	5.23	63.92

Values in parentheses are the per-bus THDv (%) used to compute the statistics. Mean and standard deviation are computed across buses (spatial dispersion) for each condition. Reductions are calculated as the relative decrease of the mean with respect to the corresponding base case.

The results show substantial reductions in the mean THDv across the monitored buses when the UPQC is installed at bus 14 in both scenarios (approximately 66% without DG and 64% with DG), together with a decrease in dispersion (standard deviation). This indicates that, beyond improving the average harmonic distortion, the proposed placement also contributes to a more uniform power quality level across the affected area of the network. Because the simulations are deterministic, classical inferential tools (e.g., p-values) are not directly applicable; nonetheless, the spatial statistics reported here provide a concise and reproducible summary of robustness across buses under the tested operating conditions.

6. Discussion and Novelty of the Proposed Approach

The UPQC placement problem in this study is explicitly formulated as a constrained combinatorial optimization task, where the decision variable indicates the location of the device within the test network, subject to technical and operational limits. The optimization simultaneously considers multiple technical objectives: (i) minimization of voltage total harmonic distortion (THDv) at all monitored buses, (ii) improvement of voltage profiles across the network, and (iii) reduction of active power losses.

Unlike many existing approaches that address these objectives separately or focus solely on harmonic mitigation, the present work integrates all three performance indicators into a unified decision-making framework. The evaluation was conducted for all buses in the system, demonstrating that the most beneficial placements under nonlinear load conditions consistently yield superior results in all objectives. This comprehensive perspective ensures that the selected location not only reduces distortion but also contributes to overall network performance improvement.

The chosen exhaustive search method guarantees the identification of the exact global optimum within the reduced candidate set, which is computationally feasible in the studied system. This ensures the robustness of the conclusions without relying on stochastic or heuristic approximations. Furthermore, the methodology is scalable: for larger systems or multi-device installations, it can be adapted to incorporate metaheuristic optimization algorithms, cost constraints, and additional power quality indices.

The novelty of the proposed approach lies in:

- The simultaneous optimization of THDv, voltage profile, and loss minimization for UPQC placement under nonlinear load conditions.
- The application of the method to the complete network, with results confirming the dominance of nonlinear-load buses as optimal locations.

- The adaptability of the framework to extended scenarios, including multiple UPQC devices, economic evaluation, and larger network sizes.

This integrated methodology provides a rigorous, transparent, and technically robust foundation for decision-making in practical UPQC deployment projects.

7. Conclusions

This study presented an exhaustive search strategy to determine the optimal placement of a UPQC in distribution networks affected by nonlinear loads and distributed generation. The IEEE 33-bus distribution system was employed as the test benchmark to evaluate two representative scenarios: the first involving the inclusion of nonlinear loads at nodes 14, 19, and 27, and the second incorporating a 3.5 kW photovoltaic-based DG at node 24, simulating typical residential or small commercial installations commonly found in urban distribution networks.

The results obtained in the first case study showed that the optimal location for the UPQC is at node 14, as confirmed by the proposed optimization model based on a weighted cost function. Placing the UPQC at this node resulted in substantial improvements in the voltage total harmonic distortion (THDv), with reductions of 93.34% at node 14, 39.44% at node 19, and 54.83% at node 27, compared to the base system without compensation.

In the second case study, which included the DG, it was also confirmed that node 14 remains the optimal location for the UPQC, yielding the lowest value of the objective function (0.26). In this scenario, the THDv was improved by 93.47% at node 14, 39.64% at node 19, and 48.84% at node 27. These results highlight the effectiveness of the proposed methodology in identifying the most suitable placement for the power quality conditioner, even when the network topology changes due to the integration of distributed generation.

Although the optimal location of the UPQC did not change between the two case studies, the inclusion of the DG did introduce variations in the voltage profile during the exhaustive search process. This outcome underlines the importance of considering the presence of distributed energy sources in future studies, particularly when analyzing the interaction between active compensation devices and renewable generation units.

Overall, this research validates the use of an exhaustive search algorithm combined with a multi-objective cost function as an effective strategy for the optimal placement of power quality improvement equipment in modern distribution systems. The proposed methodology not only enhances voltage profile stability but also achieves significant reductions in harmonic distortion, ensuring compliance with power quality standards under various operating conditions.

Author Contributions: J.S.E.G. and A.A.T. contributed equally to this work. All authors have read and agreed to the published version of the manuscript.

Funding: This research received no external funding.

Data Availability Statement: The original contributions presented in this study are included in the article. Further inquiries can be directed to the corresponding authors.

Conflicts of Interest: The authors declare no conflicts of interest.

Abbreviations

AC	Alternating Current
DC	Direct Current
DG	Distributed Generation
PV	Photovoltaic
UPQC	Unified Power Quality Conditioner

APF	Active Power Filter
DVR	Dynamic Voltage Restorer
SVC	Static Var Compensator
STATCOM	Static Synchronous Compensator
D-STATCOM	Distribution Static Compensator
THD	Total Harmonic Distortion
THDv	Voltage Total Harmonic Distortion
RMS	Root Mean Square
p.u.	Per unit
IEEE	Institute of Electrical and Electronics Engineers
IEC	International Electrotechnical Commission
ANSI	American National Standards Institute
ARCERNNR	Agencia de Regulación y Control de Energía y Recursos Naturales no Renovables

References

1. Mangalapuri, S.; Sobhan, P.V. Analysis and Improvement of the Power Quality in Grid Integrated to Distributed Generation (Solar PV System) Using UPQC. In Proceedings of the 2023 International Conference on Advanced and Global Engineering Challenges, AGEC 2023, Kakinada, India, 23–24 June 2023; pp. 20–25. [CrossRef]
2. Chindris, M.; Cziker, A.; Miron, A. UPQC—The best solution to improve power quality in low voltage weak distribution networks. In Proceedings of the 2017 International Conference on Modern Power Systems, MPS 2017, Cluj-Napoca, Romania, 6–9 June 2017; pp. 1–8. [CrossRef]
3. Saliganti, K.; Papana, V.P.; Erukula, V. Performance Analysis of Solar PV-UPQC for Enhanced Power Quality. *J. Eur. Syst. Autom.* **2025**, *58*, 149–159. [CrossRef]
4. Bhosale, S.S.; Bhosale, Y.N.; Chavan, U.M.; Malvekar, S.A. Power Quality Improvement by Using UPQC: A Review. In Proceedings of the 2018 International Conference on Control, Power, Communication and Computing Technologies, ICCPCCT 2018, Kannur, India, 23–24 March 2018; pp. 375–380. [CrossRef]
5. Ghosh, A.; Ledwich, G. Unified Power Quality Conditioner. In *Power Quality Enhancement Using Custom Power Devices*; Springer Nature: New York, NY, USA, 2018; pp. 379–406. [CrossRef]
6. Pathik, B.B.; Islam, S.; Akram, N.M.; Zaman, M.S.; Al-Nahian, M.A. Total Harmonic Distortion Reduction in a Power Grid: A UPQC Based Approach. In Proceedings of the International Conference on Robotics, Electrical and Signal Processing Techniques, Dhaka, Bangladesh, 7–8 January 2023; pp. 264–268. [CrossRef]
7. Gu, J.; Xu, D.; Liu, H.; Gong, M. Unified power quality conditioner (UPQC): The principle, control and application. In Proceedings of the Power Conversion Conference-Osaka 2002, PCC-Osaka 2002, Osaka, Japan, 2–5 April 2006; Volume 1, pp. 80–85. [CrossRef]
8. Dash, S.K.; Ray, P.K.; Ray, P. Implementation of adaptive control strategies for PV system integrated unified conditioner for optimal power quality regulation. *Results Eng.* **2025**, *26*, 105122. [CrossRef]
9. Namdev, P.; Singh, R. Advances in Power Quality Enhancement: A Comprehensive Review of Custom Power Devices and Mitigation Strategies. In Proceedings of the 2023 3rd International Conference on Energy, Power and Electrical Engineering, EPEE 2023, Wuhan, China, 15–17 September 2023; pp. 761–765. [CrossRef]
10. Karelia, N.; Sant, A.V.; Pandya, V. Comparison of UPQC topologies for power quality enhancement in grid integrated renewable energy sources. In Proceedings of the 2019 IEEE 16th India Council International Conference, INDICON 2019—Symposium Proceedings, Rajkot, India, 13–15 December 2019. [CrossRef]
11. Chourasia, M.; Srivastava, S.P.; Panda, A. Control strategy for voltage sag/swell/harmonic/flicker compensation with conventional and fuzzy controller(UPQC). In Proceedings of the India International Conference on Power Electronics, IICPE, Nagercoil, India, 19–20 March 2015; pp. 1–6. [CrossRef]
12. *IEEE Std 1159*; IEEE Recommended Practice for Monitoring Electric Power Quality. IEEE: Piscataway, NJ, USA, 2009; Volume 1995, pp. 1–94.
13. Hernández-Mayoral, E.; Jiménez-Román, C.R.; Enriquez-Santiago, J.A.; López-López, A.; González-Domínguez, R.A.; Ramírez-Torres, J.A.; Rodríguez-Romero, J.D.; Jaramillo, O.A. Power Quality Analysis of a Microgrid-Based on Renewable Energy Sources: A Simulation-Based Approach. *Computation* **2024**, *12*, 226. [CrossRef]
14. Agencia de Regulación y Control de Energía y Recursos Naturales No Renovables. *Regulación Nro. ARCERNNR 002/20 (Codificada)*; Agencia de Regulación y Control de Energía y Recursos Naturales No Renovables: Guayaquil, Ecuador, 2023; Volume 20, pp. 1–35.
15. Yadav, S.K.; Patel, A.; Mathur, H.D. Comparison of power losses for different control strategies of UPQC. In Proceedings of the PIICON 2020—9th IEEE Power India International Conference, Delhi, India, 28 February–1 March 2020; pp. 3–8. [CrossRef]

16. Da Silva, S.A.; Modesto, R.A.; Sampaio, L.P.; Campanhol, L.B. Dynamic Improvement of a UPQC System Operating under Grid Voltage Sag/Swell Disturbances. *IEEE Trans. Circuits Syst. II Express Briefs* **2024**, *71*, 2844–2848. [CrossRef]
17. Kaur, S.; Kaur, N. Analysis of Designed Power System with Power Quality Improvement by Using UPQC. In Proceedings of the International Conference on Self Sustainable Artificial Intelligence Systems, ICSSAS 2023—Proceedings, Erode, India, 8–20 October 2023; pp. 1719–1724. [CrossRef]
18. Jaiswal, P.N.; Thakre, K.; Nigam, P. Comparison of DSTATCOM, DVR and UPQC for Mitigating Voltage Sag in Distribution System. In Proceedings of the 2022 IEEE 2nd International Symposium on Sustainable Energy, Signal Processing and Cyber Security, iSSSC 2022, Gunupur, India, 15–17 December 2022; pp. 7–12. [CrossRef]
19. *IEC 61000-2-4*; Electromagnetic Compatibility (EMC)—Part 2-4: Environment—Compatibility Levels in Power Distribution Systems in Industrial Locations for Low-Frequency Conducted Disturbances. International Electrotechnical Commission: Geneva, Switzerland, 2023; Volume 2023.
20. Águila Téllez, A.; López, G.; Isaac, I.; González, J.W. Optimal reactive power compensation in electrical distribution systems with distributed resources. Review. *Heliyon* **2018**, *4*, e00746. [CrossRef] [PubMed]
21. Águila, A.; Ortiz, L.; Orizondo, R.; López, G. Optimal location and dimensioning of capacitors in microgrids using a multicriteria decision algorithm. *Heliyon* **2021**, *7*, e08061. [CrossRef] [PubMed]
22. Zhang, X.P.; Yan, Z. Energy Quality: A Definition. *IEEE Open Access J. Power Energy* **2020**, *7*, 430–440. [CrossRef]
23. Chawda, G.S.; Shaik, A.G.; Mahela, O.P.; Padmanaban, S.; Holm-Nielsen, J.B. Comprehensive Review of Distributed FACTS Control Algorithms for Power Quality Enhancement in Utility Grid with Renewable Energy Penetration. *IEEE Access* **2020**, *8*, 107614–107634. [CrossRef]
24. *IEEE Standard for Harmonic Control in Electric Power Systems*; IEEE: Piscataway, NJ, USA, 2022.
25. International Electrotechnical Commission (IEC). *IEC 61000-2-12: Electromagnetic Compatibility (EMC)—Part 2-12: Environment—Compatibility Levels for Low-Frequency Conducted Disturbances and Signalling in Public Medium-Voltage Power Supply Systems*; IEC: Geneva, Switzerland, 2003.
26. International Electrotechnical Commission (IEC). *IEC 60038:2009+AMD1:2021 CSV—Standard Voltages*; IEC: Geneva, Switzerland, 2021.
27. National Electrical Manufacturers Association (NEMA). *ANSI C84.1-2020: American National Standard for Electric Power Systems and Equipment—Voltage Ratings (60 Hz)*; NEMA/ANSI: Rosslyn, VA, USA, 2020.
28. Davidrajuh, R.; Rong, C. Solving Scheduling Problems with Randomized and Parallelized Brute-Force Approach. In Proceedings of the 12th International Conference on Computer Science and Information Technologies, CSIT 2019, Barcelona, Spain, 18–20 December 2019; pp. 1–4. [CrossRef]

energies

MDPI

Article

Counteracting the Aging Process of the Aircraft's DC Power Grid in the Context of Shaping the Characteristics of the Power Source

Tomasz Tokarski *, Sławomir Michalak and Tomasz Gajewski

Air Force Institute of Technology, ul. Księcia Bolesława 6, skryt. poczt. 96, 01-494 Warszawa, Poland; slawomir.michalak@itwl.pl (S.M.); tomasz.gajewski@itwl.pl (T.G.)
* Correspondence: tomasz.tokarski@itwl.pl

Abstract: This article presents a proprietary method for measuring and adjusting the position of DC generator brushes relative to the neutral zone. This method enables shaping the characteristics of the power sources of the aircraft's DC power grids, which has a beneficial effect on counteracting the aging process. Using generators with varying operating times (flight hours), selected research results on the effect of the brush position angle α relative to the magnetically neutral zone on operating conditions are presented. In addition, examples of abnormal operation of the DC electrical network in transient states and their possible negative effects on the operation of the aircraft (A/C) power network are presented. Finally, research results are presented demonstrating the practical application of power source characteristic shaping to effectively counteract the aging process of aircraft DC power networks.

Keywords: DC generators; power quality; transients; aircraft; aging; neutral zone in generator; operational

1. Introduction

The generation of electricity by A/C power systems is paramount to their safe operation. This system is an autonomous network of components that generate, transmit, distribute, utilize, and store electrical energy. Its primary function is to generate, regulate, and distribute power throughout the A/C [1,2]. A/C operational capabilities are largely dependent on the reliability of electrical systems and subsystems. Furthermore, this dependency is expected to increase [3–6]. Generally, A/C systems use both alternating current and direct current. The requirement for high operational reliability of powered systems, components and systems in both modern and older A/C translates into the need to ensure proper power supply of on-board equipment with appropriate quality electricity. Electricity supplied to receivers must meet numerous requirements, therefore one of the fundamental tasks of A/C on-board power supply systems is to ensure the appropriate quality of electricity generated by on-board power sources, in compliance with the requirements of relevant standards [7,8].

A significant issue in A/C operation is the aging process, which significantly reduces the reliability and durability of A/C equipment, reduces the reliability of on-board power sources, and poses a potential threat to flight safety [9,10]. Therefore, the broadly understood A/C power system, through its condition, often reduced due to aging processes, reduces the durability and reliability of all other systems, which impacts the A/C's operational safety [10].

The technical condition of an A/C's power system stems from its past, and knowledge of its current condition is necessary to determine its future behavior. The processes of aging and wear of its devices and components have a decisive influence on the change in the technical condition of the SP power system. These processes lead to changes in the values of diagnostic parameters describing their technical condition. Aging is irreversible and reduces the reliability and durability of individual devices and system components. Understanding the physics of these processes, enabling an analytical approach, is crucial. Aging and wear processes occurring during the operation of avionics devices are complex and difficult to describe deterministically. Predicting the reliability and durability of A/C devices requires identifying many destructive processes that degrade their technical condition [11] in order to effectively counteract them.

2. Literature Review

Currently, with the increasing demand for electricity on board modern large aircraft, three-phase AC generators are used because they have a significantly higher power-to-weight ratio and enable easier transformation of AC voltage [12–15]. Despite these trends, commutator generators are still used in aviation on small aircraft and helicopters as the main or auxiliary power source: PZL 130TC—II ORLIK [16], W-3WA [17], M-28B/PT "Skytrack" [18], CASA-C295M [19], as well as DC starter-generators (DCSG): SW-4 [20], Mi-17 [21]. The existing trend of using DC generators as starting generators makes the starter-generator an important part of the aircraft's electrical system, as it is used to start the engine and then generate electricity [22,23]. Where a wide range and high requirements for electrical speed control are required, DC motors still play an important role due to their good speed control parameters, frequent stepless quick starting and braking [24].

Both in the past and present, the fundamental requirement in aviation is flight safety, which determines both the design and operation of aircraft. Generating electricity on board an aircraft is a priority in its operation. The requirement for high operational reliability of components and systems in both modern and older aircraft translates into the need to ensure proper power supply of devices with electricity of the required quality [7,8,25]. Of particular importance is the fact that during the operational process, the parameters of the commutator-brush assembly (CBA), changing due to damage or natural aging, result in changes in the characteristics of the power system components [26]. If the changes are significant, significant energy losses and interference may occur in the installation, increasing the probability of incorrect operation of the systems powered by it (e.g., a shortage of electricity supplied to the receivers, overheating of the generator windings and excitation circuits on the aircraft). Hence the importance of developed operational procedures that enable maintaining the aircraft in an airworthy condition despite aging processes [27–29]. Up to now, the commutation phenomenon and problems of DC motors still attract the attention of many scholars, including the establishment of a new commutation model and the method of solving the commutation magnetic field [23,30–33].

The issues presented in this article concerning the shifting of the magnetically neutral zone in commutator-type DC generators as a result of aging processes, according to [23,31], negatively affect the operation of the commutator-brush assembly (CBA) of this type of device. As stated in [23], the technical condition of the CBA largely determines the performance and efficiency of the DC generator or motor. Testing in the neutral zone allows for determining the ideal brush position, which ensures spark-free commutation, minimizes commutator and brush wear, and promotes efficient operation [26]. The method presented in this article [29,34] enables increased detection of DC generator faults and enables their rapid improvement, taking into account aging processes, by precisely adjusting the angle α of the CBA relative to the magnetically neutral zone.

Overcompensation resulting from the displacement of the magnetically neutral zone relative to the CBA in a DC machine influences the change in generator characteristics, deterioration of the dynamics of transient states during step loading, and the correctness of the self-regulation of the AC supply system [29]. The connections between these phenomena have not been considered in the literature so far. Separate considerations have been given to, among others, the reliability indicators of the brush-commutator system and the influence of compensation windings on the operation of the generator [23,26,30].

Moreover, most of the phenomena occurring in a DC generator also occur in an AC generator, in particular those related to the interaction of the generator armature (armature reaction to the load current). The influence of the transverse field in AC machines reduces the effective power delivered to the receivers [35–38]. These losses increase with increasing capacitive load of on-board receivers (increased power, e.g., LED lighting, increasing the number and power of semiconductors, increased demand for various types of electrical filters). The existing possibility of increasing the AC voltage at the end of the capacitive load line is one of the reasons why in modern large aircraft, where the primary source is a three-phase AC generator, high-voltage direct current (270 VDC, 540 VDC) is switched to [39].

Therefore, the issues discussed in the article concerning the shaping of the characteristics of DC sources can be used for issues related to AC generators, where the strong dependence of the AC characteristics on the nature of the loads (capacitive, inductive) will undoubtedly add many threads to the considerations in relation to the issues related to DC generators.

3. Background to the Problem

An example of a malfunction of the DC electrical network affecting the AC network was recorded during engine testing on an A/C and is shown in Figure 1. The malfunction of the 28 VDC network caused interference in the 115 VAC, 400 Hz AC network. It should be noted that on the A/C under test, the excitation circuits of the AC generators are powered from the DC network. After switching on a load (transient state) in the 28 VDC network, the 28 VDC voltage pulsation increased above the permissible values [29], resulting in a lack of voltage stabilization in the AC network, and the power quality from the primary DC and AC power sources was inconsistent with the requirements. This condition of the power supply networks had a direct impact on the correct operation of the A/C systems they power. During the ground test, the pilots reported malfunctions, including: the light signals on the cockpit panels and the powerplant control system.

Another significant example of a malfunction in the A/C's power grid is the damage to the power cable and excitation circuit wires of the A/C's primary power source, shown in Figure 2. This damage was caused by a malfunction in the 28 VDC network control system. The above damage caused the fuse in the excitation circuit to trip and the damaged DC power grid to be switched off and unable to be switched on again.

The generator damage presented in Figure 3 is an example of a failure of the main power source that caused a direct threat to the A/C flight safety. The consequence of the power source failure was damage to the A/C's electrical grid, including battery damage and a loss of power to the 28 VDC grid, resulting in an emergency landing during night flights without electrical power.

The presented examples of damage highlight the importance of operational procedures that enable the A/C's power grid to remain in airworthy condition despite aging processes. Malfunctions in A/C power systems that may directly impact flight safety can be detected during periodic maintenance by observing transient parameters of the on-board power system.

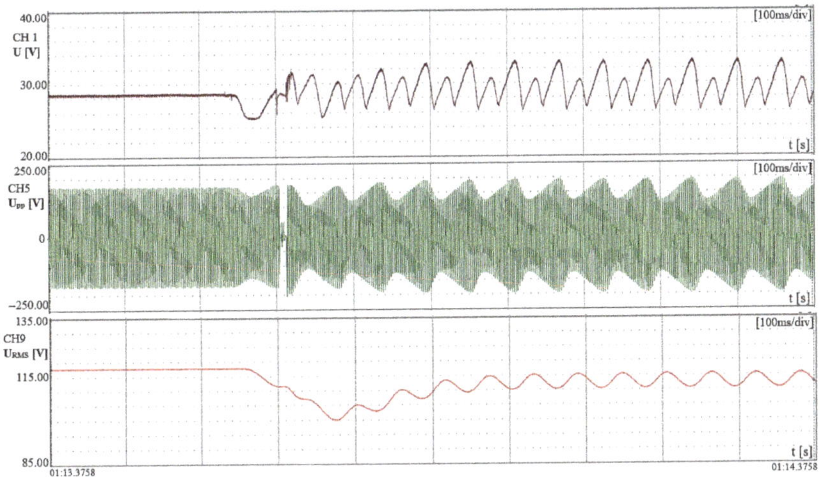

Figure 1. Voltage waveform disturbances in the 28 VDC and 115 VAC, 400 Hz AC circuits caused by a DC voltage drop pulse.

Figure 2. Damaged (melted) power cable and wires of the excitation circuit of the GS-12T 3s DC generator from the Su-22 A/C.

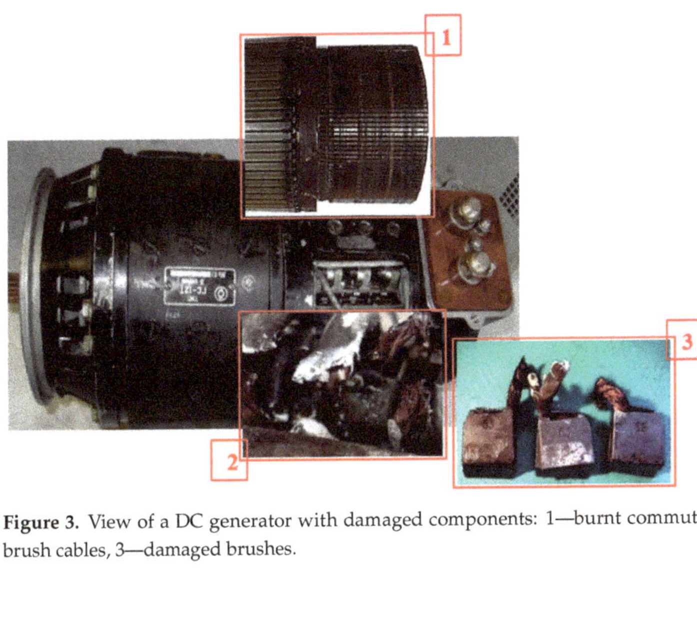

Figure 3. View of a DC generator with damaged components: 1—burnt commutator, 2—melted brush cables, 3—damaged brushes.

During operation, due to the natural aging process, the parameters of the power grids powering A/C systems change, and the quality of the electrical power from their power sources deteriorates. These changes occur in the characteristics of individual components of the power supply system and in the characteristics of its primary or auxiliary power sources [29].

During the A/C tests performed by the authors, measurements of the power quality of on-board power sources are conducted. During their implementation, tests of the DC network in a transient state are performed, among others, in order to determine the ability of the electrical power supply system to self-regulate the 28 VDC voltage [29]. To this end, the current load on their power supply system is varied, generating a pulse:

(a) a transient pulse (below 24 V)—during a sudden increase in load,
(b) a surge pulse (above 29 V)—during a sudden decrease in load.

These parameters are directly related to the aging processes occurring during the operation of the SP DC power grid, whose primary power source is DC generators. They reflect the dynamic response of the DC power grid to a sudden load or a sudden decrease in load. These dynamics are represented by the values of the reduced DC equivalent pulse time. The parameter that is directly correlated with the aging processes occurring during the operation of the DC power grid of the SP is the value of the reduced equivalent decay pulse time recorded below the voltage of 24 V. This value, in relation to the system operation time [1], clearly identifies the shelf life of the DC power grid supplying the systems of the tested SP.

If the source of electrical energy is a DC generator, then, in accordance with Standard [7], during normal operation of the electrical supply system, the transient voltages (voltage pulses) are converted into their equivalent pulses, which should be within the limit values specified in the above standard (permissible parameters). The parameter determined in the transient state is the value of the non-standard product of time and voltage, calculated as the integral of the voltage over time from the standard level to the value of the extreme deviation (beyond the standard level). These values are converted into so-called equivalent pulses, which are compared to the standard curve [2,40,41].

The core of the issue is the improper operation of the DC power grid during transient states and its faults detected during operation, which negatively impact the powered on-board systems. Figure 4, in the "Measurement waveform" tab, presents sample analyses of 28 VDC voltage waveforms during transient states, performed using the proprietary Abakus WDF v1.1 program (v1.1, Air Force Institute of Technology, Warsaw, Poland) developed in the NI LabVIEW graphical environment.

Abnormal transient characteristics (voltage reduction or increase in the supplied network exceeding the specified limits) can lead to disruptions in the operation of electrical devices and electrical drives. In the case of electronic devices, disturbances can lead to the failure of signal transmission or errors in their transmission. If changes in transient states are significant, the installation can cause significant energy losses and disruptions, which increase the likelihood of incorrect operation of the systems powered by the installation, e.g., interference with interconnecting networks, a shortage of electrical power supplied to loads, and their damage or disruption, as well as shutdowns. In extreme cases, the engine may not start or the starter system may fail. Abnormal network operation during transient states can also cause overheating of the windings of the generators serving as power sources and their excitation circuits in A/C.

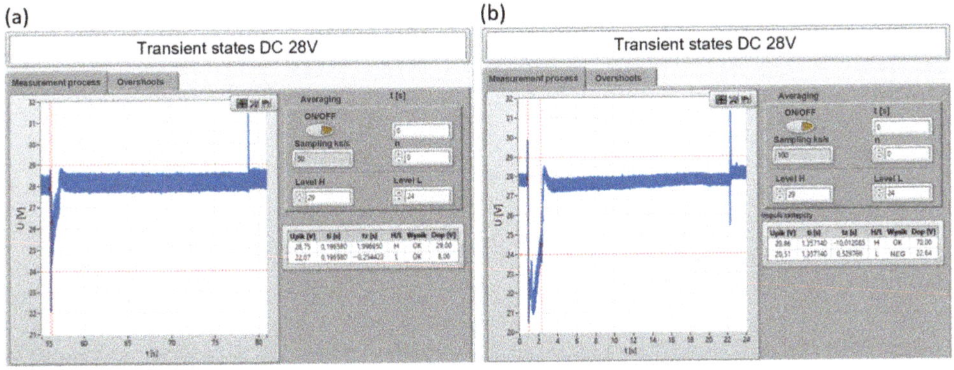

Figure 4. Examples of analysis of 28VDC voltage waveforms in transient states: H—overvoltage pulse above the upper reference level 29 V—OK (positive result), L—decay pulse below the lower reference level 24 V—exceedance (negative result), (**a**) Positive example (**b**) Negative example.

The authors' studies of the DC power systems during transient states of Su-22 A/C, conducted since 2007, provided information on the long-term power quality of the DC network under real-world operating conditions at airports. During their performance, numerous malfunctions of the electrical power installation of the tested A/C were found, among others, related to transient states in the DC network [29,42]. Example results of transient measurements of the DC network of the Su 22 A/C in subsequent months of operation are presented in Table 1. The measurement results presented in Table 1 were obtained at an aircraft turbine speed of n = 67% (minimum range), which, according to studies [43,44], corresponds to 5050 rpm of the GS-12T 3s DC generator, which is the main power source for the power grid.

Table 1. Measurement of transients in the DC power grid of the Su-22 A/C in subsequent years of operation.

No.	Number of the Tested DC Network	Number of Months of Operation on the A/C							
		I Test Initial State		II Test Condition After 31 Months of Operation		III Test Condition After 63 Months of Operation		IV Test Condition After 92 Months of Operation	
		Values of the Equivalent Pulse Parameters							
		t_z [ms]	U_{min} [V]	t_z [ms]	U_{min} [V]	t_z [ms]	U_{min} [V]	t_z [ms]	U_{min} [V]
1.	No. 1	10	21.6	39	18.89	29	19.89	40	21.54
		8	23.3	76	21.74	164	22.29	202	22.71
2.	No. 2	24	21.6	62	19.23	28	19.73	114	21.75
		20	23.1	192	20.45	315	21.98	305	21.48

The results presented in Table 1 confirm that the values of the reduced equivalent pulse time for individual DC power grids increase with their operating time. A graphical representation of the changes in the reduced equivalent pulse time for the Su-22 A/C DC power grid with respect to the number of months of operation is presented in Figure 5.

In the presented period of 92 months of operation, from test I to test IV, the value of the reduced equivalent pulse time increased for both the DC network supplied by the left and right generators. In the case of the DC network supplied by the left generator, the value of the reduced equivalent pulse time in test IV was close to the permissible limit value specified in the standard [7].

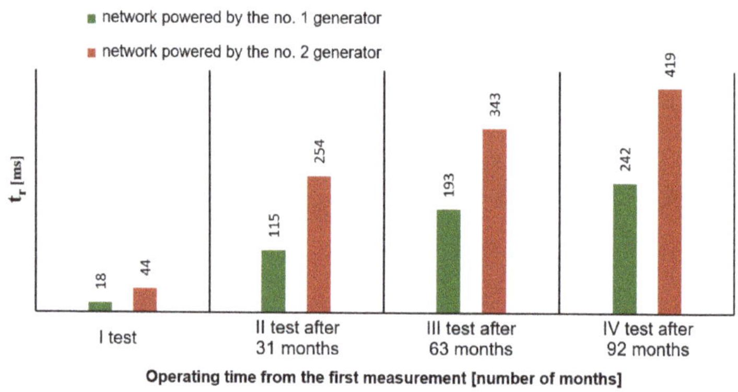

Figure 5. Values of the reduced equivalent pulse time for the Su 22 aircraft DC network powered by the left and right generators in relation to the number of months of their operation.

Analysis of the transient parameter of the on-board power system also allows for the detection of faults in the tested power grid during installation tests during overhaul procedures [7,40]. During power quality tests conducted in subsequent transient measurements for individual power grids, the permissible value of the equivalent decay pulse was exceeded. In several cases, after the generator overhaul process was completed, their use on board the A/C during engine tests resulted in exceedances of the permissible normative thresholds (reduced equivalent pulse times). Only repeated replacement of the generators (including generators and regulators) ensured that the output parameters during transient states of their DC nodes complied with the permissible normative level. Often, replacing the regulators alone did not produce significant results, and only replacing the generators resulted in a significant improvement in parameters. An example of the results of transient state measurements in the DC generator network of the Su-22 A/C in subsequent tests during the overhaul procedure is presented in Table 2 [42]. Non-conforming values are written in bold An illustration of the limit values from the measurements of transient states (equivalent decay pulses) of DC against the background of the ranges of changes equivalent to voltage step functions (permissible parameters) [7,40,41] in the electrical power supply systems supplied from 28 VDC DC generators is presented in Figure 6. The measurement results presented in Table 2 were obtained at an aircraft turbine speed of n = 67% (minimum range), which, according to studies [43,44], corresponds to a generator speed of 5050 rpm.

Table 2. Measurement of transient states of the Su-22 aircraft DC generator network in subsequent tests during the implementation of repair procedures.

No.	Number of the Tested DC Network	Measurement Results in Subsequent Tests Su-22							
		I Test		II Test		III Test		IV Test	
		Values of the Equivalent Pulse Parameters							
		t_z [ms]	U_{min} [V]	t_z [ms]	U_{min} [V]	t_z [ms]	U_{min} [V]	t_z [ms]	U_{min} [V]
1.	No. 1	**605**	**20.51**	**159**	**19.82**	**650**	**20.03**	37	20.01
			20.92	**450**	**20.82**		**21.06**	352	21.77
2.	No. 2	21	20.46	22	20.37	18	20.46	16	20.31
		139	22.87	140	22.81	158	22.83	135	22.64

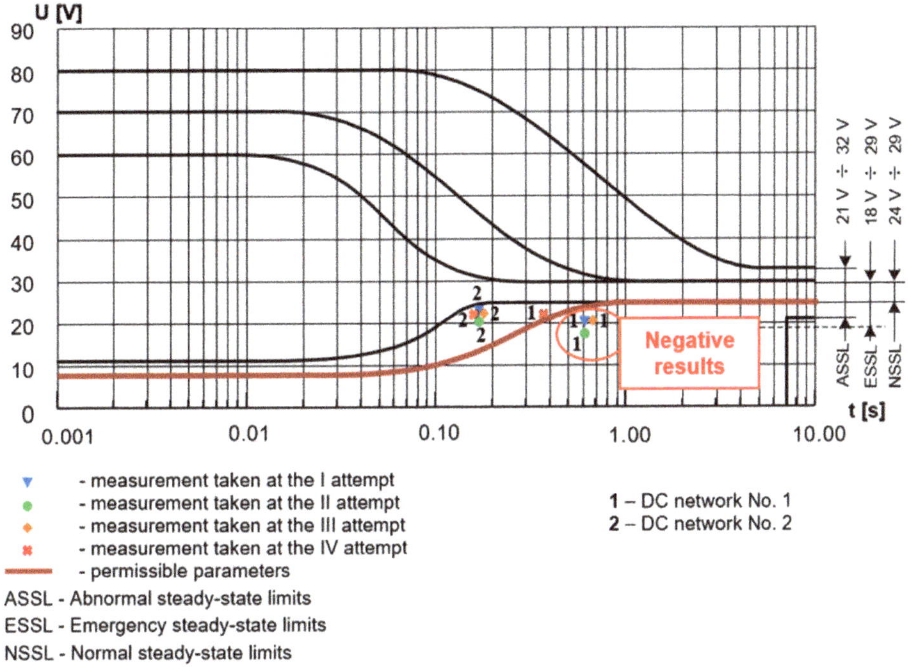

Figure 6. Limit analysis of transient state measurements (equivalent decay pulses) of the DC networks No. 1 and No. 2 in subsequent tests during the renovation procedures against the background of the ranges of changes in equivalent step functions of the DC voltage in the electrical power supply systems supplied from 28 V DC generators—permissible parameters.

The example shown in Figure 6 of the tested DC power grid No. 1 indicates that despite the completed repairs, there is a possibility of a problem with the limited self-regulation capability of its electrical power supply system, which, as mentioned earlier, can cause improper operation of A/C systems and devices powered by this system. Because the above problem of limited self-regulation capability of the electrical power supply system also occurred during engine tests on other A/C, it was necessary to take effective preventive measures. Their results [45] showed that the main cause of the lack of proper self-regulation, in the form of exceeding the permissible values of the equivalent pulse in DC networks, are aging changes resulting from the long operation time of their main power source, the GS-12T 3s generators. The equivalent pulse parameters of the aircraft's DC network deteriorated year by year, and the value of the reduced equivalent pulse time identifies the operational life of their power supply network. The inability to self-regulate the DC power supply system as required resulted from a malfunction of the GS-12T 3s DC generator. The observed malfunction, in the form of an abnormally increased duration of decay pulses, resulted from the progressive overcompensation of the GS-12T 3s generator due to aging processes. This malfunction resulted from incorrect setting of the angle α of the generator's brushes relative to the magnetically neutral zone. It turned out that with increasing generator operating time, the angle α between the magnetically neutral zone and the position of its brushes spontaneously increases, causing a significant increase in the duration of the equivalent pulse during a sudden increase in the generator's load. This confirmed the need to introduce additional procedures and checks performed on DC generators into the overhaul process, as well as additional inspections of the power grid during the A/C's post-overhaul operation. The current overhaul technology for GS-

12T 3s DC generators has been expanded to include checking the brush position relative to the neutral zone—measuring the α angle and enabling its correct adjustment. This improved the generators, thus improving the A/C's power grid's ability to self-regulate. In summary, by properly monitoring and adjusting the brush position relative to the generator's magnetic neutral zone, taking into account their service life, it is possible to shape the characteristics of the aircraft's DC power grid power source, counteracting its aging process and ensuring proper operation during subsequent operation.

4. Materials and Methods

4.1. Commutator-Based DC Generators as Electrical Energy Sources for Public Utility Power Grids

The specificity of aviation technology lies in the appropriate design and technology of parts, assemblies, and devices used in aviation, as well as their rational use, maintenance, and storage, which significantly slows down the aging process and improves the economic performance of operation. The required level of reliability and safety of modern A/C cannot be achieved without maintenance, despite ensuring compliance with the highest standards at the design and production stages. To determine the optimal operational model, it is necessary to correctly identify the processes and phenomena characteristic of aviation technology operation, study them experimentally both in the laboratory and on board the A/C and then develop countermeasures [46–49].

In both military and civil aviation, commutator-type DC generators are used as the source of electrical power for power grids in A/C. The main advantages of such power systems include easy interoperability with an on-board battery and easy parallel operation of generators, as well as the possibility of buffer operation, i.e., drawing energy from the battery during generator overload. Another advantage is that a commutator-type DC machine is reversible, meaning it can operate as a generator and then be used as a starter the next moment. Furthermore, this type of network has a simple voltage regulation system that controls only the amplitude of the generator's output voltage and lacks capacitive effects during operation under steady-state (constant) load conditions, i.e., the effects of increasing the voltage at the end of the cable under capacitive load conditions [11,37,49].

In DC commutator generators, we also encounter a negative phenomenon involving a change in the magnetic field under load. As current increases in the circuit, the magnetic neutral zone line of a DC machine shifts due to armature reaction (the creation of a magnetic field transverse to the direction of the stator magnetic field lines) relative to the geometric plane between the stator pole pieces. The stator magnetic circuit becomes locally saturated, making the generator less efficient—the external effect is a reduction in the output voltage level for systems without a voltage regulator or an increase in the excitation current for systems with an automatic voltage regulator. To counteract this phenomenon, these machines employ an automated compensation process through the installation of commutator windings and compensation windings that generate a magnetic field in the opposite direction to the armature flux reaction to the generator load. However, during long-term operation of generators, despite the use of these windings and automatic voltage regulation, due to aging processes in the electrical and magnetic circuits, the effectiveness of this automatic compensation decreases, and the machine's efficiency deteriorates. The influence of the aging process on the reduction in the efficiency of these windings has not been discussed in the known literature so far—such research was presented by the author in [29]. An effective way to counteract the phenomenon of aging of generator components is, as presented in [29], an additional precise correction of the position of the generator brushes in relation to the magnetically neutral zone. So far, in A/C designs, in operational practice, the angle of the brushes relative to the magnetically neutral zone is rarely adjusted [50,51] due to the underestimation of the impact of armature reaction (the formation of a strong transverse field in the armature when increasing the load current) on, among others:

(a) the generator's energy efficiency, especially under conditions of dynamic changes in its load level and the destruction of its commutator,
(b) the process of thermal overload of the generator,
(c) current overload of the excitation winding,
(d) the speed of the automatic voltage regulation system supplied by it.

Furthermore, operators place excessive faith in the role of the commutator and compensation windings in neutralizing the effects of transverse field influences—even though, as it turns out, their effectiveness decreases with time of operation [29].

With the development of increasingly complex voltage regulation systems for commutator DC generators (based on electromechanical and later electronic systems), the erroneous belief arose that a good voltage regulation system (i.e., with a time constant lower than the generator's time constant) is capable of freely shaping the characteristics of an A/C's power grid, even when magnetically neutral zones shift due to aging phenomena. From a formal point of view, this is true (in most operational cases observed on board the A/C), but it is related to, among others, an increase in the temperature of the generator excitation circuits (often visible in the form of overheating of the insulation of the external excitation circuit cables), non-standard voltage stabilization times after changes in the load level and, in general, an increase in operating costs resulting from premature repairs [29,52,53].

4.2. Measurement and Regulation of Commutator DC Generators in Terms of the Positioning of the Generator Brushes Relative to the Magnetically Neutral Zone

As previously discussed, DC power grid malfunctions in the form of increased decay pulses may result from incorrect adjustment of the brush position angle relative to the magnetically neutral zone in the GS-12T 3s commutator DC generator. Both measurement and regulation of this generator can be achieved using a measuring system constructed according to the proprietary solution, filed as invention [34], "System for measuring the position of DC generator brushes relative to the neutral zone."

A simplified diagram of the station for testing and regulating the brush position of the GS-12T 3s commutator DC generator relative to the magnetically neutral zone, developed based on the above solution, is shown in Figure 7 [29]. The essence of this station is that, after removing the brushes, two measuring brushes are mounted in two adjacent brush holders (closing the rotor coil circuit) of the tested generator. The measuring brushes are made of insulating material with axes of symmetry marked on their sides (from the generator terminal board side). The generator commutator is connected to the oscilloscope via measuring probes applied to the side surfaces of the commutator staves (detail "stave 1", "stave 18"), which enables the measurement and recording of voltages induced in the generator armature. The excitation circuit of the tested generator is powered from an external source for the duration of the measurement (the generator switches from shunt operation to a separately excited system)—during the measurement, the excitation windings are powered from a DC power supply with an AC pulsation component [34]. To minimize the circumferential structural clearances in the generator rotor assembly (at the splined connection between the drive shaft and the rotor), a force is applied to the generator rotor fan using a dynamometer during the measurement.

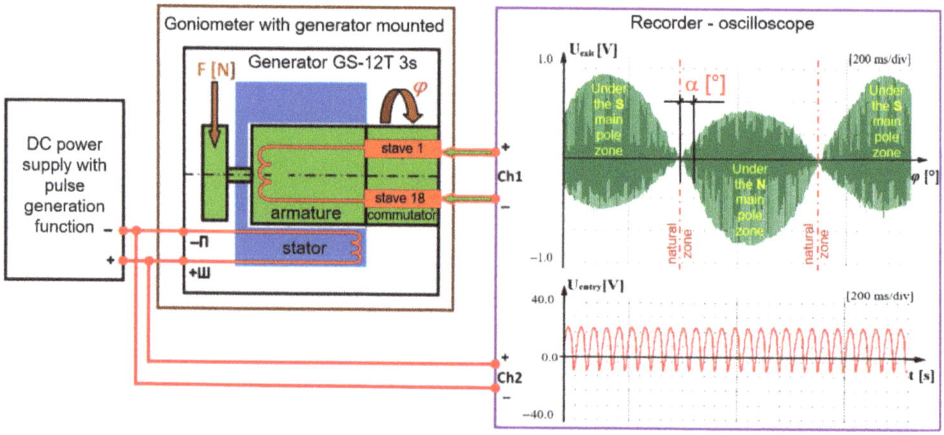

Figure 7. Schematic diagram of the stand for checking the angle α of the brush position of the GS-12T 3s DC generator in relation to the neutral zone with the elements of the measuring system: F—the force exerted on the generator rotor by means of a dynamometer, φ—the generator rotor speed, α—the angle between the brush center and the neutral axis (magnetic neutral)—defining the position of the brushes in relation to the neutral zone, the measurement of which is performed indirectly by means of a depth gauge.

The basic element of the stand for testing and adjusting the brush position of the GS-12T 3s commutator DC generator relative to the magnetically neutral zone (proprietary design solution) is the goniometer described in utility model [54]. Using the goniometer in the above system allows for precise determination of the brush position angle relative to the neutral zone. Appropriate configuration of the goniometer in the measurement system also enables precise adjustment of the brush position angle α of the generator relative to its neutral zone, as required for both generator and motor operation.

In experimental tests and during the overhaul process, the stand shown in Figure 8 is used to measure and adjust the brush position angle α relative to the neutral zone of GS-12T 3s generators.

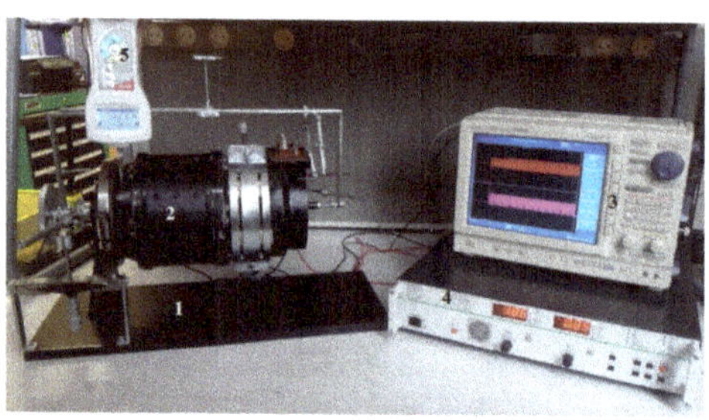

Figure 8. View of the stand for measuring and adjusting the angle α of the brush position of the GS-12T 3s DC generator in relation to the neutral zone: 1—goniometer for checking and adjusting the brush position of the GS-12T 3s DC generator in relation to the neutral zone, 2—tested GS-12T 3s generator, 3—oscilloscope, 4—DC power supply, 5—dynamometer.

4.3. *Shaping the Characteristics of a Commutator DC Generator*

This article presents selected test results of the characteristics of a commutator-type DC generator after changes in the brush angle (α) relative to the neutral zone. Examples include a GS-12T 3s DC generator No. 1, which has had 0 h of operation on the primary line since its inception, and a generator No. 2, which has had 1677 h of operation on the primary line. Tests determining the effect of the brush angle (α) relative to the neutral zone on selected characteristics of a commutator-type DC generator were performed on a VR 600 drive station, shown in Figure 9.

Figure 9. View of the VR-600 test stand for testing GS-12T 3s DC generators: 1—VR-600 drive stand, 2—GS-12T 3s generator under test, 3—EL 9080—600 HP thyristor load, 4—oscilloscope, 5—DC power supply.

Before proceeding to the next stages of testing selected characteristics of the tested generators, on the stand shown in Figure 8, the current angle α of the generator brushes position relative to the neutral zone was determined, and then the brush position was adjusted to the value of the angle α required for the performed stage of testing.

4.3.1. The Influence of the Brush Position Angle Relative to the Magnetically Neutral Zone on the Idle Operating State of the Generator

To determine the effect of the brush position angle α relative to the magnetically neutral zone on the no-load operating state of the tested GS-12T 3s DC generators, characteristic tests were performed [37,55,56], i.e., the dependence of the generator's electromotive force E at an open external circuit (no load) on its excitation current I_e, at a constant rotational speed and load current $I_{load} = 0$, i.e., at an open external circuit (n = 4200 rpm = const = n_N) according to:

$$U_o \approx E = f(I_e)$$

Example test results of no-load characteristics, measured values of the electromotive force E at the terminals of GS-12T 3s DC generators No. 1 and No. 2, obtained for the generators with their brushes positioned relative to the neutral zone for angles α: 0°, 1.0°, respectively. 1.6°, 2.2° and 3.4° are presented in Table 3. Figure 10 shows the no-load characteristics for generators No. 1 and No. 2 obtained after their adjustments to the individual values of the α angle. The generator no-load test results presented in Table 3 were performed at a rotational speed of 4200 rpm, which, according to [57], corresponds to the minimum rated rotational speed of the GS-12T 3s DC generator, and the limitation of the excitation current range to 20 A resulted from the use of such protection in the generator excitation circuit on the aircraft.

Table 3. Comparison of the electromotive force E (n = const = 4200 rpm and I_{load} = 0 A) of the GS-12T 3s DC generators No. 1 and No. 2 with different run-on time and different positions of the generator brushes relative to the neutral zone.

No.	Excitation Current	The Angle α of the Brush Position Relative to the Neutral Zone of the Generator [°]									
		0		1.0		1.6		2.2		3.4	
		The Value of Electromotive Force E/Generator Number									
	I_e [A]	E_o [V]		E_1 [V]		$E_{1.6}$ [V]		$E_{2.2}$ [V]		$E_{3.4}$ [V]	
		No. 1	No. 2	No. 1	No. 2	No. 1	No. 2	No. 1	No. 2	No. 1	No. 2
1.	0	0.95	0.91	0.88	0.86	0.92	0.85	0.87	0.88	1.04	0.83
2.	2	8.05	9.51	9.71	10.11	9.05	11.06	8.85	10.03	6.25	8.02
3.	4	16.21	19.32	19.25	20.02	18.15	20.95	17.65	20.21	11.21	15.17
4.	6	23.80	27.36	27.12	27.81	26.12	28.85	25.65	28.07	17.01	22.51
5.	8	30.20	32.49	32.80	33.02	32.02	34.13	31.61	33.25	23.51	29.42
6.	10	35.11	35.83	36.50	36.51	36.21	37.29	35.85	36.39	30.23	34.91
7.	12	38.70	37.92	39.02	38.72	39.02	39.53	38.62	38.61	36.52	38.26
8.	14	41.24	39.62	40.95	40.08	41.25	41.12	40.35	40.40	40.61	40.42
9.	16	43.32	41.24	42.51	41.31	43.03	42.33	42.02	42.03	42.73	41.82
10.	18	44.75	42.19	44.02	42.41	44.25	43.28	43.42	43.02	43.81	42.74
11.	20	45.80	43.08	45.11	43.17	45.19	43.98	44.53	43.91	44.24	43.14

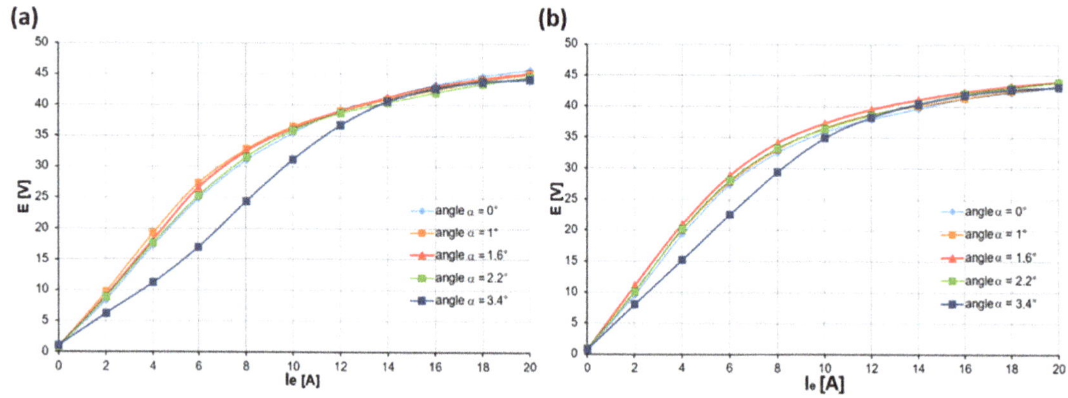

Figure 10. No-load characteristics of GS-12T 3s DC generators (n = const = 4200 rpm and I_{load} = 0 A) with different run-on time and different positions of the generator brushes relative to the neutral zone of the α angle: 0°, 1.0°, 1.6°, 2.2° and 3.4°: (**a**) generator No. 1, (**b**) generator No. 2.

Test results for GS 12 T 3s DC generators showed that each change in the brush position in the generator (angle α: 0°, 1.0°, 1.6°, 2.2°, and 3.4°) resulted in a change in the no-load characteristics. The obtained characteristics, presented in Figure 10, for example, generators No. 1 and No. 2, determine their dependence on both the angle α of the brush position in the generator relative to the magnetically neutral zone and on the tested generator's accretion.

To illustrate the established relationship, based on the no-load characteristics presented above, the no-load excitation current I_{0e} was determined, i.e., the excitation current at which the electromotive force E is equal to the rated voltage of the generator U_N = 28.5 V. The obtained values of the rated excitation current I_{0e} at no-load and a specific angle α of the brushes setting in the generator are presented in Table 4, and their characteristics in Figure 11.

Table 4. Values of the rated excitation current I_{0e} at no-load speed of 4200 rpm and a voltage of 28.5 V for GS 12 T 3s DC generators with a specific angle α of the generator brushes position relative to the neutral zone.

No.	Generator GS-12T 3s	Generator Hours Since the Beginning of Operation on the A/C	The Value of the Electromotive Force on the Generator Armature at n = 4200 rpm	The Angle α of the Brush Position Relative to the Neutral Zone of the Generator [°]				
				0	1.0	1.6	2.2	3.4
				Rated No-Load Excitation Current				
		[hours]	E [V]	I_{0e} [A]				
1.	No. 1	0	28.5	7.12	6.36	6.57	7.15	9.18
2.	No. 2	1677		6.37	6.11	5.90	6.14	7.21

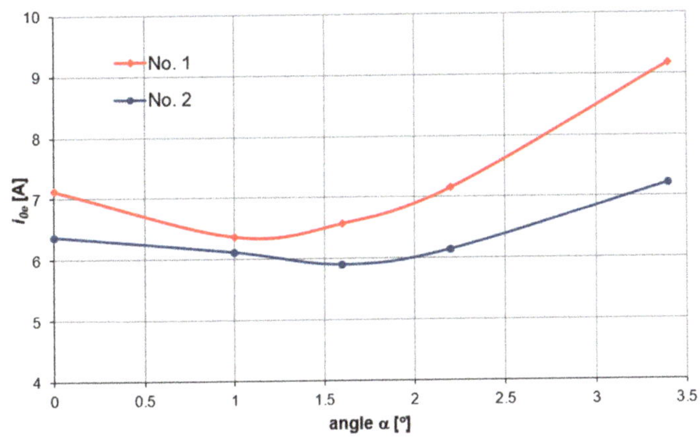

Figure 11. Influence of the operating time of DC generators No. 1 and No. 2 on the values of the rated excitation current I0W at no-load n = 4200 rpm and voltage 28.5 V depending on the angle α of the position of the generator brushes relative to the neutral zone.

Comparing the characteristics of generators No. 1 and No. 2 shown in Figure 11, it can be concluded that with increasing operating time, the value of the excitation current I_{0e} at which the generator reaches its rated voltage decreases. This indirectly confirms the hypothesis that with increasing operating time (mainly due to increasing armature winding resistance), the generator systematically overcompensates [29]. At the same time, the minimum I_{0e} value shifts to larger α angles with flight time (operating time on the SP). This results, among other things, from [11,56] due to:

(a) aging changes in the rotor magnetization characteristics—the rotor's magnetic permeability characteristics become lower and flatter over time—these changes are directly proportional to the operating time and temperature,
(b) reduced commutator diameter (due to periodic rewinding)—the brush shorts a larger number of rotor turns per unit of time,
(c) increasing the effective air gap—due to increased clearances in the rotor assembly bearing.

4.3.2. The Influence of the Brush Position Angle Relative to the Magnetically Neutral Zone on the Generator's Control Properties

To determine the effect of the brush position angle α relative to the magnetically neutral zone on the control properties of the GS-12T 3s DC generators, tests of control characteristics were performed [37,55], i.e., the dependence of the excitation current I_e on the load current I_{load} at a constant rated voltage U_r and a constant rotational speed equal to the rated n_r, in accordance with:

$$I_e = f(I_{load})$$

Example test results of the generator control characteristics determined on the GS-12T 3s DC generators No. 1 and No. 2, in which the position of the generator brushes relative to the neutral zone was regulated for the angle α: 0°, 1.0°, 1.6°, 2.2°, and 3.4° are presented in Table 5. The control characteristics obtained on their basis, defining their dependence on the position of the generator brushes at a specific value of the angle α, are presented in Table 5. in Figure 12. The results of the generator control characteristics tests presented in Table 5 were performed at a rotational speed of 9000 rpm (corresponding to the maximum rotational speed of the generator) and within the range of current changes in the excitation circuit I_e ($\Delta I_e = I_{emax} - I_{emin}$) necessary to maintain the output voltage of 28.5 V obtained with a (slow or step-by-step 50 A) load change from 0 A to the rated current (I_{Nload} = 400 A), which values were determined based on the study [57].

Table 5. Comparison of the excitation current values I_e (n = const = 9000 rpm and I_{load} = 0 ÷ 400 A and U = 28.5 V) in the excitation circuit of GS-12T 3s generators No. 1 and No. 2 with different accumulator run time and different positions of the generator brushes relative to the neutral zone.

No.	Generator Load Current I_{load} [A]	Angle α of the Brush Position Relative to the Neutral Zone [°]									
		0		1.0		1.6		2.2		3.4	
		Average Value of Excitation Current I_e/Generator Number									
		I_{e0} [A]		I_{e1} [A]		$I_{e1.6}$ [A]		$I_{e2.2}$ [A]		$I_{e3.4}$ [A]	
		No. 1	No. 2	No. 1	No. 2	No. 1	No. 2	No. 1	No. 2	No. 1	No. 2
1.	0	2.60	2.60	2.59	2.60	2.60	2.59	2.59	2.57	2.60	2.57
2.	50	2.75	2.52	2.74	2.61	2.94	2.71	2.96	2.69	3.05	2.97
3.	100	2.87	2.41	2.91	2.63	3.27	2.75	3.38	2.78	3.55	3.34
4.	150	2.95	2.34	3.07	2.80	3.58	2.85	3.77	2.89	4.0	3.63
5.	200	2.98	2.37	3.25	2.85	3.78	3.05	4.15	3.04	4.43	4.05
6.	250	3.06	2.40	3.43	2.88	4.07	3.27	4.53	3.30	4.79	4.50
7.	300	3.05	2.42	3.70	2.92	4.22	3.39	4.80	3.45	5.02	4.78
8.	350	3.04	2.34	3.91	2.93	4.40	3.62	5.12	3.72	5.39	4.98
9.	400	3.03	2.27	4.05	3.01	4.52	3.82	5.37	3.90	5.78	5.38
10.	$\Delta I_e = I_{400A} - I_{0A}$	0.46	−0.33	1.46	0.41	1.92	1.23	2.78	1.33	3.18	2.81

The results of tests on GS-12 T 3s DC generators, performed with different brush positions relative to the neutral zone, showed that any change in brush position (change in α angle) in the generator affects the shape of its control characteristic $I_e = f(I_{load})$. As the α angle increases from 0° to 3.4°, the difference between the minimum and maximum current values in its excitation circuit increases, the angle of inclination of the resulting control characteristic increases, and its shape changes.

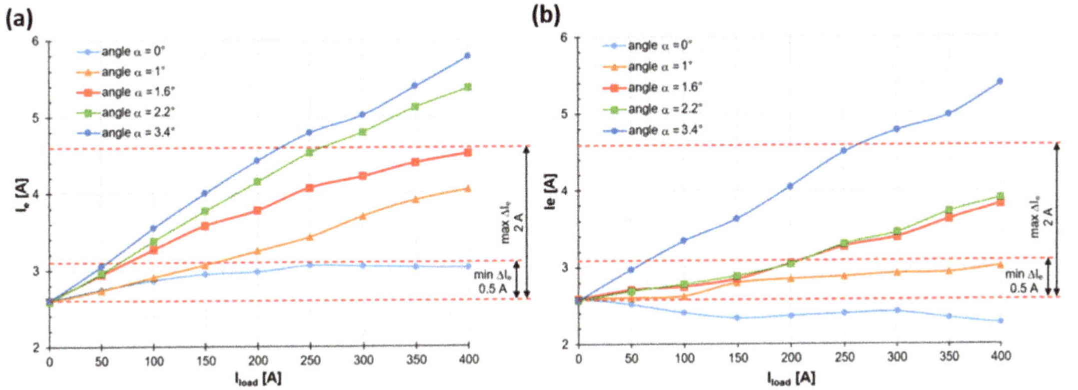

Figure 12. Control characteristics of GS-12T 3s DC generators with brushes shifted relative to the neutral zone in the direction of generator rotor rotation by angle α: 0°, 1°, 1.6°, 2.2°, and 3.4° against the background of the required values: (**a**) generator No. 1, (**b**) generator No. 2.

The changes in current intensity in the excitation circuit of GS-12 T 3s DC generators with different run-on times ($\Delta I_e = I_{400A} - I_{0A}$), obtained at different brush position angles relative to the neutral zone, presented in Table 5, indicate that the shape of the control characteristics of individual generators is influenced by aging changes resulting from the operating time of these generators. An example of the influence of aging changes is given by the characteristics of changes in the current intensity in the excitation circuit of generators No. 1 and No. 2 (with different accumulator run-time) depending on the angle α shown in Figure 13.

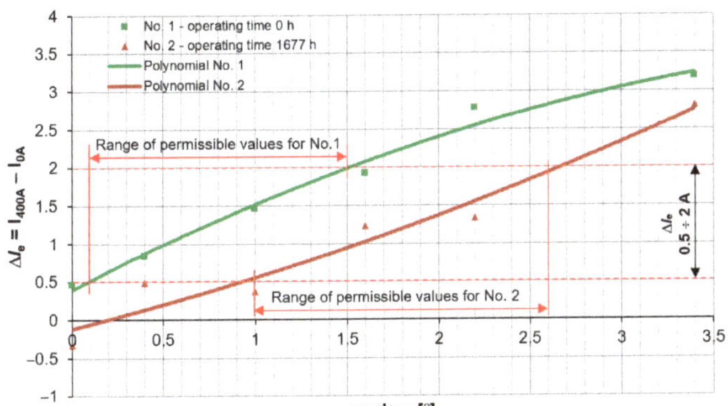

Figure 13. The influence of the operating time of DC generators No. 1 and No. 2 on the characteristics of current changes in the excitation circuit of the generators depending on the angle α of the position of the generator brushes relative to the neutral zone.

The characteristics shown in Figure 13 indicate that:

(a) for a generator with 0 h of operation on the primary winding, the transverse armature reaction flux Φ_q is compensated by the magnetic flux of the compensating winding Φ_{kom}, which is opposite in direction but equal in absolute value, in accordance with the requirements [33] for angle settings Φ in the range from 0.1° to 1.5°,

(b) for a generator with a significant operating time (1677 h of operation on the primary winding), due to aging changes, the weakening transverse armature reaction flux Φ_q is overcompensated in the range from 0° to 1.0°—this flux is compensated by the magnetic flux of the compensating winding Φ_{kom}, which is opposite in direction but equal in absolute value, in accordance with the requirements [57] only for angle settings Φ in the range from 1.0° to 2.6°.

4.3.3. The Influence of the Brush Position Angle Relative to the Magnetically Neutral Zone on the Generator Properties in Transient States

Studies of the DC characteristics, which determine the dependence of the brush position relative to the magnetically neutral zone (angle α) on the value of the equivalent decay pulse S_{iz} generated during step changes in the generator load from 10% I_r to 170% I_r (at n = const) [40,41], allow us to determine the effect of the α angle on the transient operating state of GS-12T 3s generators. The value of the equivalent substitute impulse S_{iz} is calculated according to the formula:

$$S_{iz} = 24 - U_{min} \cdot t_z$$

U_{min}—minimum voltage value
t_z—reduced time
The reduced time t_z of the equivalent pulse is calculated according to formula:

$$t_z = S/(24 - U_{min})$$

S—area under the lower reference level of 24 V of the actual voltage waveform

Example test results of GS-12T 3s DC generators No. 1 and No. 2 in transient states (caused by a step load on the generator from 10% I_z to 170% I_z) performed at three generator speeds and rotational speeds, obtained for the generators with their brushes positioned relative to the neutral zone for angles α: 0°, 1.0°, 1.6°, 2.2°, and 3.4°, respectively, are presented in Table 6.

Table 6. Average values of equivalent decay pulses S_{iz} of the GS-12T 3s DC generator No. 1 and No. 2 obtained for different angles α of the brush position relative to the neutral zone in the generator.

No.	Generator Speed [rpm]	Angle α of the Brush Position Relative to the Neutral Zone [°]									
		0		1.0		1.6		2.2		3.4	
		Equivalent substitute Impulse S_{iz} [Vms] /Generator Number									
		S_{iz0} [A]		S_{iz1} [A]		$S_{iz1.6}$ [A]		$S_{iz2.2}$ [A]		$S_{iz3.4}$ [A]	
		No. 1	No. 2	No. 1	No. 2	No. 1	No. 2	No. 1	No. 2	No. 1	No. 2
1.	5050	58.6	43.4	84.6	67.1	134.5	98.5	229.2	148.1	462.8	441.2
2.	6633	44.3	36.7	67.8	49.8	106.9	65.4	139.1	87.0	200.9	175.1
3.	9000	41.2	32.5	62.7	43.9	87.2	54.2	112.7	79.7	158.0	167.7

The transient state characteristics derived from these results, i.e., the dependence of the equivalent decay pulse S_{ir} on the angle α of the generator brushes positioned relative to the magnetically neutral zone, as a response of the generator control system cooperating with the voltage regulator to a single step loads in Figure 14. The results of equivalent decay pulse tests presented in Table 6 were performed at rotational speeds of 5050 rpm and 6633 rpm, which, according to studies [43,44], reflect the actual operating conditions of the GS-12T 3s DC generator on the Su-22 aircraft (turbine speed n = 67%—minimum range and n = 88%) and additionally at the maximum rated rotational speed of the generator of 9000 rpm [57].

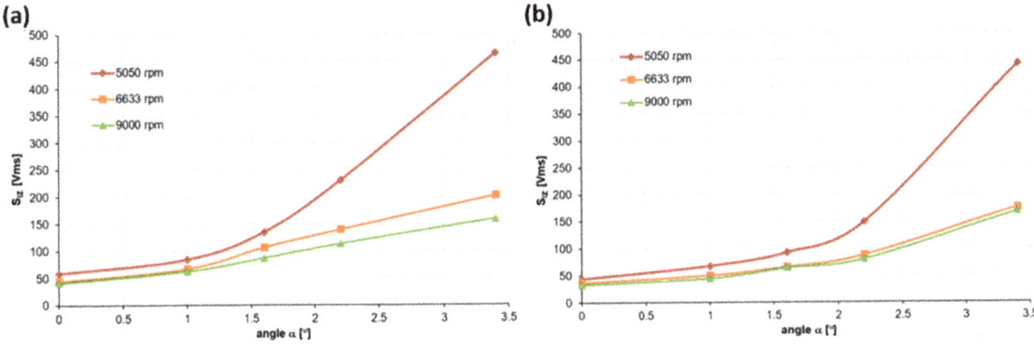

Figure 14. Values of equivalent decay pulses of the GS-12T 3s DC generator No. 1 and No. 2 as a function of changes in the brush position angle relative to the magnetically neutral zone (pulses were generated with load changes from 10% to 170% of the rated current I_r of the generator). (**a**) generator No. 1; (**b**) generator No. 2.

Tests of GS-12 T 3s DC generators have shown that changing the brush position (angle α: 0°, 1.0°, 1.6°, 2.2°, and 3.4°) in the generator affects the obtained equivalent decay pulse values (S_{iz}), which is particularly noticeable at lower generator speeds. Increasing the brush position angle (α) relative to the generator's neutral zone (from 0° to 3.4°) increases the value of the equivalent decay pulse generated during transient states.

The characteristics of GS-12T 3s generators No. 1 and No. 2 with different flight times, presented in Table 6, show the effect of aging changes on the obtained equivalent decay pulse values (S_{iz}) for all brush position angle values (α) relative to the generator's neutral zone. Figure 15 shows examples of obtained S_{iz} values for the presented transient tests of DC generators GS-12T 3s No. 1 and No. 2 (with different run-in time), with load changes from 10% to 170% of the rated current I_r of the generator and rotational speed n_{pr1} = 5050 ± 50 rpm.

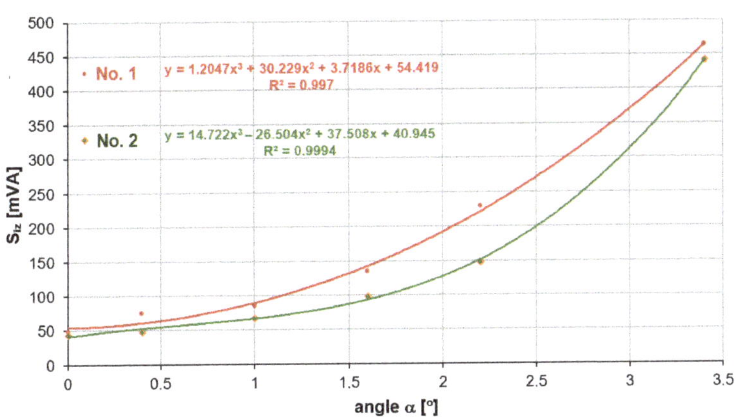

Figure 15. Influence of the operating time of DC generators No. 1 and No. 2 on the characteristics of equivalent decay pulses of generators with brushes shifted relative to the neutral zone in the direction of generator rotor rotation by the angle α: 0°, 1°, 1.6°, 2.2° and 3.4°, with load changes from 10% to 170% of the rated current I_r of the generator and rotational speed n_{pr1} = 5050 ± 50 rpm.

The trend line of the tested parameter S_{iz} for generator No. 1 (except for the angle α = 0°) observed in Figure 15 deviates from the trend line for generator No. 2. The illustrated shift in the trend line for generator No. 2 with a long flight time compared to generator No. 1 with zero flight time indicates aging changes occurring in the generators, which affect their proper self-regulation during operation. The characteristics presented in Figure 15 show that in generators with a longer flight time, the angle α can be increased to a greater extent than in generators with zero flight time without causing a significant increase in the equivalent decay pulse.

5. Mathematical Model of a Commutator-Type DC Generator

5.1. Mathematical Model

The research materials presented in this article illustrate numerous phenomena related to aging processes occurring in GS-12T 3s commutator-type DC generators. These measurements were performed over many years during the operation of these generators on board aircraft during their normal operation. For this reason, not all experiments could be conducted on these generators and their power grids. Therefore, a mathematical model based on the MATLAB R2021b environment was developed to model the static and dynamic processes occurring in the DC power node of the Su 22 aircraft [29].

Data for the model were obtained from available literature sources and from direct measurements of the characteristics of a representative generator and regulator [1,37,42,45,49,57]. Data that were impossible to obtain (e.g., the coefficient of viscous friction in the moving parts of the voltage regulator) were estimated using heuristic methods, while maintaining the correct external characteristics and boundary conditions.

The mathematical model of the GS-12T 3s generator with the RN-180M voltage regulator was developed based on the diagram presented in [57], which was used to create the equivalent circuit diagram of the generator with voltage regulator shown in Figure 16.

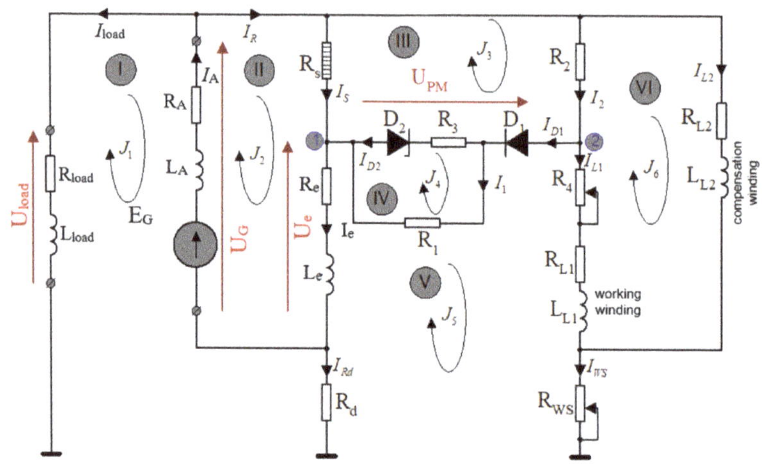

Figure 16. Equivalent diagram of the electrical circuits of the GS-12T 3s generator with the RN 180M voltage regulator.

The above diagram (Figure 16) was used to determine the voltage balance equations of the next six loops [41] using the loop method [1,55], from which the following were determined:

Laplace transform of the armature current I_A:

$$I_A\ (s) = J_2\ (s) - J_1\ (s)$$

Laplace transform of the generator voltage U_G:

$$U_G(s) = E_G(s) - I_A(s)\,[R_A + s\,L_A] = E_G(s) - [J_2(s) - J_1(s)]\,[R_A + s\,L_A]$$

Laplace transform of the generator excitation current I_e:

$$I_e(s) = J_2(s) - J_5(s)$$

Laplace transform of the generator excitation voltage U_e:

$$U_e(s) = I_e(s)\,[R_e + s\,L_e] = I_e(s)\,R_e + s\,I_e(s)\,L_e$$

Laplace transform of the current flowing through the carbon column of the regulator I_s:

$$I_s(s) = J_2(s) - J_3(s)$$

Laplace transform of the voltage across the carbon column of the regulator U_s:

$$U_s(s) = I_s(s)\,R_s$$

Laplace transform of the current in the additional resistance I_{Rd}:

$$I_{Rd}(s) = J_1(s) - J_5(s)$$

Laplace transform of the voltage across the additional resistance U_{Rd}:

$$U_{Rd}(s) = I_{Rd}(s)\,R_d$$

Laplace transform of the current in the coil of the working winding I_{L1}:

$$I_{L1}(s) = J_5(s) - J_6(s)$$

Laplace transform of the current in the coil of the compensation winding I_{L2}:

$$I_{L2}(s) = J_6(s)$$

U_{PM} bridge diagonal voltage in operator form:

$$U_{PM}(s) = U_{P2}(s)\,U_{P1}(s)$$

If the bridge diagonal voltage is greater than zero, then:

$$U_{P2}(s) > U_{P1}(s)$$

$$\text{or}\quad I_2\,R_2 < I_s\,R_s$$

In addition to the effect of temperature, the developed mathematical model takes into account the dynamics of the moving elements of the RN-180M voltage regulator, where the moving element is a core compressing a carbon pillar—the greater the compressive force, the lower the resistance of the carbon pillar [7,10,11,45]. The compressive force on the carbon pillar is reduced by the attractive force of the electromagnet in the regulator's coil L1 (Figure 16). The input to the model is therefore the current in coil L1. An auxiliary element is coil L2, which performs the task of temperature compensation of the voltage stabilization system in the generator. This model therefore includes control from the side of coil L2 (Figure 16). The input current to coil L2 is the second control input in the model. The output in the model is the resistance of the carbon post of the voltage regulator.

The equations of the mathematical model of the moving elements of the voltage regulator can be presented in the form:

$$F_{L1} = f\ (I_{L1})$$

$$F_{L2} = f\ (I_{L2})$$

$$F_{el} = F_{L1} - F_{L2}$$

where F_{L1}—electromagnet force as a function of coil L1 current,

F_{L2}—*electromagnet force as a function of coil L2 current,*

$f(I_{L1})$—nonlinear force characteristic depending on the current in coil L1,

$f(I_{L2})$—nonlinear force characteristic depending on the current in coil L2,

I_{L1}—current in coil L1,

I_{L2}—current in coil L2,

F_{el}—resultant electromagnet force.

The equation for the dynamics of the moving elements of the voltage regulator (all moving elements move simultaneously in one axis) is as follows:

$$m \cdot \frac{d^2x}{dt^2} + k_{TL} \cdot \frac{dx}{dt} + F_x(x) = F_{el}$$

where m—mass of all moving parts of the RN-180M voltage regulator,

x—displacement of moving parts relative to the regulator housing,

k_{TL}—*coefficient of viscous friction,*

F_x (x)—nonlinear force characteristic of the diaphragm spring (compressing the carbon column).

For the purpose of solving the equation in the simulation model, this equation was transformed into the following form:

$$\frac{d^2x}{dt^2} = \frac{1}{m} \cdot F_{el} - \frac{k_{TL}}{m} \cdot \frac{dx}{dt} - \frac{x}{m} \cdot f(x)$$

The equation relating the resistance of the carbon column to the contact force is as follows:

$$R_s\ (F_x) = f\ [F_x\ (x)]$$

where R_s *(Fx)*—resistance of the carbon column,

$f\ [F_x\ (x)]$—nonlinear characteristic of the change in the resistance of the carbon column.

Furthermore, the mathematical model of the electromotive force (EMF) generation of the GS 12T 3s generator assumes that the EMF value depend on the following variable input parameters:

I_e—generator excitation current,

n—generator rotor speed,

α—brush angle relative to the stator's magnetically neutral zone.

With these assumptions, the electromotive force (EMF) of the generator depends on three variables:

$$E = f\ (I_e, n, \alpha)$$

The generator's EMF is included in the voltage balance loop equations for the electrical circuits of the generator-voltage regulator system.

The characteristics of the change in EMF when individual variables are changed were measured on a real generator [29] for specific values, which served as interpolation nodes for determining the EMF value between nodes.

5.2. Examples of Simulation Test Results

To demonstrate the value of the research, example results of simulation tests of static characteristics of the GS-12T 3s generator model with a regulator obtained for no-load operation are presented below and compared with the experimental results.

Simulation studies were performed using a simulation model of the electrical circuits of a GS-12T 3s generator with an RN-180M voltage regulator, developed using the Matlab-Simulink programming environment [1,2]. The developed model contains an internal structure whose main element is the "Electrical Circuit Differential Equations Module." This module solves the system of six differential equations of electric circuits shown in Figure 16 for six loops (mesh currents: J1, J2, J3, J4, J5, J6). The outputs of the mesh currents J1 to J6 and the derivatives of the mesh currents are coupled to the inputs using memory modules to store the parameter values from the previous integration step. An additional element in the internal structure is the "Logic Module" for commutating algorithms for solving nonlinear differential equations depending on the parameters of the electrical circuits.

Example results of simulation tests of the no-load running of the GS 12T-3s generator (obtained for the rotor rotational speed n = 4200 rpm and various angles of the generator brushes position relative to the neutral zone α: 0°, 1.0°, 1.6°, 2.2° and 3.4°) are shown in Figure 17.

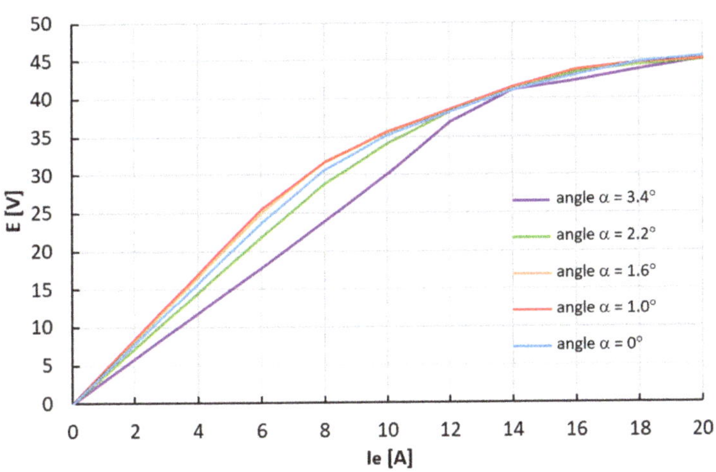

Figure 17. Simulated no-load characteristics of the GS-12T-3s generator for the rotor speed n = 4200 rpm and different positions of the generator brushes relative to the neutral zone of the angle α: 0°, 1.0°, 1.6°, 2.2° and 3.4°.

In order to assess the consistency of the simulation model (Figure 17) with the experimental data (Figure 10a), regression lines were fitted [29] using the no-load characteristics of the GS-12T-3s generator for the results obtained at different angles α. An example scatter plot of the simulation model and experimental results obtained for α: 0°, 1.0°, 1.6°, 2.2° and 3.4° is shown in Figure 18.

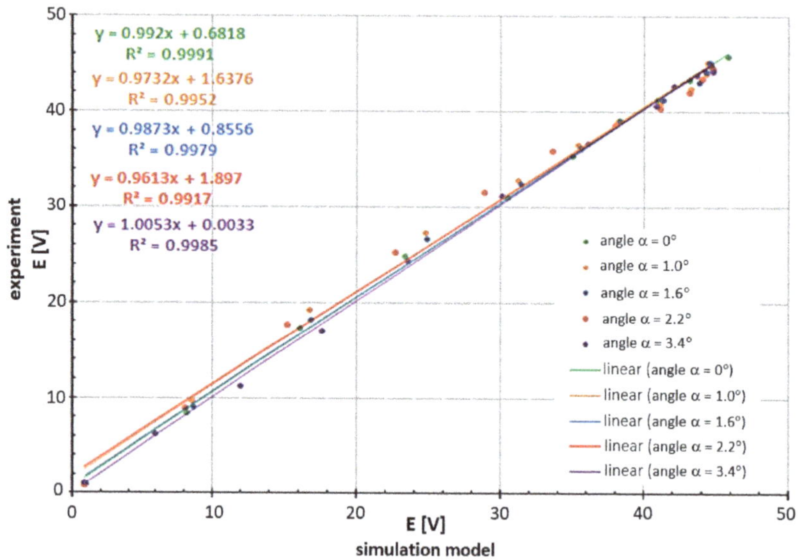

Figure 18. Comparison of simulated and experimental no-load characteristics of the GS-12T 3s DC generator (n = const = 4200 rpm and I_{obc} = 0 A).

The presented example comparison of the results of simulated and experimental tests of the no-load characteristics of the GS-12T 3s DC generator (for which the regression equations and R2 determination coefficients were determined) showed that the fit of the developed simulation model to the obtained experimental data is correct.

The presented mathematical and simulation model (described in more detail in [1]) should therefore be treated as an approximation of the description of physical phenomena that can be further developed and supplemented in order to adapt them to other electrical machines currently operated both on aircraft and in industry.

6. An Example of the Practical Use of Power Source Characteristics Shaping to Effectively Counteract the Aging Process of an Aircraft DC Power Grid

The presented possibilities of shaping the characteristics of the A/C power source by appropriately regulating the DC generators (changing the angle α of the generator brushes position relative to the magnetically neutral zone) in practice create the possibility of effectively counteracting the aging process of the aircraft DC power network. Below are sample test results of the DC power network of the Su-22 A/C with GS 12T 3s generators No. 3 and No. 4, performed during engine tests in transient states, which confirm the possibilities of practical use of shaping the power source characteristics to effectively counteract the aging process. The tests were performed before and after adjusting the angle α of the brushes position of generators No. 3 and No. 4 relative to the magnetically neutral zone (according to the method patented by the author) [34]. Data on the operating time of generators No. 3 and No. 4, which are the main power sources of the tested electrical power grid of the Su-22 A/C, as well as the value of the angle α in the generators before the angle adjustment (1st engine test) and after the angle adjustment in the generators (2nd engine test) are presented in Table 7.

Table 7. Data of generators No. 3 and No. 4 constituting the main power sources of the tested electrical network of the Su-22 A/C.

No.	Generator Number GS-12T 3s	Operation Time of Generators on the Su-22 A/C	The Value of the α Angle in the Generator (Before Adjustment) During the First Engine Test	The Value of the α Angle in the Generator (After Adjustment) During the Second Engine Test
		[hours]	[°]	[°]
1.	No. 3	1744	3.1	1.6
2.	No. 4	1168	2.6	1.4

Example results of measurements in transient states of DC power networks during engine tests of the Su-22 A/C performed with DC generators GS 12T 3s No. 3 and No. 4 are presented in Table 8. Non-conforming values are written in bold. The measurement results presented in Table 8 were obtained at an aircraft turbine speed of n = 67% (minimum range), which, according to studies [43,44], corresponds to a generator speed of 5050 rpm.

Table 8. Measurement results of transient parameters of the Su 22 A/C power grid with GS-12T 3s DC generators No. 3 and No. 4 before and after the α angle adjustment.

No.	Network Power Source	Transient Parameters of the Su-22 A/C Power Grid					
		I Test Before Adjusting the α Angle			II Test After Adjusting the α Angle		
		t_z [ms]	U_{min} [V]	S_{iz} [Vms]	t_z [ms]	U_{min} [V]	S_{iz} [Vms]
1.	Generator No. 3	115	20.26	**430.1**	20	20.16	76.8
		895	20.17	**3427.9**	108	22.52	159.8
2.	Generator No. 4	36	19.86	**149.0**	16	20.50	56.0
		702	21.34	**1867.3**	95	22.87	107.4
3.	Generators No. 1 and No. 2 operating in parallel	13	20.97	39.4	7	21.07	21.5

Non-conforming values are written in bold

As presented in the introduction, the parameter that is directly correlated with the aging processes occurring during the operation of aircraft DC power networks powered by GS-12T 3s DC generators is the value of the reduced DC equivalent pulse time recorded during transients. This value, given the known minimum voltage value of this pulse, allows for the determination of the equivalent DC voltage decay pulse S_{iz}. These values, after comparison with the ranges of changes in equivalent DC voltage functions in electrical power supply systems powered by 28 V DC generators specified in the standard [7], clearly identify the period of usability of the DC power networks of the tested A/C. An illustration of such analyses comparing the obtained measurement results from the first test of the Su-22 A/C engines (before adjusting the α angle in the generators) and from the second test (after adjusting the α angle in the generators) with the ranges of changes in the equivalent DC voltage functions in the electrical power supply systems powered by 28 V DC generators is presented in Figure 19.

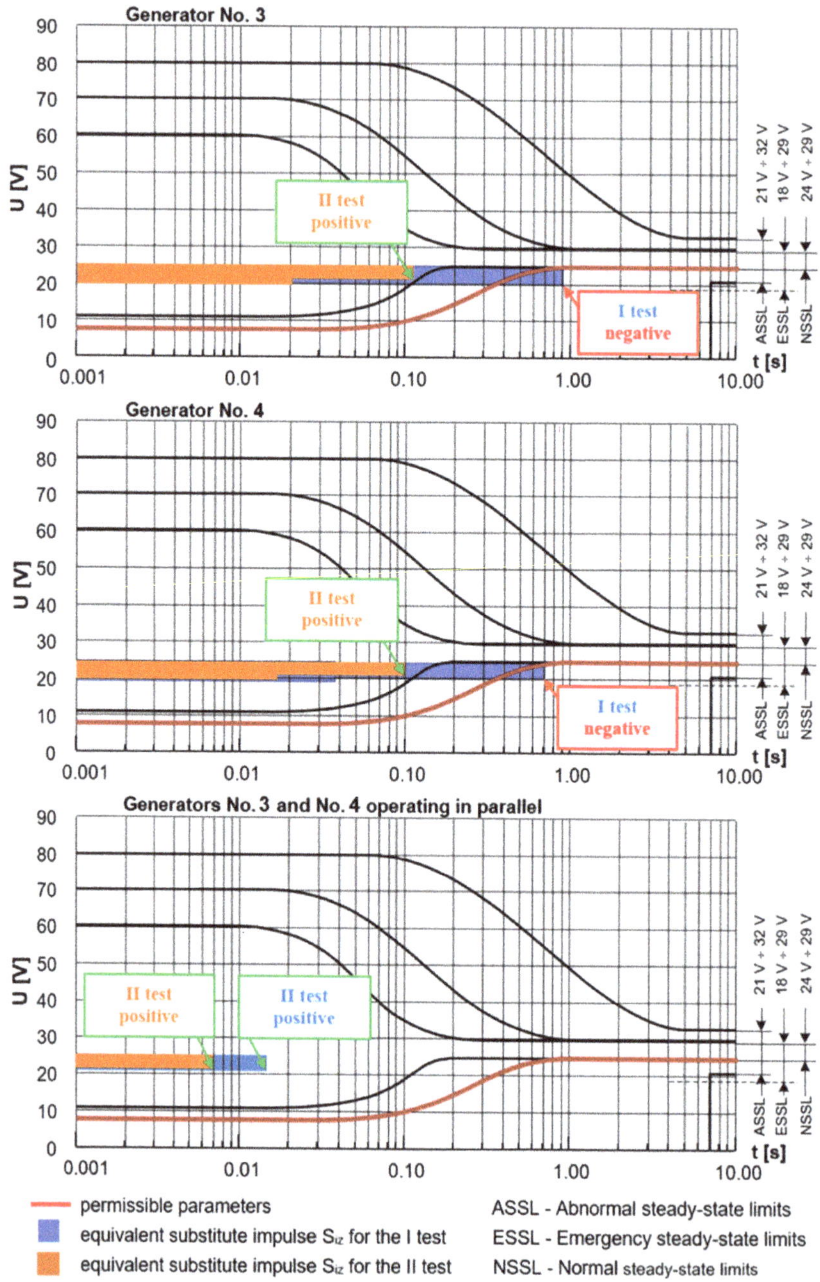

Figure 19. Comparison of the analyses of the measurements of the transient parameters of the Su-22 aircraft DC network against the background of the ranges of changes in the equivalent DC voltage functions in the electric power supply systems powered by DC generators No. 3 and No. 4.

The test results presented in Table 8 and the analyses illustrated in Figure 19 are an example confirming the effectiveness of using power source characteristic shaping to effectively counteract the aging process of an aircraft's DC power grid. By adjusting the α angle in the GS-12T 3s DC generators [34] while taking into account the aging processes of the generators, it is possible to obtain the correct parameters of the Su-22 aircraft's DC network when operating single generators, which confirms the need to shape the power source characteristics before installation on the A/C. Precise brush positioning relative to the magnetic neutral line effectively counteracts the aging process of the aircraft DC power grid by achieving the required network parameters in steady and transient states, and also influences the approximation of the time constants of the control systems, which has a positive effect on the approximation of the dynamics of the generator systems during parallel operation.

7. Conclusions

The topic of this article stems from many years of operational research (including those conducted by the authors) related to the quality of electrical power in A/C on-board power grids, conducted to improve flight safety. The analysis of the Su-22 aircraft's DC power grid demonstrated that aging processes occurring during its operation affect the properties of the power grid during transient states, including the correctness of the system's self-regulation. It was found that the parameters directly correlated with aging processes occurring during the operation of the DC power grid are the amplitude and duration of the equivalent pulse during transient states of the DC network. Analysis of the results of the Su-22 aircraft's DC power grid tests, obtained over the years and after overhaul, indicated that the direct cause of the malfunctions is the long-term operation of the grid's primary sources, the GS-12T 3s generators. Aging changes in the generators cause their overcompensation and change their characteristics. By shaping the characteristics of DC generators, compensating for the shifting of the magnetically neutral zone relative to the geometric plane between the stator pole pieces due to aging processes, it is possible to counteract the aging processes occurring in DC networks during their long-term operation. The author's method [28] of precise measurement and precise adjustment of the brush position angle α in DC generators, after taking into account the changes in the characteristics occurring in the generators during long-term operation as a result of the aging process, allows for further operation of these generators on A/C as a power source for their power networks while maintaining the required quality of electricity in both steady and transient states.

The research results and the developed proprietary solution for regulating the α angle presented in this article are implemented for operation on the Su-22 A/C as part of the bulletin, and the applied methodology ensures effective improvement of the characteristics of the GS 12T 3s DC generators and the parameters of the obtained DC power grids of the above A/C. This has reduced the failure rate of the DC network and has a direct impact on flight safety and reduced A/C operating costs. Many of the research topics covered in this article can be transferred to modern A/C in the future. The presented solution can also be applied to other A/C, including civil aviation and the national economy.

Author Contributions: Conceptualization, T.T. and S.M.; methodology, T.T.; software, T.G.; validation, T.T. and T.G.; formal analysis, T.T. and S.M.; investigation, T.T.; resources, T.T. and T.G.; data curation, T.T. and T.G.; writing—original draft preparation, T.T.; writing—review and editing, T.T. and S.M.; visualization, T.T.; supervision, T.T. and S.M. All authors have read and agreed to the published version of the manuscript.

Funding: This research received no external funding.

Data Availability Statement: The original contributions presented in this study are included in the article. Further inquiries can be directed to the corresponding author.

Conflicts of Interest: The authors declare no conflicts of interest.

Abbreviations

The following abbreviations are used in this manuscript:

A/C	Aircraft
ASSL	Abnormal steady-state limits
DC	Direct Current
ESSL	Emergency steady-state limits
NSSL	Normal steady-state limits
CBA	commutator-brush assembly

References

1. Abu-Rub, H.; Malinowski, M.; Al-Haddad, K. (Eds.) *Power Electronics for Renewable Energy Systems, Transportation, and Industrial Applications*, 1st ed.; John Wiley & Sons Ltd.: Hoboken, NJ, USA, 2014.
2. Gębura, A.; Setlak, L.; Tokarski, T.; Borowski, J. (Eds.) *Lotnicze Urządzenia Elektryczne: Monografia*; Wydawnictwo Instytutu Technicznego Wojsk Lotniczych: Warszawa, Poland, 2023; ISBN 978-83-61021-59-9.
3. Brombach, J.; Lucken, A.; Nya, B.; Johannsen, M.; Schulz, D. Comparison of different electrical HVDC-architectures for aircraft application. In *Electrical Systems for Aircraft, Railway and Ship Propulsion (ESARS)*; IEEE: Piscataway, NJ, USA, 2012; pp. 1–6.
4. Sarlioglu, B.; Morris, C.T. More Electric Aircraft: Review, Challenges, and Opportunities for Commercial Transport Aircraft. *IEEE Trans. Transp. Electrif.* **2015**, *1*, 54–64. [CrossRef]
5. Sciascera, C.; Giangrande, P.; Brunson, C.; Galea, M.; Gerada, C. Optimal design of an electro-mechanical actuator for aerospace application. In Proceedings of the Annual Conference of the IEEE Indus-Trial Electronics Society, Yokohama, Japan, 9–12 November 2015.
6. Zhao, X.; Guerrero, J.M.; Wu, X. Review of Aircraft Electric Power Systems and Architectures. In Proceedings of the International Energy Conference (ENERGYCON), Cavtat, Croatia, 13–16 May 2014; IEEE: Piscataway, NJ, USA, 2014.
7. *NO-17-A206:2019*; Wojskowe Statki Powietrzne—Naziemne Układy Zasilania Elektroenergetycznego. Podstawowe Parametry, Wymagania i Badania. MINISTRA OBRONY NARODOWE: Warsaw, Poland, 2019.
8. *MIL-STD-704F (W/Change-1)*; Department of Defense INTERFACE Standard: Aircraft Electric Power Characteristics. United States Department of Defense: Washington, DC, USA, 2016.
9. Cur Krzysztof, Gołda Paweł, Jemielniak Mirosław, Sarnowski Witold: Analysis of selected problems of the operation of means of air transport. *Aviat. Secur. Issues* **2023**, *4*, 405–421. [CrossRef]
10. Zieja, M. Prognozowanie Trwałości Wybranych Urządzeń Osprzętu Lotniczego w Aspekcie Procesów Starzenia. Ph.D. Thesis, Instytut Techniczny Wojsk Lotniczych, Warsaw, Poland, 2008.
11. Gębura, A.; Radoń, T. Pulsacje napięcia wyjściowego jako źródło informacji diagnostycznych o komutatorowych prądnicach prądu stałego. *Pr. Nauk. Inst. Tech. Wojsk Lotniczych* **2013**, *33*, 57–74. [CrossRef]
12. Wang, Y.; Nuzzo, S.; Zhang, H.; Zhao, W.; Gerada, C.; Galea, M. Challenges and opportunities for wound field synchronous generators in future more electric aircraft. *IEEE Trans. Transp. Electrif.* **2020**, *6*, 1466–1477. [CrossRef]
13. Setlak, L.; Ruda, E. Review, analysis and simulation of advanced technology solutions in power electronics systems (PES) of more electric aircraft. *World Acad. Sci. Eng. Technol.* **2015**, 9.
14. Gębura, A.; Szelmanowski, A.; Jacyna-Gołda, I.; Gołda, P.; Kalbarczyk, M.; Tomaszewska, J. Concepts and Experiments on More Electric Aircraft Power Systems. *Energies* **2025**, *18*, 1653. [CrossRef]
15. Aguida, M.E.; Che, Y.; Yang, J. Technological Advancements and Future Prospects of Electrical Power Systems for Sustainable More Electric Aircraft. *Propuls. Power Res.* **2024**, *13*, 475–486. [CrossRef]
16. Samolot PZL-130 TC—II ORLIK. *Opis Techniczny Część II Osprzęt*; EADS PZL Warszawa-Okęcie S.A.: Warszawa, Poland, 2004.
17. PZL-Świdnik, S.A. *Śmigłowiec PZL W-3WA—Instrukcja Obsługi Technicznej*; Wyd. PZL-Świdnik S.A.: Świdnik, Poland, 2012.
18. Samolot M-28B/PT-GC „Skytruck" z awioniką Glass Cockpit. *Album Schematów Elektrycznych Samolotu „M28B/PT" z Awioniką Glass Cockpit T2/2. Część 2. Osprzęt/Uzbrojenie*; Nr PRG/Z/15/2010; PZL-Mielec: Mielec, Poland, 2010. Available online: https://pzlmielec.pl/ (accessed on 17 February 2025).
19. *CASA C-295M*; C-295 Electrical System Description and Operation. MI-295-3-056-I. Construccions Aeronauticas S.A.: Madrid, Spain, 2012.
20. *Instrukcja Obsługi Śmigłowca SW-4*; PZL Świdnik: Świdnik, Polonia, 2015.
21. Petreski, D.; Iliev, A.; Trajanovska, E.; Gjurov, L.; Petrovski, A. Improved technical characteristics of the MI-8/17 in the Army of FYR Macedonia. *Mil. Tech. Cour.* **2018**, *66*, 122–142. [CrossRef]
22. Lee, D.; Choi, D. Analysis of the reliability of a starter-generator using a dynamic Bayesian network. *Reliab. Eng. Syst. Saf.* **2020**, *195*, 106628. [CrossRef]

23. Wawrzyński, W.; Zieja, M.; Tomaszewska, J.; Michalski, M. Reliability Assessment of Aircraft Commutators. *Energies* **2021**, *14*, 7404. [CrossRef]
24. Zhuo, L.; Yang, D.; Sun, R.; Shi, D.; Zhou, J. Research on Ignition Starting Characteristics of Aeronautical Starter-Generator with Turbine Engine Based on Reduced Order Model. In Proceedings of the 2023 26th International Conference on Electrical Machines and Systems (ICEMS), Zhuhai, China, 5–8 November 2023. [CrossRef]
25. *ISO 1540:2006*; Aerospace—Characteristics of Aircraft Electrical Systems. ISO Copyright Office: Cham, Switzerland, 2006.
26. Filina, O.A.; Martyushev, N.V.; Malozyomov, B.V.; Tynchenko, V.S.; Kukartsev, V.A.; Bashmur, K.A.; Pavlov, P.P.; Panfilova, T.A. Increasing the Efficiency of Diagnostics in the Brush-Commutator Assembly of a Direct Current Electric Motor. *Energie* **2024**, *17*, 17. [CrossRef]
27. Żurek, J.; Zieja, M. The aging-processes-based forecasting of service lives of selected items of air craft equipment. *J. KONBiN* **2008**, *4*, 287–305. Available online: http://worldcat.org/issn/189582814 (accessed on 2 March 2025).
28. Dub, M.; Drlik, P. Modelling of aircraft DC generator voltage regulation In Proceedings of the MECHATRONIKA, 15th International Symposium, Prague, Czech Republic, 5–7 December 2012.
29. Tokarski, T. Kształtowanie Charakterystyk Komutatorowej Prądnicy Prądu Stałego Statku Powietrznego. Ph.D. Thesis, 561/506 Instytut Techniczny Wojsk Lotniczych: Warszawa, Poland, 2023.
30. Sun, R.; Peng, H.; Shi, D.; Huang, G. Research on Optimal Design of Commutation Performance of Starter-Generator Used in Aero-Engine. In Proceedings of the 2021 IEEE International Magnetic Conference (INTERMAG), Lyon, France, 26–30 April 2021. [CrossRef]
31. Shantarenko, S.; Shantarenko, S.; Vaganov, A. Performance Control of the Commutator-and-brush Assembly of the Traction Motor. *Transp. Res. Procedia* **2021**, *54*, 854–861. [CrossRef]
32. Sawa, K.; Isato, M.; Ueno, T.; Nakano, K. Commutation characteristics and brush wear of DC motor at high rotation speed. In Proceedings of the 2017 IEEE Holm Conference on Electrical Contacts, Denver, CO, USA, 10–13 September 2017. [CrossRef]
33. Zhang, Q.; Ueno, T.; Morita, N. The new commutation theory to the optimization technique of the large dc machine commutation design. *IEEJ Trans. Ind. Appl.* **2011**, *131*, 1518–1521. [CrossRef]
34. Tokarski, T.; Kalisiak, M. Układ do pomiaru położenia szczotek prądnicy prądu stałego względem strefy neutralnej. Opis patentowy PL 238723 B1 ogłoszono 27.09.2021 r. WUP 26/21. Available online: https://ewyszukiwarka.pue.uprp.gov.pl/search/pwp-details/P.430838 (accessed on 2 March 2025).
35. Kozłowski, J. *Laboratorium Elektrotechniki Teoretycznej*; Wydawnictwo Politechniki Poznańskiej: Poznań, Poland, 1998.
36. Wardach, M. Magnetic wedges influence of pulsation in electrical machine with permanent magnets. *Przegląd Elektrotechniczny* **2010**, *86*, 163–166. Available online: https://www.researchgate.net/publication/272143434_Magnetic_wedges_influence_of_pulsation_in_electrical_machine_with_permanent_magnets (accessed on 2 March 2025).
37. Sołbut, A. *Maszyny Elektryczne 2—Maszyny Prądu Stałego, Maszyny Synchroniczne*; Oficyna Wyd. Politechniki Białostockiej: Białystok, Poland, 2019; ISBN 978-83-66391-10-9.
38. Wasiluk, W. *Poradnik Inżyniera Elektryka—Tom 2*; Wydawnictwo Naukowo Ttechniczne: Warszawa, Poland, 2007. Available online: https://www.ksiegarnia-fachowa.pl/wiecej.php?id=ni-471 (accessed on 1 March 2025); ISBN 978-83-204-3292-3.
39. Setlak, L.; Ruda, E. Współczesne rozwiązania technologiczne, analiza i symulacja wybranych komponentów architektury HVDC samolotów zgodnych z koncepcją MEA/AEA. *Przegląd Elektrotechniczny* **2017**, *2*. [CrossRef]
40. Tokarski, T.; Michalak, S. The effect of the operation time of the aircraft power system on power quality in transient states. *J. KONBiN* **2024**, *54*, 47–64. [CrossRef]
41. Tokarski, T. Evaluation of direct current electric power systems of the aircraft based on characteristics of a transient state. *Diagnostyka* **2019**, *20*, 81–91. [CrossRef]
42. Leśniczak, A. *i Inni: Badania i Ocena Stanu Technicznego Samolotu Su-22 M4 nr 30304 w Czasie Remontu Weryfikacyjnegosprawozdanie nr SP-147/31/2015*; Instytut Techniczny Wojsk Lotniczych: Warszawa, Poland, 2016.
43. Silnik 89—Opis techniczny; Dowództwo Wojsk Lotniczych: Poznań, Poland, 1988; Lot. 2560/86.
44. Samolot Su-22M4—Aerodynamika Praktyczna; Dowództwo Wojsk Lotniczych: Poznań, Poland, 1987; DWL Lot. 2596/87.
45. Tokarski, T.; Gębura, A. *Analiza Możliwości Doprowadzenia Układu Zasilania Elektrycznego Prądu Stałego Samolotów Su-22 Do Samoregulacji Napięcia Zgodnie z Wymaganiami Normy Obronnej NO-15-A200: 2007 z Uwzględnieniem Wyników Badań Wykonanych w Czasie Remontu Weryfikacyjnego. Sprawozdanie Nr 177/OP/43/16—BT ITWL 8623/50*; Instytut Techniczny Wojsk Lotniczych: Warszawa, Poland, 2016.
46. Dzik, M. Bezpieczeństwo lotów w lotnictwie wojskowym—Czynniki i okoliczności wpływające na powstawanie wypadków I incydentów lotniczych. *Obron.—Zesz. Nauk. Wydziału Zarządzania I Dowodzenia Akad. Obrony Nar. Nr* **2013**, *1*, 73–91.
47. Lewitowicz, J. *Podstawy Eksploatacji Statków Powietrznych—Eksploatacyjne Problemy w Projektowaniu i Modernizacji Statków Powietrznych*; Wydawnictwo Instytutu Technicznego Wojsk Lotniczych: Warszawa, Poland, 2012.
48. Zadencki, R. Analiza Uszkodzeń Samolotów Wielozadaniowych F-16 Block 52+ Podczas ich Eksploatacji. Ph.D. Thesis, Politechnika Poznańska, Poznań, Poland, 2020.

49. Glinka, T. *Maszyny Elektryczne i Transformatory—Podstawy Teoretyczne, Eksploatacja i Diagnostyka*; Instytut Napędów I Maszyn Elektrycznych Komel: Katowice, Poland, 2015.
50. Hockmeyer, I.O. The generation and regulation of electric power in aircraft: A survey of design features of generators and their control. *J. Inst. Electr. Eng.-Part II Power Eng.* **1946**, *93*, 2. [CrossRef]
51. Milewski, Z. *Aerodynamika Konstrukcja i Systemy Statku Powietrznego*; Wydawnictwo WAT: Warszawa, Poland, 2008; ISBN 978-83-89399-82-3.
52. Leśniczak, A. *i Inni: Kompleksowa Ocena Stanu Technicznego Samolotów Su-22 na Podstawie Wyników Badań i Weryfikacji Wykonanych w Procesie Remontu Weryfikacyjnego Oraz Opracowanie Zaleceń Profilaktycznych Do Eksploatacji Samolotów po Remoncie*; BT ITWL 1830/I; Instytut Techniczny Wojsk Lotniczych: Warszawa, Poland, 2018.
53. Lewitowicz, J.; Szelmanowski, A.; Gębura, A.; Pazur, A.; Franczuk, E.; Tokarski, T. Badanie zjawiska zniekształceń stref magnetycznie obojętnych w lotniczych komutatorowych prądnicach prądu stałego. *Przegląd Elektrotechniczny* **2019**, *95*, 78–81. [CrossRef]
54. Tokarski, T.; Kalisiak, M. Stanowisko do badania i regulacji położenia szczotek komutatorowej prądnicy prądu stałego GS-12T 3s względem strefy neutralnej Wzór użytkowy W.128337. Numer prawa wyłącznego Ru.072444 ogłoszono 07.03.2022. Available online: https://uprp.gov.pl/pl/uslugi-online/e-wyszukiwarka (accessed on 1 March 2025).
55. Bolkowski, S. Elektrotechnika. Wydawnictwo Szkolne i Pedagogiczne, S.A.: Warszawa, Poland, 2005, ISBN 978-83-02.
56. Jaworski, B.; Dietłaf, A.; Miłkowska, L.; Siergiejew, G. *Kurs Fizyki. Elektryczność i Magnetyzm. Tom 2*; PWN: Warszawa, Poland, 1984; ISBN 83-01-01265-X.
57. Samolot S-32MK—Opis Techniczny i Eksploatacja Oddzielnych Przyrządów i Agregatów Osprzętu; Ministerstwo Obrony Narodowej—Dowództwo Wojsk Lotniczych: Poznań, Poland, 1976; Lot.1758/76.

MDPI AG
Grosspeteranlage 5
4052 Basel
Switzerland
Tel.: +41 61 683 77 34

Energies Editorial Office
E-mail: energies@mdpi.com
www.mdpi.com/journal/energies

www.ingramcontent.com/pod-product-compliance
Lightning Source LLC
Chambersburg PA
CBHW061831210926
51850CB00006B/854